D1571665

Fuels and Engines

TECHNOLOGY ▪ ENERGY ▪ ENVIRONMENT

COVER ILLUSTRATION
Institut Français du Pétrole document

FROM THE SAME PUBLISHER

- A New Generation of Two-Stroke Engines for the Future?
  P. Duret
- Multi-Dimensional Simulation of Engine Internal Flows
  T. Baritaud, Ed.
- Direct Numerical Simulation for Turbulent Reacting Flows
  T. Baritaud, T. Poinsot, T. Baum
- Physical Chemistry of Lubricants
  J. Denis, J.C. Hipeaux, J. Briant
- Rheological Properties of Lubricants
  J. Briant, J. Denis, G. Parc
- Combustion and Flames. Chemical and Physical Principles
  R. Borghi, M. Destriau

Institut Français du Pétrole Publications

Jean-Claude GUIBET
Ph.D. in Science from the University of Louvain
Graduate Engineer from IFP School
Fuels coordinator at the Institut Français du Pétrole
Professor at IFP School

with the assistance of

Emmanuelle FAURE-BIRCHEM
Graduate Engineer from ENSCP and IFP School
Chevron Chemical S.A., Oronite Division, Fuel Additive Specialist

Foreword by
Raymond H. LÉVY
Honorable President of Renault

Fuels and Engines

TECHNOLOGY ▪ ENERGY ▪ ENVIRONMENT

Volume 1

Translated from the French by
Frank Carr, Mots Corporation

REVISED EDITION

1999

Editions TECHNIP 27 RUE GINOUX 75737 PARIS CEDEX 15 FRANCE

Translation of
Carburants et moteurs
Technologies • Énergie • Environnement
J.C. Guibet

ISBN 2-7108-0753-X (volume 1)
ISBN 2-7108-0751-3 (complete edition)

Foreword

After a long and distinguished career as a researcher, engineer, and teacher, Jean-Claude Guibet, assisted by Emmanuelle Faure-Birchem, has compiled an all-encompassing scientific and technological achievement. As well as an all-encompassing work, this book is also a state of the art review of the subject.

And what a subject it is! To measure its importance, one need only think of the place engines and fuels hold in our daily lives, in the economies of our countries, and in the destiny that our developments are preparing for future generations.

In our daily lives, fuels and engines are a key to our personal freedom. The freedom to come and go, the freedom to meet and discuss things face-to-face, a freedom that even the most sophisticated communication technology cannot totally replace. It is the liberty that all young people aspire to when they approach the legal age for obtaining a driver's license and the prospect of personal motorized transport. It is the liberty that people in countries with state-managed economies have aspired to: remember the extraordinary purchases of automobiles by the new German länder after the fall of the Berlin wall.

The preceding paragraphs undoubtedly explain the role that fuels and engines play in the economy of our countries. In France, transportation represents 10% of all employment. On a global scale, General Motors, Ford, Toyota, Exxon, and Royal-Dutch Shell regularly appear among the world's top ten businesses; Shell and General Motors are first-place competitors in profitability, and Shell, Exxon, and Toyota are rivals for the top spot in total assets. Automobiles also figure prominently in the operations of the top ten Japanese companies.

In France, which is a major consumer of nuclear energy, motor fuels represent one third of all fossil fuel consumption. This figure is somewhat lower for other nations with less emphasis on nuclear power, but it will no doubt increase as less-developed countries industrialize. Hydro-electric, nuclear, and other renewable energy sources can substitute for fossil fuels for various uses, but not for transportation: energy supply, the key to the modern economy, is closely tied to motorized transport; this is also true for pollution, which is closely related to the products of combustion from fossil fuels.

The exhaustive treatment the subject demanded of the authors is the result of the interplays I have just described. Based on the expertise and experience of the research teams at the Institut Français du Pétrole (French Petroleum Research Institute), this book provides in-depth coverage of the questions related to fuel formulation and engine behavior. The book begins with a general explanation of the physical and chemical characteristics of fuels and their refining methods, and proceeds to address the problems caused by each type of product. The environmental and energy consumption effects of the two most important products (gasoline and diesel fuel), which are often market rivals, are described in detail. The book concludes by examining the new refining techniques and engine designs that are currently at the research and development stage, and which could appear in the fuels and engines of tomorrow.

Jean-Claude Guibet is not satisfied with just a state-of-the-art treatment of the subject matter. His work is directed to the future, which is evident in his treatment of economic and ecological problems, and their resolution. Solving these problems involves satisfying our irrepressible need for personal liberty, especially freedom of movement, and preserving the future of our planet, whose deterioration results from these very activities.

Therefore, global energy consumption—and the ways of reducing it—receive extensive treatment. This important subject is examined in detail, especially the latest research on spark-ignition engines.

Extensive treatment is also devoted to the pollution caused by fuels, to the current and future regulations intended to control fuel-related emissions, and to the various techniques that will meet these regulations. Concerning the sensitive topic of pollution in urban areas, it is important to differentiate, as the author indicates, between the pollutants derived from engines and the green-house effect caused by the CO_2 *produced by the combustion of coal, natural gas, and other petroleum products.*

This last point is an important one in industrialized countries, particularly in France where nuclear power is extensively *used—instead of fossil fuels—to produce electricity, thereby ensuring that the share of the CO_2 contributed by engines is considered to be significant. However, the effect of green-house gases should be considered on a global scale, where the reduction in the use of fossil fuels for combustion can be applied more effectively using other forms of energy, especially nuclear power. The need to reduce the growth in the consumption of fossil fuels, for economic and environmental reasons, can best be accomplished by using easily replaceable substitutes. Reducing the fuel consumption of engines should not be neglected, because the ease of transporting stored fuel is an essential requirement for vehicle propulsion systems.*

One cannot analyze energy consumption and emissions without comparing diesel engines and spark-ignition engines. This is a delicate subject and the truth often clashes with the interests of particular manufacturers or suppliers. Despite his remarkable objectivity, Jean-Claude Guibet does not completely escape from this difficulty, notably in matters of fuel consumption wherein a particular publication that is undoubtedly oriented to a particular interest is referenced. Also, noise pollution seems to have been neglected as well. The technological progress currently underway on both of these engine types may, in the very-near future, lead to some more definitive conclusions.

As mentioned earlier, petroleum products should be used primarily for engines, but the early developments in electric vehicles should not be ignored; perhaps the author has dealt with them too lightly. The handicaps of electric propulsion are apparent at our current state of the understanding of energy storage, but the book barely mentions hybrid propulsion systems. I would not be surprised if, in a few years time, an addendum to this work includes hybrid propulsion systems that will refer to motorized adaptations that are not mentioned herein.

The state of art is evolving. This work is a new, and completely revised, edition of a work that first appeared in French about ten years ago. Those of us that will still be active in the field of engines and fuels in the year 2009 will benefit no doubt from a new edition of this work. That will be due to the efforts of Jean-Claude Guibet and Emmanuelle Faure-Birchem who have, with all the passion and rigor of their discipline, provided us with an up-to-date work that will be retained in the reference libraries of engineers—engineers who need to understand the characteristics and the operation of heat engines used for propulsion, the properties of the fuels required by these engines, and how to prepare these fuels. "Fuels and Engines: technology, energy, and the environment" will remain a valuable reference for many years.

How many years? Given the impetus being applied to this field by rational progress, where each advance in engine technology requires an advance in fuels, and each advance in fuel technology enables a new advance in engine technology, one cannot improve on the author's choice of a concluding quotation from Claude Lévi-Strauss, who wrote: "Each advance provides a new hope, which depends on the solution of a new problem. The case is never closed."

Shall we resume this discussion in ten years time?

Raymond H. LÉVY

Preface

La perfection de l'invention confine ainsi à l'absence d'invention. De même que dans l'instrument, toute mécanique apparente s'est peu à peu effacée et qu'il nous est livré un objet aussi naturel qu'un galet poli par la mer, il est également admirable que, dans son usage même, la machine se fasse progressivement oublier.

The perfection of invention verges on the absence of invention. It is the same way with the evolution of machines. All the apparent mechanical developments gradually disappear and we are left with an object that is as natural as a pebble polished by the sea. It is also admirable that through the very use of the machines themselves, the developments are gradually forgotten.

A. de Saint-Exupéry
Terre des hommes *(Man and his world)* (1939)

This work is an extensive revision of the first French-language edition that was published in 1987.

All the data have been updated and extended to include both European and global perspectives. Our analysis of the field of fuels and engines is focused on three main themes:

- the progress of technology
- reducing energy consumption
- protecting the environment

One cannot justify having precise fuel-quality criteria without taking into account the evolution of the technologies used to produce these fuels in the refinery and the technologies required for their optimal use in engines.

As to the combined objectives of mastering both energy consumption and environmental concerns, we know that these concerns are currently of paramount importance and that they will remain so, no doubt, for the rest of the decade.

The information in this book is destined primarily for students in the energy field, especially the students at *École Nationale Supérieure du Pétrole et des Moteurs* who have been an attentive and receptive audience year after year.

In addition to this purely academic group, we are also addressing a much larger audience who, by virtue of their work and professional responsibilities, or their interests and curiosity, are interested in engines, automobiles, aviation, and the transportation world of today and tomorrow.

I was able to complete this work with the help of Emmanuelle Faure-Birchem, who holds engineering degrees from *École Nationale Supérieure de Chimie de Paris* and *École Nationale Supérieure du Pétrole et des Moteurs*. She has my warmest thanks and I commend her for showing, at the beginning of her career, remarkable qualities of scientific rigor, determination, and organization at every stage of the work—from the initial manuscript through to the final edit.

I would also like to thank my colleagues: Brigitte Martin, André Douaud, Xavier Montagne, and Philippe Pinchon, who provided their comments, advice, and suggestions.

I am also very thankful to the numerous specialists at the *Institut Français du Pétrole (IFP)* and the other organizations that have provided vital information. Also, I extend my thanks to the representatives of *Éditions Technip* and the *Direction de la Documentation* of the *IFP*, who provided their experience and efficient collaboration.

Mister Raymond H. Lévy, the Honorable President of Renault, agreed to write the foreword and to provide his recommendation. He has my deepest gratitude for that expression of interest and esteem.

J. C. G

Our sincere thanks to the representatives of the following organizations:

Aérospatiale.
Agence de l'Environnement et de la Maîtrise de l'Énergie (ADEME).
Agence Internationale de l'Énergie (AIE).
Airbus Industrie.
Airparif.
Arco Chimie France, SNC.
Association Française des Techniciens et Professionnels du Pétrole (AFTP).
Association Française du Gaz Naturel pour Véhicules (AFGNV).
Association pour le Développement des Carburants Agricoles (ADECA).
Association Technique de l'Industrie du Gaz en France (ATG).
Beru.
BK-Gas.
British Gas.
Bureau de Normalisation du Pétrole (BNPé).
Comité Professionnel Du Pétrole (CPDP).
Comité Professionnel du Butane et du Propane (CPBP).
Concawe.
Deltec.
Elf Antar France.
Ethyl Petroleum Additives International.
Exxon Chemical International INC, (Paramins).
Garret.
Gaz de France.
Institut National de Recherche sur les Transports et leur Sécurité (INRETS).
Lucas Diesel.
Ministère Français de l'Environnement.
Ministère Français de l'Industrie, Direction des Hydrocarbures (DHYCA).
Proléa.
PSA Peugeot Citroën.
Renault SA.
Renault Véhicules Industriels (RVI).
Robert Bosch (France) SA.
Shell Recherche SA.
Société d'Applications Générales Électriques et Mécaniques (SAGEM).
Société d'Études de Machines Thermiques (SEMT-PIELSTICK).
Société des Ingénieurs de l'Automobile (SIA).
Société Nationale d'Étude et de Construction de Moteurs d'Avion (SNECMA).
Sulzer Burckhardt.

Syndicat National des Producteurs d'Alcool Agricole (SNPAA).

The Associated Octel Company Limited.

TNO Road-Vehicles Research Institute.

Total Raffinage Distribution.

Ullit.

Union Française des Industries Pétrolières (UFIP).

Union Technique de l'Automobile et de Cycle (UTAC).

Vialle Autogas Systems.

If we have accidentally forgotten to include some organizations, please forgive our oversight.

Table of contents

(Volumes 1 and 2)

The Table of Contents for Volumes 1 and 2 is included at the front of each volume. The Bibliography and the Index for both Volumes 1 and 2 are included at the back of each volume.

Volume 1

(p. I to XXXVI and 1 to 386)

Chapter 1

Chapter 3

Volume 2

(p. I to XLII and 387 to 773)

Chapter 8

Annex 1

Annex 2

Abbreviations and acronyms

ACEA	*Association des Constructeurs Européens d'Automobiles* (European automakers association)
AFQRJOS	Aviation Fuel Quality Requirements for Jointly Operated Systems
AGP	Aniline Gravity Product
AGRICE	Agriculture-Chemistry-Energy
AKI	Anti-Knock Index
AP	Aniline Point
API	American Petroleum Institute
AQIRP	Air Quality Improvement Research Program
ATAC	Active Thermo-Atmosphere Combustion
ATR	Atmospheric Residue
AVTUR	Aviation Turbine
BOCLE	Ball On Cylinder Lubricity Evaluator
CAA	Clean Air Act
CAAA	Clean Air Act Amendments
CAFE	Corporate Average Fuel Economy
CAP	Common Agricultural Policy
CARB	California Air Resources Board
CCAI	Calculated Carbon Aromaticity Index
CCI	Calculated Cetane Index
CEC	Coordinating European Council
CEP	Car Efficiency Parameter
CFC	Chloro-Fluoro-Carbons
CFPP	Cold Filter Plugging Point
CFR	Cooperative Fuel Research (committee and engine)
CI	Cetane Index
CII	Calculated Ignition Index
CNG	Compressed Natural Gas
CORC	Coordinating Octane Requirement Committee
CP	Cloud Point

CRC	Coordinating Research Council (United States)
CVS	Constant Volume Sampler
DAO	Deasphalted Oil
DBE	Dibromocthane
DCE	Dichloroethane
DERD	Directorate, Engine Research and Development
DI	Direct Injection
DI	Driveability Index
DM	Detonation Meter
DMA (B, C, or X)	Fuels of the type Distillates, Marine
DME	Dimethylether
DNPH	Di-Nitro-2,4-Phenyl Hydrazine
DON	Distribution Octane Number
ΔR	Delta RON (when accounting for volatile fractions)
DTBP	Di-tertio-Butyl Peroxide
E0, E5, … , E85	Fuel with 0%, 5%, … , 85% ethanol
E70-100-180-210 …	Volume fraction evaporated at 70-100-180-210 … °C
ECE	Economic Commission for Europe
ECVM	Energy Content of the Vaporized Mixture
EFEG	Environmental Fuels Experts Group
EGR	Exhaust Gas Recirculation
EHN	Ethyl-2 Hexyl Nitrate
EKMA	Empirical Kinetic Modelic Approach
EPA	Environmental Protection Agency
EPEFE	European Program on Emissions, Fuels, and Engine Technologies
EPN dB	Effective Perceived Noise (decibels)
ESFC	Effective Specific Fuel Consumption
ETBE	Ethyl tert butyl ether
EUDC	Extra-Urban Driving Cycle
EUROPIA	European Petroleum Industry Association
FAP	Friction Average Pressure
FCC	Fluid Catalytic Cracking
FFV	Flexible Fueled Vehicles
FIA	*Fédération Internationale Automobile*
FIA	Fluorescent Indicator Adsorption
FID	Flame Ionization Detector
FOD	Fuel Oil, Domestic
FP	Flash Point
FP	Final Point (distillation)
FTP	Federal Test Procedure (US standard test cycle)

FVI	Fuel Volatility Index
GATT	General Agreement on Tariffs and Trade
GC	Gas Chromatography
GFC	*Groupement Français de Coordination* (French coordinating group)
GFI	Gaseous Fuel Injection
GHSV	Gas Hourly Space Velocity (m^3 feedstock/m^3 of catalyst · hour)
GHV	Gross Heating Value
GWP	Global Warming Potential
HC	Hydrocarbons (as emissions)
HCO	Heavy Cycle Oil
HDM	Hydrodemetalyzation
HDN	Hydrodenitrification
HDS	Hydrodesulfurization
HDT	Hydrotreatment
HFRR	High Frequency Reciprocating Rig
HMN	Hepta Methyl Nonane
HPLC	High Performance Liquid Chromatography
IAP	Indicated Average Pressure
IAPAC (engine)	*Injection Assistée par Air Comprimé* (two-stroke)
IATA	International Air Transportation Association
ICAO	International Civil Aviation Organization
IP	Initial Point (distillation)
IPCC	Intergovernmental Panel on Climate Change
ISFC	Indicated Specific Fuel Consumption
JFTOT	Jet Fuel Thermal Oxidation Tester
JP4, JP5, JP7, JP8, JP9, JP10	Jet Propelled (aviation fuels)
KLSA	Knock Limited Spark Advance
LA-4	Los Angeles 4 (driving cycle)
LAFY	Los Angeles Freeway (driving cycle)
LANF	Los Angeles Non Freeway (driving cycle)
LCO	Light Cycle Oil
LEV	Low Emission Vehicles
LF	Luminosity factor
LNG	Liquefied Natural Gas
LPG	Liquefied Petroleum Gas
LTFT	Low Temperature Flow Test
M0, M5, … , M85	Fuels containing 0%, 5%, … , 85% methanol
MAN	*Machinenfabrik Augsburg Nürnberg* (corporation)
MAV	Maleic Anhydride Value

MEP	Mean Effective Pressure
MIP	Mean Indicated Pressure
MIR	Maximum Incremental Reactivity
MMT	Methylcyclopentadienyl Manganese Tricarbonyl
MON	Motor Octane Number
MOR	Maximum Ozone Reactivity
MSEP	Micro Separometer rating
MTBE	Methyltertiobutylether
MTG	Methanol To Gasoline (Mobil process)
MVEC	Mitsubishi Innovative Valve timing and lift Electronic Control
MVEG	Motor Vehicles Emissions Group
NAAQS	National Ambient Air Quality Standards
NDIR	Non Dispersive Infra Red
NGV	Natural Gas for Vehicles
NHV	Net Heating Value
NMHC	Non Methane Hydrocarbons
NMOG	Non Methane Organic Gases
NMR	Nuclear Magnetic Resonance
NO_x	Oxides of Nitrogen (emissions)
NPRM	Notice of Proposed Rulemaking (reformulated gasolines)
NYNF	New York Non Freeway (driving cycle)
OBD	On Board Diagnostics
OFP	Ozone Forming Potential
ORI	Octane Requirement Increase
OTL	Operability Temperature Limit
PAH	Polynuclear Aromatic Hydrocarbons
PAN	Peroxy Acetyl Nitrate
PP	Pour Point
PRF	Primary Reference Fuels
psi	pounds per square inch (units of pressure)
RAT	Residue, Atmospheric
RJ4, RJ5, RJ6	Ram Jet (aviation fuels)
RMA (… L)	Fuels, Residue, Marine
RME	Rapeseed Methyl Ester
RON	Research Octane Number
RUFIT	Rational Utilization of Fuels In Private Transport
RVP	Reid Vapor Pressure
SE	Specific Energy
SFPP	Simulated Filter Plugging Point

SHED	Sealed Housing for Evaporative Determination
SOF	Soluble Oil Fraction (particulates)
SP	Smoke Point
SR	Specific Reactivity
SR	Straight Run (output from direct distillation)
T10-50-90-95- ...	Temperature at which 10%-50%-90%-95% ... evaporates
TAME	Tertioamylmethylether
TAP	Toxic Air Pollutants
TBA	Tertiobutylalcohol
TBN	Total Basic Number
TEL	Tetra Ethyl Lead
TLEV	Transitional Low Emissions Vehicle
TML	Tetra Methyl Lead
toe	Tonne of Oil Equivalent
TR0, TR4, TR5	Turbo-Reactor fuels (aviation)
TS	Thermally Stable (aviation fuels)
UAM	Urban Airshed Model
ULEV	Ultra Low Emission Vehicles
UTAC	Union Technique de l'Automobile, du Motocycle et du Cycle (French industrial assoc.)
UTM	Union Test Method
V/L	Vapor-Liquid ratio (gasoline volatility)
VD	Vacuum Distillate
VFV	Variable Fueled Vehicles
VGO	Vacuum Gas Oil (vacuum distillate)
VOC	Volatile Organic Compound
VR	Vacuum Residue
VTEC	Variable valve Timing and lift Electronic Control system
WPI	Wax Precipitation Index
ZEV	Zero Emission Vehicle

Introduction

In this book, the word "**fuels**" refers to a range of products—which are usually liquids and occasionally gases—that burn in the presence of air and enable the operation of heat engines, either piston (gasoline or diesel) or turbine (aircraft engines). The term "**fuel oil**" refers to the fuels burned in boilers, ovens, furnaces, and similar devices.

Petroleum is now the primary source of fuels and this situation is not expected to change for several years. To be more precise, natural gas, coal, and biofuels currently represent less than 1% of the global consumption of fuels used in engines. Moreover, the latter are mainly used in the field of transportation, where road transport predominates to a great extent over air, rail, or water transport.

Therefore, this book deals primarily with fuels refined from crude oil and used in **road vehicles** (passenger cars and heavy trucks). The book also covers **alternative fuels** (natural gas, biofuels, ...) and **special** fuels (jet fuels, heavy fuels, ...) to provide the reader with an overall perspective and a preview of the subjects.

For each type of fuel, the book begins by describing the current state of the technology applied to engines to optimize performance and reduce emissions. The sections that follow describe the fuel characteristics that are required and the processes that are needed to prepare these fuels in sufficient quality and quantity. New constraints, which are imposed by the essential objectives of reducing energy consumption and protecting the environment, are studied in depth.

Before exploring these themes, it may be useful to review some of the present circumstances that are leading us to the year 2000. The reader is invited to examine more closely the events and trends that developed throughout the world during the 1970-2000 period, and how these occurrences have affected all forms of transportation, the proportional demand for petroleum products, and environmental protection.

Transportation on the advent of the year 2000

During the 60 or so years between 1938 and 2000, the world vehicle fleet has gone from 43 million vehicles to over 700 million. The fleet should realistically top the one billion mark by the year 2010. Fig. 1 shows the projections to the year 2010 for various regions of the world. The highest growth rates are projected for the developing countries, which includes China. In China, where the population represents about one quarter of the world total, the vehicle fleet is expected to reach 20 million vehicles by the year 2010.

Longer-range estimates have been produced (Dessus, 1993), which project that the world fleet will be 2.5 billion vehicles by 2060, of which 70% will be in regions where the current level of engine use is very low (countries in central Africa, India, China, etc.).

The ratio of the number of commercial vehicles (CV) to private vehicles (PV) is an interesting indicator, because the former is usually diesel powered and consumes diesel fuel, whereas the latter type is primarily equipped with gasoline engines. As shown in Fig. 2, the distribution of these types of vehicles varies strongly from one region of the world to another. In the heavily indus-

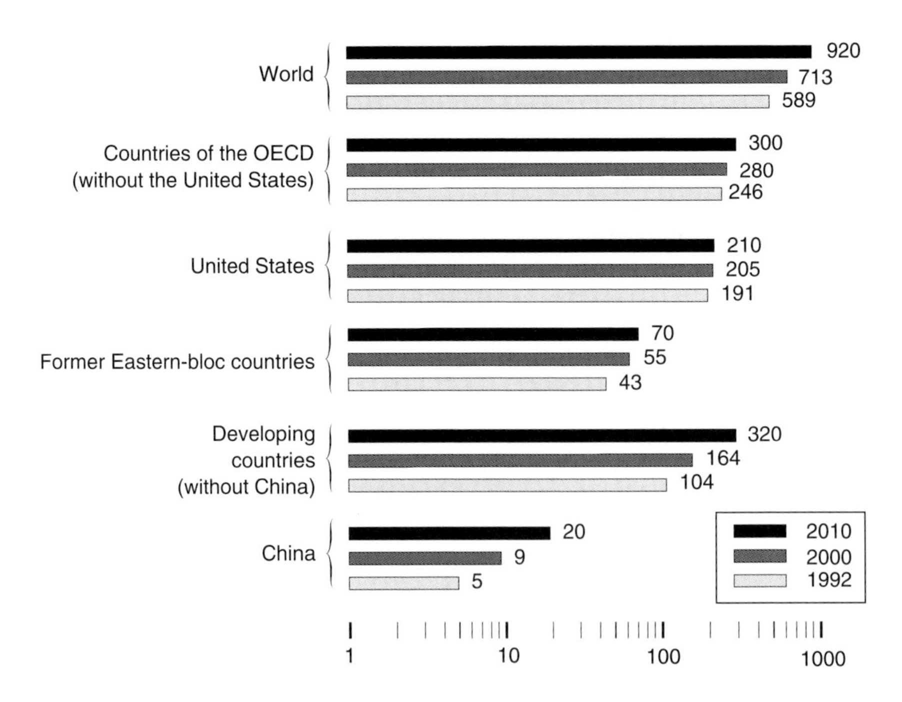

Figure 1 *Predicted trends in the world vehicle fleet by region (in millions of vehicles).*

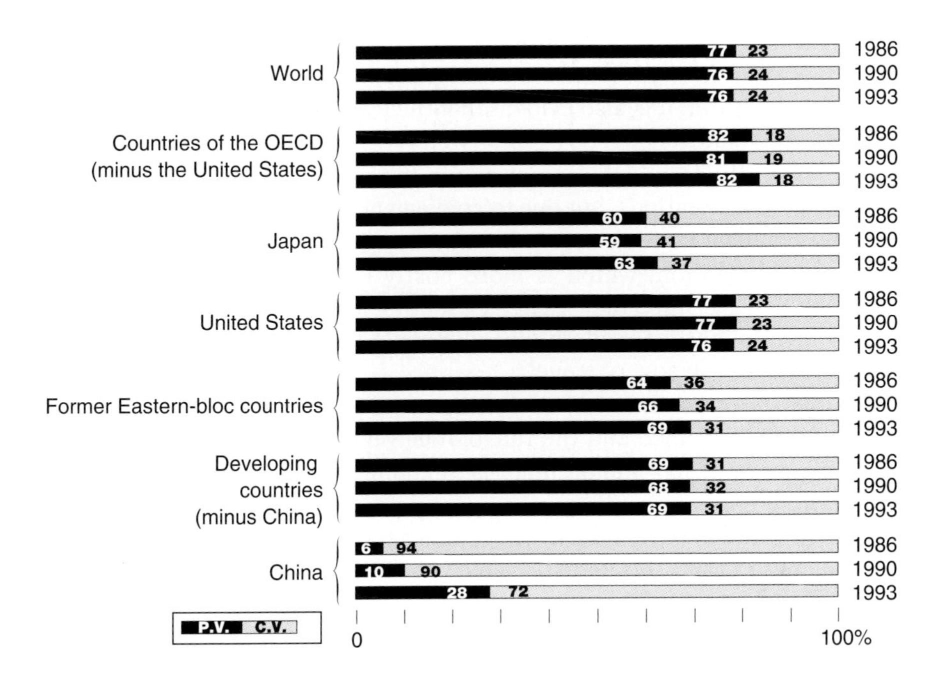

Figure 2 *Distribution of private versus commercial vehicles in the world vehicle fleet (in percent).*

trialized areas (Europe, USA, Japan), the distribution of these two large categories of vehicles will change very little in the near future. The distribution will level off in favor of private vehicles in the countries with the greatest promise of economic growth. For example, China, which in 1998 had more CVs than PVs, will see the situation reversed—likely before the year 2000.

Two-wheeled motorized vehicles also represent an important part of the world vehicle fleet. There are, in fact, almost 100 million two-wheelers in operation worldwide. Japan is the leader with about 18 million two-wheelers, which places it ahead of both Italy and the United States.

In some countries such as Thailand, Indonesia, India, Taiwan…, two-wheelers are the most common type of vehicle (60 to 70% of the total). In China, where almost 250 million bicycles were in operation in 1996, the prospect of a large increase in the use of motorcycles is evident, because the number has already reached 5 million.

The worldwide fuel consumption of two-wheelers is already approaching 20 million tonnes per year, which effectively represents the total automobile consumption for a European country like France, Italy, Germany, or the United Kingdom.

The road transport of people and goods is evidently not optimal on a **energy efficiency** basis, which is usually expressed as passenger-km (or tonne-km for goods) per kg equivalent of petroleum. The data shown in Fig. 3 provides some information on the subject, which is tempered with inherent reservations about the applicability of this type of overall estimate. Note that a private vehicle consumes more fuel, per passenger and per kilometer, than public transport by road or by water. It is also evident that the transport of goods by truck uses more energy than transport by rail or water.

Therefore, the formidable expansion in the use of road vehicles is not due to rational considerations, but due to convenience, speed, attractiveness, comfort, and personal liberty.

Air transportation, which is more costly on an energy basis, has also expanded to a considerable extent. No doubt it will continue to grow because it is an integral part of modern life, whether it is used for business travel or for vacation and discovery travel.

Frequency of use and **driving conditions** are important factors that affect the total fuel consumed and the rate-of-renewal of the vehicle fleet.

Although the total annual mileage accumulation of vehicles varies from country to country according to geography, lifestyle, or economic standards, it remains within a fairly narrow range of 9000 to 16 000 km per year (5593 to 9944 miles per year). This statistic, which has increased slightly over the years, should stabilize, particularly in areas of high automobile density.

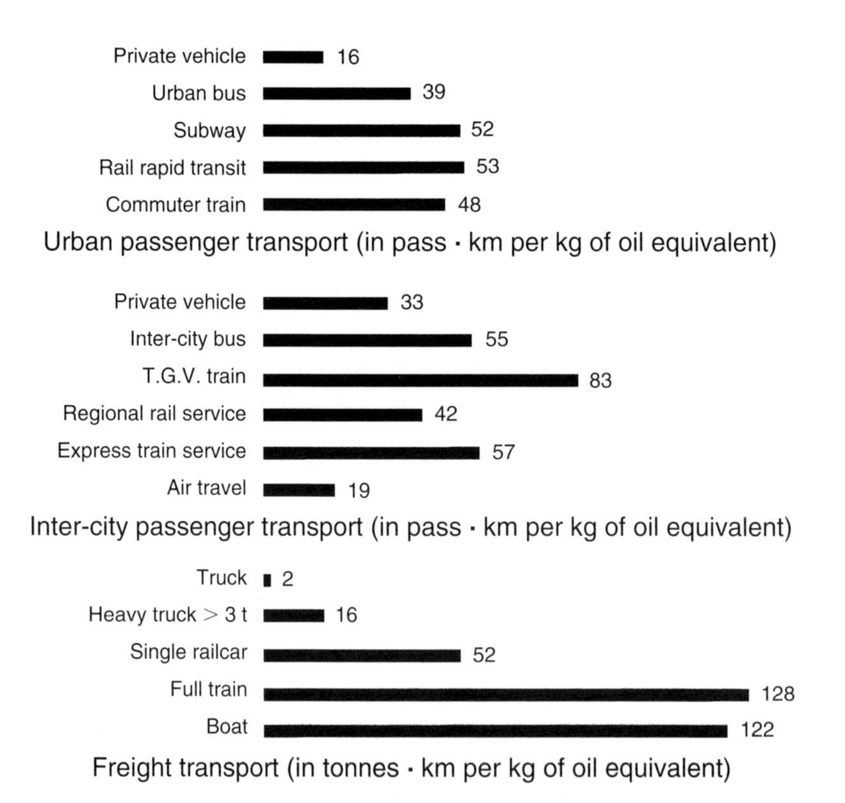

Figure 3 *Energy efficiency of various modes of transport.*

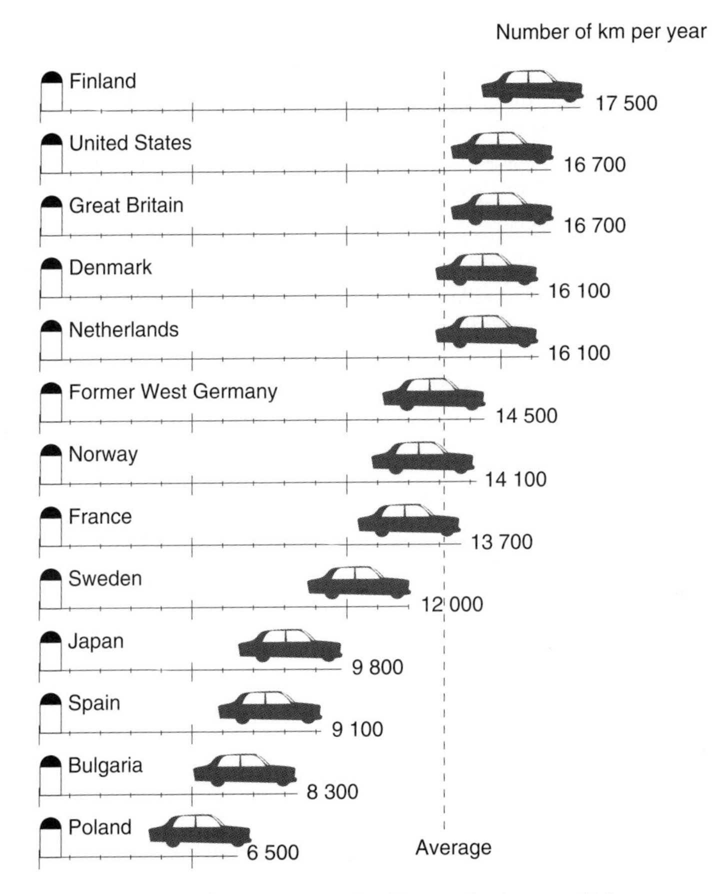

Figure 4 *Average annual mileage of private vehicles.*

The average annual mileage of commercial vehicles is usually very high and it often exceeds 100 000 km per year. The road transport of goods, expressed as tonnes-per-km per year, is expanding throughout the world, particularly in the developing countries.

A private vehicle will accumulate between 100 000 and 200 000 km in its lifetime, but a truck can travel over one million km. The average lifetime of both types of vehicles is usually more than 10 years, which means that the renewal of the vehicle fleet occurs in a **slow** and **progressive** manner. Throughout the book, we will often mention this situation, particularly in relation to all the activities undertaken to reduce pollution and the other nuisances related to transportation.

Driving conditions have changed considerably in 25 years, particularly in Europe. Highway driving (80 to 100 km/h) has given way to urban/suburban driving (between 20 and 60 km/h) (Lamure 1995) and freeway driving (over 100 km/h). We estimate that the average European private vehicle consumes 35% of its fuel in urban areas, 40% on freeways, and 25% on highways. Light duty commercial vehicles are used primarily on the periphery of cities and heavy vehicles are used mostly on the freeways.

Although these statistics are essential to establish the energy and environmental balance for transportation, other statistics such as speed limits should be added to the list. Speed limits are much lower in the US and Japan (from 90 to 110 km/h) than they are in Europe (130 km/h or more). Fuel consumption increases by 30 to 40% for the same vehicle when its speed increases from 90 to 130 km/h.

The size and weight of vehicles, and the resulting engine size, are important parameters. Fuel consumption can be doubled or tripled when a small vehicle with a 1000 cm^3 engine is compared to a larger vehicle with an engine of over 2000 cm^3. This latter type of vehicle is representative of the US vehicle fleet (over 80% of the total), but it is rarely found in Europe (less than 10%).

Matching engines and fuels: the priority considerations

Until 1970, the most sought after qualities of fuels were the ones that delivered the **best performance** in terms of starting, lack of stalling, acceleration performance, and maximum speed. All possible means were used to improve the physical qualities of gasoline (volatility, vapor pressure) and to increase its octane rating, which included the addition of lead alkyls (up to 0.8 g Pb/liter).

The energy crisis of 1973 was the beginning of major research and development efforts on the part of manufacturers—primarily European—to reduce the **specific fuel consumption** of vehicles. Gains of 20 to 30% were achieved in 20 years, along with other improvements that were related to appearance, reliability, and safety.

At first, these activities were directed to the engine itself (fuel system, combustion), and then, to the overall vehicle design (power plant, transmission, aerodynamics ...).

At the end of the 1970s, a parallel requirement appeared with the need to quickly reduce **atmospheric pollution of automotive origin**. At the same time, the lead content of gasoline in the United States and Japan was reduced to permit the use of catalytic converters. The same situation ensued in Europe, where increasingly tighter emission controls were mandated during the 1980s.

The period from 1985 to 1995 was undoubtedly identified as the time when a series of initiatives were introduced to reduce the tailpipe emissions (in g/km) from vehicles. Significant developments included the worldwide adoption of unleaded gasoline and low-sulfur diesel fuels, the emergence of alternative fuels (natural gas), the introduction of reformulated gasoline in the United States, and the development of vehicles (especially in Japan) that meet very-low emission standards——the California standards for Ultra Low Emission Vehicles (ULEV).

The period that is just beginning, which will undoubtedly extend beyond the year 2000, will bring new preoccupations with it. These will include the **greenhouse effect**, which is linked in an unexclusive but significant way to the atmospheric release of CO_2 formed by the combustion of fossil fuels (petroleum, natural gas, coal).

The worldwide CO_2 emissions that are shown in Fig. 5. are expected to double between 1970 and 2010. One estimate, which is based on various Organization for Economic Cooperation and Development (OECD) countries, indicates that transportation is responsible for about 30% of the emissions from the various sources of CO_2.

These considerations led to important efforts that were aimed at reducing **transportation-related energy consumption** in all its forms, and to the development of energy sources that were not based on fossil fuels (biofuels).

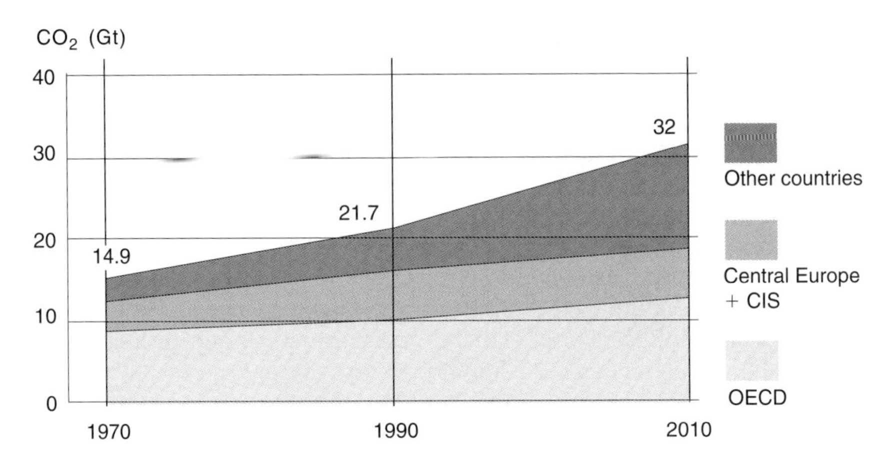

Figure 5 *Worldwide emissions of CO_2. (Source: AIE)*

On the eve of the year 2000, attention is equally focused on the ease of use fuels and on environmental protection at the local level—mainly in urban areas. This situation provides a new challenge for fuels, engines, and conventional vehicles, which must now compete with alternative solutions (the use of natural gas and electricity, and the development of public transportation).

The production of fuels

The major types of fuels—gasoline, jet fuel, and diesel fuel—are primarily petroleum-based, as they have been for the past 30 years. However, some energy products—usually heavy fuels—have faced competition from and, in some cases, have been replaced by new sources. Domestic fuel oil, for example, is being progressively replaced by natural gas and electricity. The use of heavy fuel oils has also dropped off sharply in some countries like France, since the advent of nuclear power.

Since 1970, there has been a notable increase in the petroleum fractions devoted to the formulation of motor fuels. This evolution, which is occurring in various regions of the world as shown in Fig. 6, will certainly continue through the early years of the next millennium.

This situation has a powerful effect on refinery schemas. Although **conversion procedures** are complex and expensive to implement, their implementation in the operation of refineries is increasingly necessary to convert

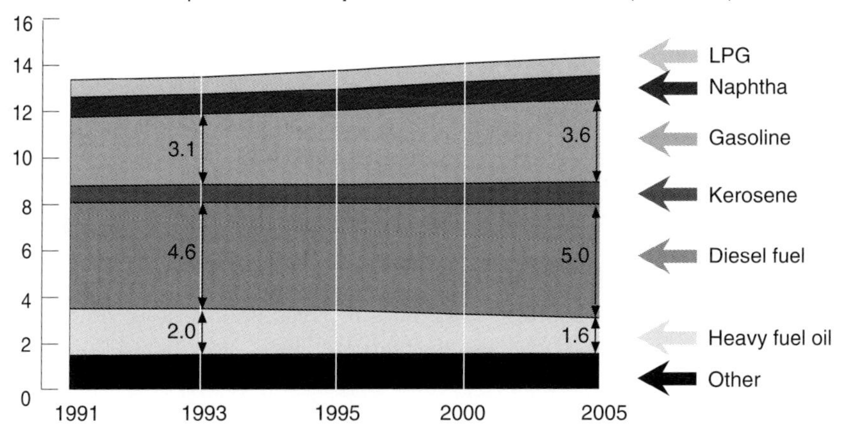

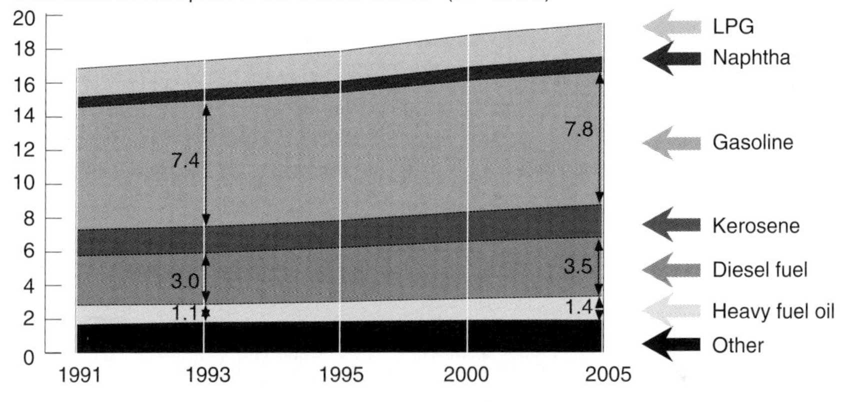

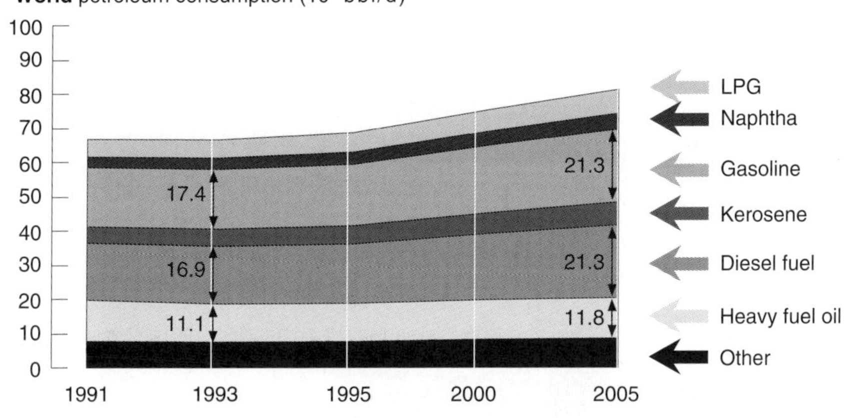

Figure 6 *The foreseeable evolution of the balance of petroleum production in various regions of the world. (Source: IEA)*

heavy feedstocks into medium and lighter fractions. These changes to refinery schemas have an effect on **product quality**. This effect is mentioned several times throughout the book.

To a certain extent, the diesel fuel/gasoline ratio within the motor fuels market is a parameter that defines the structure of refining. Typical ratios for various countries are shown in Fig. 7 and Fig. 8. The differences are important, since they vary considerably from 0.47 in the United States to 1.32 in France and to 1.40 in Belgium, where the use of diesel powered automobiles is the highest in the world.

During the 1970s, the formulation of products primarily involved **separation**, which was usually done by distillation. Formulation now requires the increased use of chemical procedures (hydrogenation, desulfurization, alkylation, isomerization, ...), which fall more and more into the realm of petrochemistry.

In a modern refinery, there are multiple interconnections between the production processes for gasoline and diesel fuel. For instance, the hydrogen resulting from the reforming of gasoline can be put to profitable use in the hydrodesulfurization of diesel fuel.

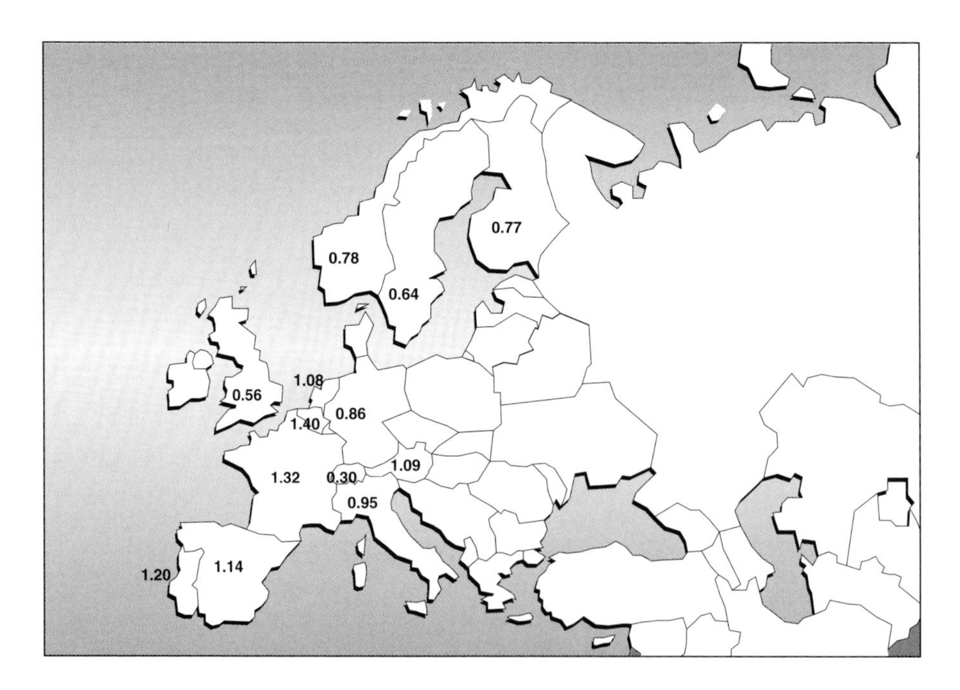

Figure 7 *Diesel fuel/gasoline ratios (by mass) for various countries in Europe.*

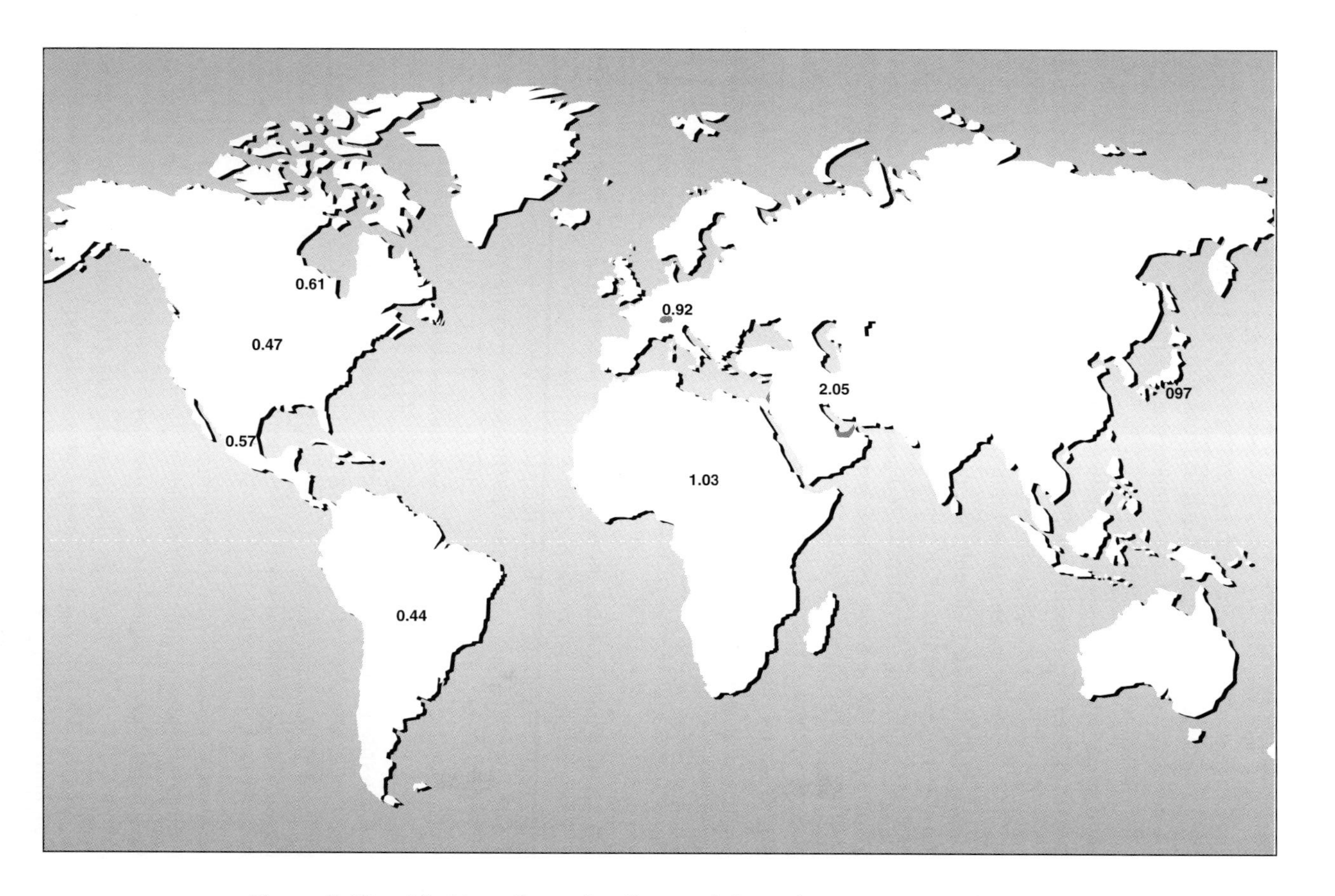

Figure 8 *Diesel fuel/gasoline ratios (by mass) for various countries in the world.*

Environmental protection

All the important changes made to engines and fuels between 1970 and 2000, have been mandated to reduce pollution from the use of automobiles. These changes have evolved over the years due to the requirement to meet ever more stringent regulations. The following changes, which affect the production of fuels, are costly and constraining for the refining industries:

- **Reduced lead-content** in gasoline will be in effect throughout the world soon after 2000. The use of lead content in gasoline had reached its peak in 1975 at 375 Mt/yr and it had fallen to less than 100 Mt/yr by 1998.
- The **desulfurization** of fuels, especially diesel, has reduced sulfur content by a factor of ten in 20 years.
- The introduction of **reformulated gasoline** in the United States will be followed in other areas of high automobile density, such as South-East Asia. These products, as described later, are subject to very strict rules governing their formulation and characteristics.
- **Additives** will be used to keep fuel systems and combustion chambers clean, which avoids a performance degradation of antipollution devices.

Likewise, all the engine research and development over the last 30 years has been focused on obtaining the lowest level of pollutants at the source—the engine exhaust.

Other more-sophisticated techniques have simultaneously been put into place. These techniques include the post-treatment of emissions from gasoline engines using catalytic converters, the study of devices that reduce the emission of oxides of nitrogen (NO_x) and particulates from diesel engines, and the recent work on new lean-burn engines; these techniques demonstrate some of the efforts and the progress made to date in this field.

During the 1990s, we became more aware of the effects of two-wheeled vehicles on atmospheric pollution, especially in South-East Asia. These vehicles, when they are equipped with older-generation two-stroke engines, emit an average of 20 g of pollutants (carbon monoxide, hydrocarbons, and NO_x) per km, which contrasts with the 1.5 g/km or less emitted by automobiles constructed to 1996 European standards.

This awareness led to the development of new motorcycle engines (**two-stroke direct-injection**) that deliver better performance and lower emissions. In some cases, these engines are equipped with catalytic converters.

Environmental protection was also an objective that led to the development of new directions for the use of alternative fuels such as LPG, natural gas, methanol, and biofuels.

These efforts will certainly help to bring about improvements, but they will not ensure a complete and perfect protection of the environment. Therefore, during the years between 2000 and 2010, we must again design, develop, and put in place new measures to conserve and extend the progress we have already achieved.

This brief introduction demonstrates that the following themes are important: reduced energy consumption, environmental protection, and improved comfort and driveability. These criteria are used in this book to establish the level of technical advancement and the current state of engines and fuels.

Chapter 1

The physical properties and chemical characteristics of fuels

Crude oil, and the petroleum products refined from it, contain mainly atoms of **carbon** and **hydrogen** combined in chemical compounds called hydrocarbons. Other elements such as sulfur, nitrogen, oxygen, and certain metals are also present in low concentrations or as trace elements in the heavier fractions.

Some of the basics of hydrocarbon chemistry are discussed in this chapter. The nomenclature and the rules of classification are introduced, as well as some of the structures of the various families of organic chemistry. These families include the oxygenated, sulfurized, and nitrogenated products that are frequently used along with hydrocarbons in the various applications of fuels. The chapter concludes with some information about the composition and the general properties of commercial fuels.

1.1 A review of the chemistry of hydrocarbons

This section contains only general guidelines. More complete information can be found in the references on organic chemistry (Niviere 1993; McMurry 1994).

The carbon and hydrogen atoms that form hydrocarbons establish chemical bonds between themselves by using electrons. The simplest bond is an association of two electrons. Carbon is **quadrivalent** and hydrogen is **monovalent**. This means that an atom of carbon can create four bonds with adjacent elements, either carbon or hydrogen, while an atom of hydrogen can form only one bond with carbon.

There can be one, two, or three bonds between carbon atoms. These bonds can be single, double, or triple bonds, as indicated by the following symbols:

$$-\overset{|}{\underset{|}{C}}-\overset{|}{\underset{|}{C}}- \qquad >C=C< \qquad -C\equiv C-$$

Each bond involves two, four, or six electrons respectively. Hydrocarbons are classified according to the types of carbon bonds in their structure. Each class has particular physical and chemical properties.

1.1.1 General classification

Three major families of hydrocarbon compounds are presented in Table 1.1:

- **Saturated hydrocarbons**, which are made up of single carbon-to-carbon bonds, are divided into two categories:
  - **Paraffins** or alkanes, which have open carbon bonds.
  - **Naphthenes** or cyclanes, which have closed bonds formed in a ring.
- **Unsaturated hydrocarbons** divide themselves into two distinct groups:
  - **Olefins**, which have one or more double bonds, are referred to in official terms as alkenes or cyclenes, depending on their molecular configuration as either chains or rings.
  - **Acetylenics** or alkynes are identified by at least one triple bond. In this family, ring structures, which are referred to as cyclenes, are very rare.
- **Aromatic hydrocarbons** consist of one or several unsaturated cyclic configurations that contain six carbon atoms—like benzene.

Table 1.1 *General classification of hydrocarbons.*

Family type	Common name	Official name	Structure	General formula
Saturated	Paraffins	Alkanes	$-\overset{|}{\underset{|}{C}}-\overset{|}{\underset{|}{C}}-$	C_nH_{2n+2}
	Naphthenes	Cyclanes	Ring of 3, 4, 5 or 6 carbon atoms	C_nH_{2n}*
Unsaturated	Olefins**	Alkenes	$>C=C<$	C_nH_{2n}***
	Acetylenic	Alkynes	$-C\equiv C-$	C_nH_{2n-2}
Aromatics	Aromatics	–	(benzene ring)	C_nH_{2n-6}****

* Formula valid only for compounds with a single ring.
** Cycloolefins, diolefins, and polyolefins are classed in this category.
*** Formula valid only for non-cyclic monoolefins.
**** Formula valid only for compounds with a single aromatic ring and a laterally saturated bond.

Another older system of classification consists of distinguishing between hydrocarbons that are aliphatic, alicyclic, or aromatic. Aliphatic compounds are all the structures with open chains, which are saturated or unsaturated (paraffins, olefins, acetylenics). This designation can be extended to other families of organic products such as alcohols, aldehydes, ketones, and acids. Alicyclic hydrocarbons include products such as cyclanes, cycloolefins, and cyclodiolefins.

The International Union of Pure and Applied Chemistry (IUPAC) has established a nomenclature (Panico 1994) that provides for the systematic designation of organic compounds. This system of nomenclature is used in the following sections.

1.1.2 Chemical formulas and nomenclature

1.1.2.1 Paraffins or alkanes

Paraffin and alkane hydrocarbons have the following general formula: C_nH_{2n+2}.

These hydrocarbons can be

- **straight** chains: C—C—C—C
- **branch** chains:

```
C—C—C—C—C—C—C
      |  |
      C  C
      |
      C
```

The straight chains are referred to as **normal paraffins** (*n*-paraffins) and the branch chains are referred to as **isoparaffins**. Two hydrocarbons with the same general formula, but which differ in arrangement, are called isomers. Each isomer can have distinct physical and chemical properties.

The first four members of the paraffin family are known by their usual names: methane, ethane, propane, and butane. The formulas for these compounds are shown in Table 1.2.

Branching is possible after butane. The two isomers with four carbon atoms (C_4) are normal butane (*n*-butane) and isobutane.

To designate straight chain hydrocarbons with more than four atoms, a numbered prefix based on the number of atoms is used (penta, hexa, deca, ...), along with the suffix "ane" to indicate that the compounds belong to the alkane group.

The designation thus obtained is preceded by the term "normal" or the letter "*n*".

For example

- normal-octane or *n*-octane: C_8H_{18}
- *n*-undecane: $C_{11}H_{24}$
- *n*-pentadecane: $C_{15}H_{32}$
- *n*-eicosane: $C_{20}H_{42}$

Moving to a higher order requires the addition of a $—CH_2—$ link.

The nomenclature for **branched paraffins** follows certain rules that are explained here.

Table 1.2 *Terms used for the first members of the paraffin group.*

<table>
<tr><th rowspan="2">Name</th><th colspan="3">Type of structure</th></tr>
<tr><th>Developed</th><th>Semi-developed</th><th>General</th></tr>
<tr><td>Methane</td><td>II
|
H—C—H
|
H</td><td>-</td><td>CH_4</td></tr>
<tr><td>Ethane</td><td>H H
| |
H—C—C—H
| |
H H</td><td>$CH_3 - CH_3$</td><td>C_2H_6</td></tr>
<tr><td>Propane</td><td>H H H
| | |
H—C—C—C—H
| | |
H H H</td><td>$CH_3 - CH_2 - CH_3$</td><td>C_3H_8</td></tr>
<tr><td>Butane</td><td>H H H H
| | | |
H—C—C—C—C—H
| | | |
H H H H</td><td>$CH_3 - CH_2 - CH_2 - CH_3$</td><td>$C_4H_{10}$</td></tr>
<tr><td>Isobutane</td><td>H H H
| | |
H—C—C—C—H
| | |
H | H
H—C—H
|
H</td><td>$CH_3 - CH - CH_3$
|
CH_3</td><td>C_4H_{10}</td></tr>
</table>

The principal chain of branched paraffins is, by definition, the longest and carries the name of the corresponding normal hydrocarbon. The lateral chains, or alkyls, are referred to by specific terms—ethyl, propyl, butyl, etc.—where the root of the term refers to the alkane with the same number of carbon atoms; the root is followed by the suffix "yl". When alkyls are not described in detail, they are referred to as **radicals** and denoted by the letter R. Because the carbon atom attached to the principal chain is also linked to one, two, or three other carbon atoms, the radical is referred to as **primary**, **secondary**, or **tertiary**:

- **primary**: $R-CH_2-$
- **secondary**: R—CH—R'
|
- **tertiary**: R'
|
R—C—R''
|

Table 1.3 shows the names of some frequently encountered alkyl radicals. These radicals are considered to be entities and use the suffix "yl".

Table 1.3 *Nomenclature of some radicals alkyl.*

Name	Formula	Name	Formula
Methyl	CH_3-	*n*-Pentyl (or *n*-amyl)	$CH_3-CH_2-CH_2-CH_2-CH_2-$
Ethyl	CH_3-CH_2-	Isopentyl (or iso-amyl)	CH_3 \\ $CH-CH_2-CH_2-$ / CH_3
Propyl	$CH_3-CH_2-CH_2-$	Neopentyl	CH_3 | CH_3-C-CH_2- | CH_3
Isopropyl	CH_3 \\ $CH-$ / CH_3	*tert*-Pentyl (or *tert*-amyl) (or *t*-amyl)	CH_3 | CH_3-CH_2-C- | CH_3
n-Butyl	$CH_3-CH_2-CH_2-CH_2-$		
sec-Butyl (or secondary butyl)	CH_3-CH_2 \\ $CH-$ / CH_3		
tert-Butyl (or *t*-butyl)	CH_3 | CH_3-C- | CH_3		

To designate a branched paraffin, each carbon atom in the principal chain is given a number that begins with the end of the chain that is the closest to a branch. The reference number for each branching point is the number of the carbon atom at that point on the principal chain.

For example:

```
 1     2     3     4     5     6     7
CH3 — CH — CH2 — CH2 — CH — CH2 — CH3
       |                 |
      CH3               CH2
                         |
                        CH3
```

2-methyl-5-ethylheptane

As a general rule, the number of possible branches, or isomers, increases very quickly with the number of carbon atoms. Therefore, there are three isomers for C_5, five for C_6, eighteen for C_8, seventy-five for C_{10}, and theoretically over $6.2 \cdot 10^{13}$ for C_{40}.

The prefix "iso", which is often used to indicate a branched paraffin, is insufficient to describe the structure of a compound. Therefore, 2,2,4-trimethylpentane

```
       CH3
        |
CH3 — C — CH2 — CH — CH3
        |          |
       CH3        CH3
```

is only one of the many isoparaffins of C_8; it is also one of the products used as a standard reference to determine octane ratings and it is commonly referred to as "isooctane."

1.1.2.2 Naphthenes or cycloparaffins

Naphthenes are ring structures where each carbon atom is linked to two hydrogen atoms and to two carbon atoms. This arrangement applies to single-cycle structures. The general formula is C_nH_{2n}.

The prefix "cyclo" applies to naphthenes and it is followed by the name of the alkane with the appropriate number of carbon atoms.

For example:

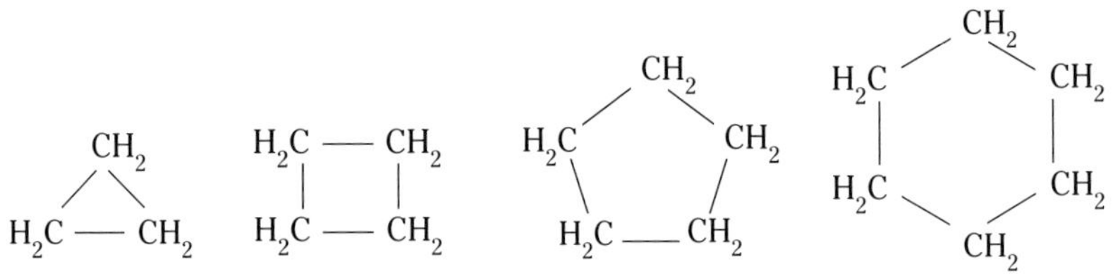

cyclopropane cyclobutane cyclopentane cyclohexane

Naphthenes are usually represented symbolically by using neither the cyclic carbon atoms nor the hydrogen atoms. Therefore, the following symbols

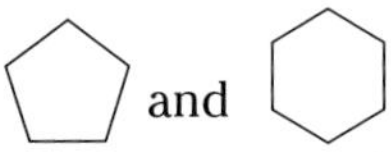

and

represent cyclopentane and cyclohexane respectively.

Naphthenes can also consist of alkyl chains bonded to a ring. In that case, the cyclic carbon atoms are numbered in a way that describes the various substituents by using the lowest possible values.

For example:

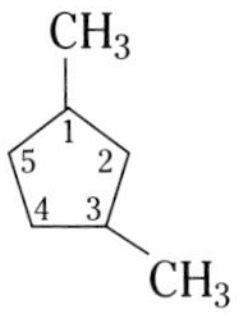

1,3-dimethylcyclopentane (not 1,4-dimethylcyclopentane)

There are also naphthenes that consist of several condensed rings. The most common configuration is a compound with two C_6 rings.

This compound is **decaline**,

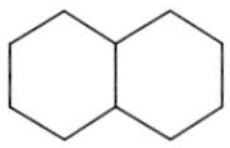

which has the general formula $C_{10}H_{18}$.

1.1.2.3 Olefins or alkenes

The olefin group consists of two types of compounds, one with double bonds (**monoolefins**) and the other with multiple bonds (**polyolefins**).

A. Monoolefins

These compounds have two less hydrogen atoms than the corresponding saturated compounds. The general formula is C_nH_{2n} for open chains (alkenes) and C_nH_{2n-2} for ring compounds (cyclenes). The following explanation is directed primarily at alkenes, since they are more widely distributed than cyclenes.

The simplest olefins are:

- ethylene or ethene: $H_2C{=}CH_2$
- propylene or propene: $CH_3{-}CH{=}CH_2$

Olefins use the same nomenclature as paraffins, from C_4 onwards, and replace the "ane" suffix with "ene": pentene, hexene, octene, etc.

Olefin branches are described in the same way as alkanes. However, the position of the double bonds in the principal chain must be identified. This is indicated by preceding the "ene" suffix with the number of the first carbon atom with a double bond.

The numbering system is applied in a way that ensures that the lowest number is used:

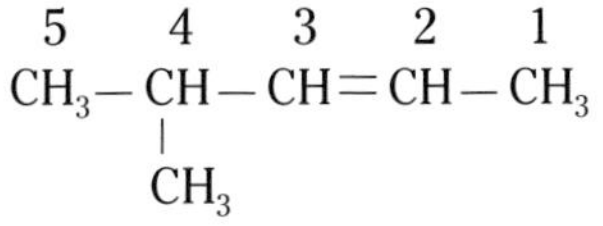

```
 5     4     3     2     1
CH3—CH—CH=CH—CH3
       |
      CH3
```

4-methyl-pent-2-ene

Therefore, for the same carbon chain, several isomers can occur depending on the location of the unsaturation.

However, **geometric isomers** can exist because double bonds do not permit the free rotation of hydrogen or substituent alkyls. These isomers are found in the same plane and they are referenced by their position as either "cis" (same side) or "trans" (opposite sides).

For example:

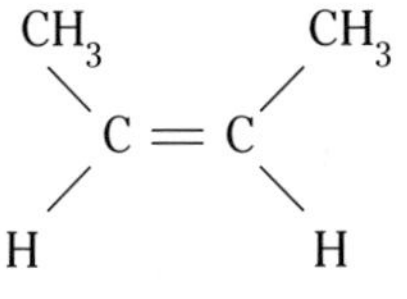

```
CH3        CH3          CH3        H
   \      /                \      /
    C = C                   C = C
   /      \                /      \
  H        H              H        CH3
```

(*Z*)-but-2-ene (or *cis*-but-2-ene) — (*E*)-but-2-ene (or *trans*-but-2-ene)

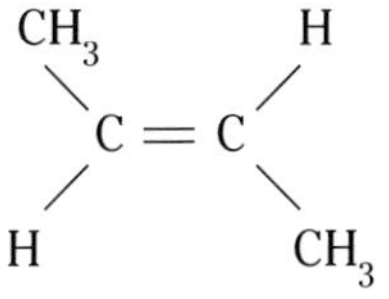

These two products are isomers of

- but-1-ene: $CH_3{-}CH_2{-}CH{=}CH_2$
- isobutene: $CH_3{-}C(CH_3){=}CH_2$

```
isobutene: CH3—C=CH2
               |
              CH3
```

B. Polyolefins

Polyolefin hydrocarbons are named according to their double bonds, such as **dienes**, **trienes**, etc. These double bonds can be adjacent ones, which are referred to as "allenic":

$$-\overset{|}{\underset{|}{C}}-\underset{|}{C}=C=\underset{|}{C}-\overset{|}{\underset{|}{C}}-$$

or they can be arranged alternately with single bonds, which are referred to as "conjugated":

$$-\overset{|}{\underset{|}{C}}-\underset{|}{C}=\underset{|}{C}-\underset{|}{C}=\underset{|}{C}-\overset{|}{\underset{|}{C}}-$$

The products most frequently encountered or quoted in petrochemistry are

- allene (or propadiene) $H_2C=C=CH_2$
- buta-1,3-diene $CH_2=CH-CH=CH_2$
- Isoprene (2-methylbuta-1,3-diene) $CH_2=\underset{\substack{|\\CH_3}}{C}-CH=CH_2$
- cyclopentadiene

The compound buta-1,3-diene,[1] which is emitted in small quantities in exhaust gas, is monitored very closely because of its toxicity.

1.1.2.4 Acetylenics

Acetylenic hydrocarbons, which have triple bonds, are represented by the general formula C_nH_{2n-2}. They are referred to in the nomenclature by the suffix "yne" and they follow the same general rules as the preceding families of compounds. The first compound in the series has retained its historical name of acetylene and it is represented as $H-C\equiv C-H$.

There are two types of acetylenics:

- "real" or **monosubstitute** acetylenics: $R-C\equiv C-H$
- **disubstituted** acetylenics: $R-C\equiv C-R'$

Most of these compounds rarely appear in fuels. However, small amounts of acetylene are found in exhaust gas.

1.1.2.5 Aromatics

The first compound in the aromatics group is **benzene**, which has the general formula C_6H_6.

The six carbon atoms are apparently joined by either single bonds or double bonds. In fact, the bonds are arranged as a six–member ring with three

1. Henceforth, this hydrocarbon will be referred to by its common name of 1,3-butadiene or simply butadiene. In fact, buta-1,3-diene has an isomer known as buta-1,2-diene and represented by the formula $CH_3-CH=C=CH_2$.

conjugated double bonds. That is why the following symbols are used to represent the compound:

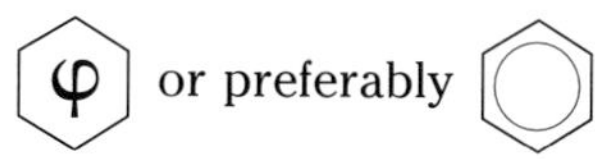

There are no precise rules for designating the nomenclature of aromatics, which have names that are already in general use. Table 1.4 provides a list of simple compounds that consist of one or several aliphatic groups related to the benzene ring.

Depending on the location of the attachment of the two CH_3 groups, the following three isomers of benzene occur: orthoxylene, metaxylene, and paraxylene. The prefixes "ortho", "meta", and "para" are used in organic chemistry to designate the relative locations or the various possible substituents of benzene.

The term "phenyl", which is given to the C_6H_5 radical derived from benzene, is sometimes used to refer to aromatic hydrocarbons with several rings.

For example:

CH_2

biphenyl   biphenylmethane

The benzene ring can also be fused to another aromatic or naphthalenic ring:

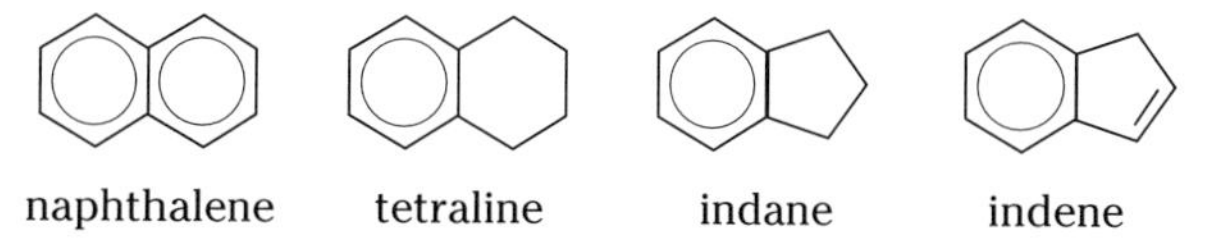

naphthalene   tetraline   indane   indene

There is also a group of very-complex aromatic hydrocarbons that consist of several rings of various diverse forms. These compounds are know as **polynuclear aromatic hydrocarbons** (PAH). Table 1.5 provides a list of some PAHs, which occur in very small amounts in the combustion products of fuels, and they deserve special attention because of their toxicity (Degobert 1995).

The hydrogen content of these compounds diminishes with the growth in the number of rings. The most complex structures are almost exclusively carbon and they begin to approach the composition of graphite.

1.1.3 Physical properties

At ambient temperature, **paraffins** are gases up to C_4, liquids between C_4 and C_{16}, and solids above C_{16}. However, even for the long–chain compounds, the fusion point does not exceed 100°C. As shown in Fig. 1.1, the boiling point and the density of the liquids increases along with the number of carbon atoms.

The melting and the boiling points are reduced by the introduction of branches to the principal chain; for example, *n*-pentane boils at 36°C, isopentane at 28°C, and neopentane at 9.5°C.

Table 1.4 *A listing of some simple aromatic hydrocarbons.*

Name	Formula	Chemical structure (carbon backbone)
Benzene	C_6H_6	
Toluene	C_7H_8	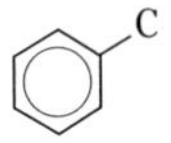 C
Ethylbenzene	C_8H_{10}	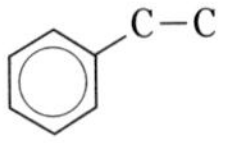 C–C
o-Xylene	C_8H_{10}	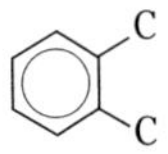 C C
m-Xylene	C_8H_{10}	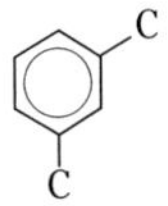 C C
p-Xylene	C_8H_{10}	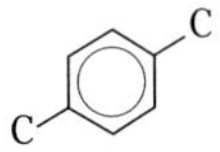 C C
Styrene (vinylbenzene)	C_8H_8	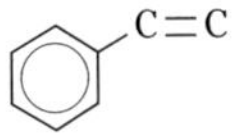 C=C
n-Propylbenzene	C_9H_{12}	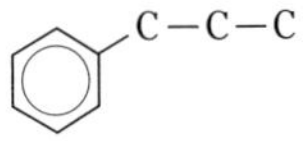 C–C–C
Cumene (isopropylbenzene)	C_9H_{12}	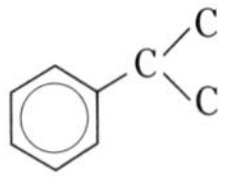 C C C
n-Butylbenzene	$C_{10}H_{14}$	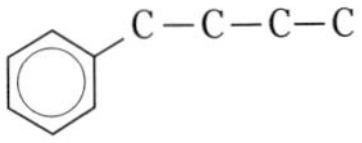 C–C–C–C
Durene (1,2,4,5-tetramethylbenzene)	$C_{10}H_{14}$	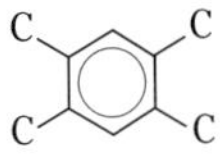 C C C C
Naphtalene	$C_{10}H_8$	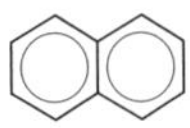
1-Methylnaphthalene (or α-methylnaphthalene)	$C_{11}H_{10}$	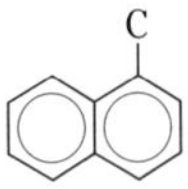 C
2-Methylnaphthalene (or *β*-methylnaphthalene)	$C_{11}H_{10}$	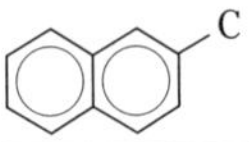 C

Table 1.5 *A partial list of polynuclear aromatic hydrocarbons.*

Name	Formula
Anthracene	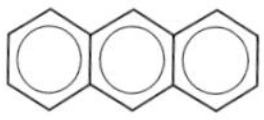
Phenantrene	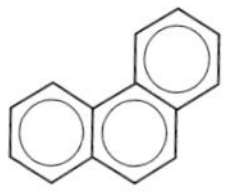
Acenaphtene	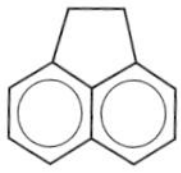
Fluorene	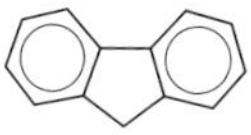
Chrysene	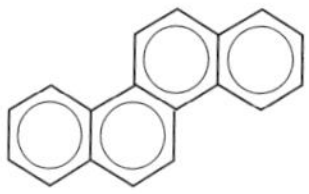
Benzo[a] anthracene	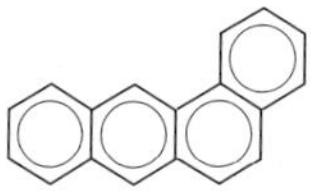
Cholanthrene	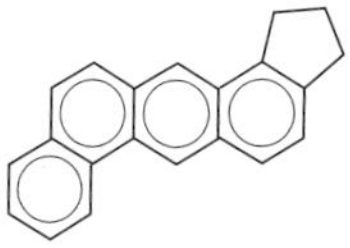
Pyrene	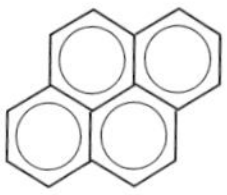
Fluoranthene	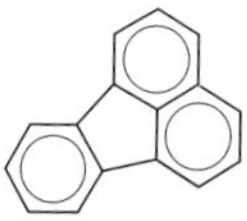
Benzo[k] fluoranthene	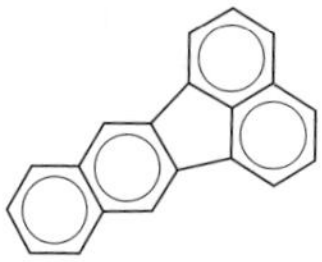
Naphtacene	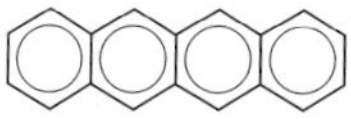
Benzo[j] fluoranthene	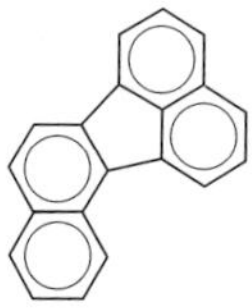
Benzo[b] fluoranthene	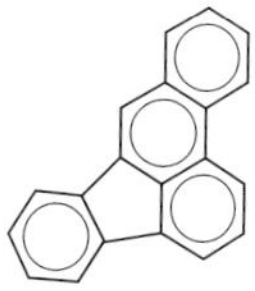
Benzo[a] pyrene	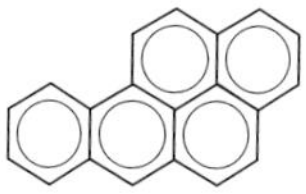
Benzo[e] pyrene	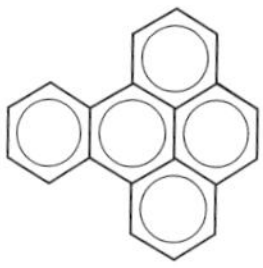
Perylene	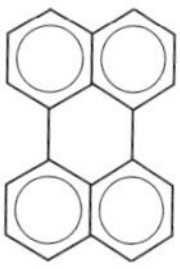
Benzo[g, h, i] perylene	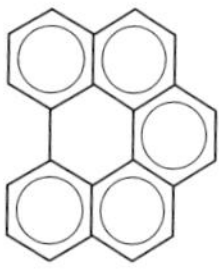
Dibenzo [a, h] anthracene	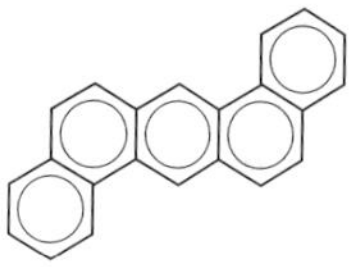
Indenopyrene	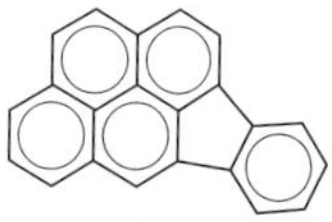
Coronene	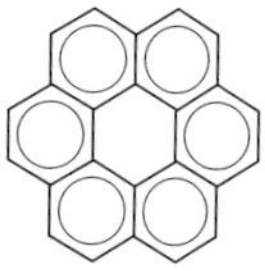
Anthranthene	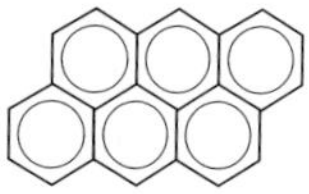

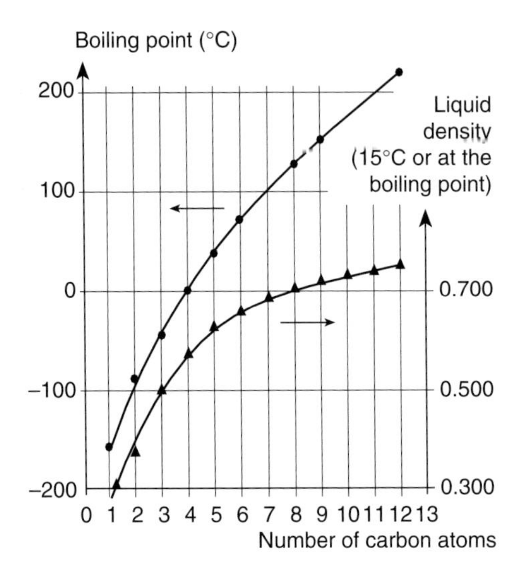

Figure 1.1
The boiling point and density of normal paraffin liquids.

Paraffins have a low density of around 0.700. Paraffins are insoluble in water but they are soluble in a number of organic products. However, simple alcohols are not readily miscible in alkanes.

For the same number of carbon atoms, the movement from paraffins to **naphthenes** tends to increase density, and to raise the fusion and boiling points. Cyclohexane boils at 81°C and *n*-hexane boils at 69°C; their densities at 15°C are 0.778 and 0.659 respectively.

For a given form of carbon structure, **olefins** have physical properties that are very close to those of paraffins. However, unsaturation tends to reduce the boiling point somewhat and to increase the density. Their solubility in water and alcohols is very low, but it is markedly superior to those of alkanes.

The double bond $\gt C=C \lt$ has an absorption characteristic in the far ultra-violet (about 18.5 nm) and infra-red (1620 to 1680 cm^{-1}) spectra.

Aromatic hydrocarbons are identified by their high density, which is greater than 0.870. Benzene has a higher density than toluene, xylene, or cumene. The boiling temperature of benzene is about 80°C. The boiling temperature of the other aromatics exceeds 100°C. Compounds with several benzene rings have a density that approaches 1.000 and they are frequently solid at ambient temperature. PAHs are found in the smoke and particulates emitted with exhaust gases.

Although the water solubility of aromatics is very low, it is better than that of other hydrocarbons. However, aromatics have a high solubility in alcohols.

Certain physical particularities of aromatics are also used in analytical techniques. For example, these products have a particularly high refractive index and they exhibit a very-high absorption in the ultra-violet range.

1.1.4 Chemical properties

Hydrocarbons are the primary material for most high-consumption chemical products. Evidently, it is impossible to describe here, even in summary, all the various processes used in petrochemistry. However, this section does provide some description of the major types of chemical reactions that can occur in the preparation and use of fuels.

1.1.4.1 Chemical reactivity

In a general sense, **paraffins are not very reactive**, which means that they do not readily participate in chemical transformations. This results from the fact that the C—C and C—H bonds in these compounds are very strong and difficult to break. The bonding energy is 19.9 kJ/mol for C—C and it varies from 21 to 24 kJ/mol for C—H.

As a result, paraffins bonds can only be broken by powerful reactants at high temperatures. Because the molecule is saturated, new elements cannot be added; for example, hydrogenation is not possible.

Naphthenes have characteristics that are similar to paraffins, which include the additional possibility of an opening in the ring structure and a certain tendency to transform into aromatics.

The **olefin** double bond is easily transformed into a single bond by adding about 14.3 kJ/mol of energy. Under these conditions, the free-valence carbon atoms can form a bond with the radicals provided by the reactant. This provides multiple opportunities for **addition reactions**. The double bond is also a vulnerable spot in the carbon chain because of its reactivity and the chain can break at that point during certain reactions—especially oxidation.

The special structure of the **benzene ring** provides a **very stable chemical structure**. Compared to a system with normal bonds, the delocalization of benzene electrons results in an 8.6 kJ/mol surplus in the energy of formation. This difference is called "resonance energy".

Aromatic compounds do not behave like traditional unsaturated hydrocarbons. Addition reactions are possible with this structure, but they are very difficult to accomplish because they cause the aromatic characteristics to disappear along with the corresponding stability. This also explains why the ring structure is not easily broken and why it is very resistant to the action of oxidants.

Aromatics frequently behave like saturated compounds and they are conducive to **substitution reactions**—the replacement of hydrogen with an element or chemical group from a reactant—either with the ring or with the lateral chains.

1.1.4.2 Major types of reactions

This section describes the transformations of hydrocarbons and leaves out the numerous petrochemical reactions that involve elements such as oxygen, nitrogen, sulfur, chlorine, etc.

Table 1.6 contains descriptions and schematics for the main reactions that enable the modification of hydrocarbon structures. Explanations are provided for some types of reactions. Industrial procedures of major importance, such as cracking, reforming, alkylation, isomerization, and hydrogenation, are described in more detail in Chapter 2, which is devoted to refining techniques.

A. *Isomerization*

Isomerization is the process of transforming from one isomer to another. It can be applied to all groups of hydrocarbons, but in the preparation of fuels, it is applied in particular to *n*-paraffins:

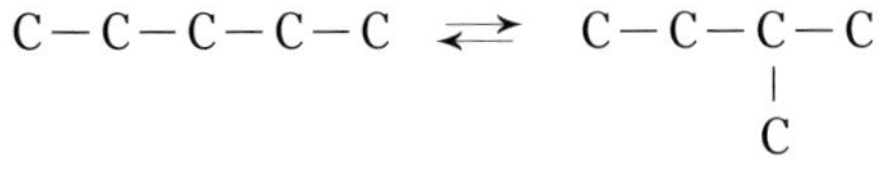

$$\mathrm{C-C-C-C-C} \rightleftarrows \begin{array}{l}\mathrm{C-C-C-C}\\ \qquad\;\; | \\ \qquad\;\; \mathrm{C}\end{array}$$

These are **balanced** reactions that lead to a complex range of products where several isomers often coexist. Isomerization can be a purely thermal process, but catalysts that operate between 100 and 200°C are favored.

B. *Alkylation*

Alkylation is the reaction of an olefin with a paraffinic or aromatic hydrocarbon. The **alkylation of isobutane by light olefins** is obtained in a refinery by using propylene or butenes that result in an isoparaffin mix of C_7 and C_8. For example:

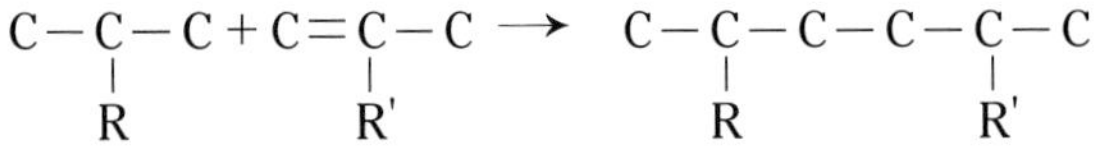

$$\begin{array}{ccccccccccccccccc}\mathrm{C} & - & \mathrm{C} & - & \mathrm{C} & + & \mathrm{C} & = & \mathrm{C} & - & \mathrm{C} & \rightarrow & \mathrm{C} & - & \mathrm{C} & - & \mathrm{C} - \mathrm{C} - \mathrm{C} - \mathrm{C}\\ & & | & & & & & & | & & & & & & | & & \quad\;\;\;\; |\\ & & \mathrm{R} & & & & & & \mathrm{R'} & & & & & & \mathrm{R} & & \quad\;\;\;\; \mathrm{R'}\end{array}$$

(or other isomers)

Alkylates are always complex mixtures, even if the reactants are pure hydrocarbons. The reaction is thermodynamically enhanced by reducing the temperature and increasing the pressure. Industrial operations use acidic catalysts (sulfuric or hydrofluoric acid) at less than 100°C and at a pressure of approximately 50 bar.

C. *Polymerization and oligomerization*

These reactions mainly involve olefins whose molecules can be assembled according to the following schema:

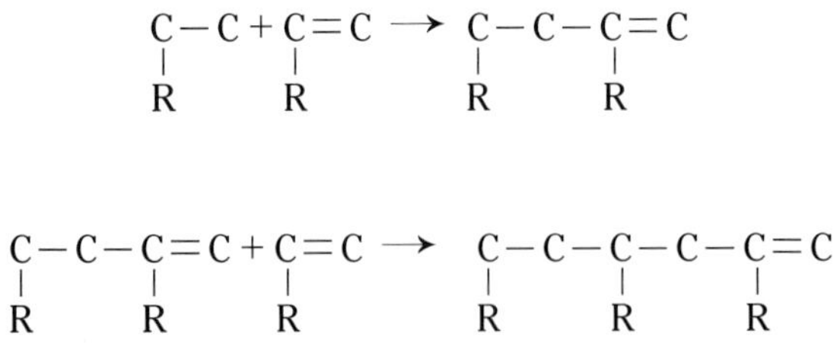

$$\begin{array}{ccccccccccccccc}\mathrm{C} & - & \mathrm{C} & + & \mathrm{C} & = & \mathrm{C} & \rightarrow & \mathrm{C} & - & \mathrm{C} & - & \mathrm{C} & = & \mathrm{C}\\ | & & & & | & & & & | & & & & | & & \\ \mathrm{R} & & & & \mathrm{R} & & & & \mathrm{R} & & & & \mathrm{R} & & \end{array} \qquad \text{dimer}$$

$$\begin{array}{ccccccccccccccccccccccc}\mathrm{C} & - & \mathrm{C} & - & \mathrm{C} & = & \mathrm{C} & + & \mathrm{C} & = & \mathrm{C} & \rightarrow & \mathrm{C} & - & \mathrm{C} & - & \mathrm{C} & - & \mathrm{C} & - & \mathrm{C} & = & \mathrm{C}\\ | & & & & | & & & & | & & & & | & & & & | & & & & | & & \\ \mathrm{R} & & & & \mathrm{R} & & & & \mathrm{R} & & & & \mathrm{R} & & & & \mathrm{R} & & & & \mathrm{R} & & \end{array} \qquad \text{trimer, ... etc.}$$

Table 1.6 *The main types of chemical reactions between hydrocarbons.*

Type of reaction	Preferred reactant	Products	Reaction schematic
Isomerization	*n*-Paraffin	Isoparaffins	$C-C-C-C-C \rightleftarrows C-C(C)-C-C$
Alkylation	Paraffin + olefin	Isoparaffins	$C-C(C)-C + C=C(C) \rightarrow C-C-C(C)-C-C(C)$
Polymerization (oligomerization)	Olefins	Olefins	$C(R)=C + C(R)=C \rightarrow C(R)-C-C(R)=C$
Cracking	Paraffins, aromatics*, naphthenes	Olefins + other hydrocarbons	$R-CH_2-CH_2-CH_2-R' \rightarrow R-CH=CH_2 + R'-CH_3$
Hydrocracking	Paraffins, aromatics*	Paraffins and eventually aromatics	$R-CH_2-CH_2-R' + H_2 \rightarrow R-CH_3 + R'-CH_3$ $C_6H_5-CH_2-R + H_2 \rightarrow C_6H_6 + R-CH_3$
Hydrogenation	Olefins, aromatics	Paraffins, naphthenes	$R-CH=CH_2 + H_2 \rightarrow R-CH_2-CH_3$ $C_6H_5-R + 3\,H_2 \rightarrow C_6H_{11}-R$
Dehydrogenation (aromatization)	Naphthenes, paraffins	Aromatics	$C_6H_{11}-R \rightarrow C_6H_5-R + 3\,H_2$ $R-CH_2-CH_2-CH_2-CH_2-CH_2-CH_3 \rightarrow C_6H_5-R + 4\,H_2$

* Straight chain

This is the principal of macromolecule synthesis that is so important in petrochemistry (polyethylene, polypropylene). Refining provides for the controlled dimerization and trimerization of light olefins into C_3 and C_4, which yields a mixture of branched olefins from C_6 to C_9. This transformation is called **oligomerization**.

D. Cracking

As the name suggests, cracking consists of breaking some carbon-carbon bonds to obtain **light hydrocarbons** from a heavier feedstock (feed). In reality, cracking induces very complex reactions, especially condensations, where the light-product output is accompanied by products that are heavier than the initial reactants. There are multiple industrial uses for this type of transformation:

- **pure thermal cracking**
- **steam cracking (in the presence of water vapor)**
- **catalytic cracking**
- **visbreaking**
- **coking**

These processes are examined in depth in subsequent sections.

Cracking produces olefins and it is represented by the following equation,

$$R-CH_2-CH_2-CH_2-R' \rightarrow R-CH=CH_2 + R'-CH_3$$

where paraffin is the reactant. In reality, all the chemical families of hydrocarbons can be subjected to cracking.

Cracking is a very **endothermic** reaction that occurs at high temperature, between 400 and 1 000°C, depending on the nature of the feedstock and the characteristics of the procedure (thermal or catalytic). The reaction always produces **coke** as a sub-product, which results from the complete scission of the C — C and C — H bonds. Reduced pressure improves cracking. Using water vapor to dilute reactants, which is done in steam cracking, is beneficial and brings the added advantage of directing the reaction toward the formation of olefins instead of secondary products or parasites.

E. Hydrocracking

Hydrocracking is a cracking process that occurs under hydrogen pressure, which produces products that are **light and saturated**. The simplest schematic is the following one:

$$R-CH_2-CH_2-R' + H_2 \rightarrow R-CH_3 + R'-CH_3$$

The hydrodealkylation of an aromatic hydrocarbon is also feasible:

$$C_6H_5-CH_2-R + H_2 \rightarrow C_6H_6 + R-CH_3$$

Thermodynamically speaking, this is an **exothermic** reaction without molecular expansion. This differs considerably from cracking, which is an endothermic reaction that results in an increased number of molecules.

Industrial hydrocracking occurs under high hydrogen pressure (50 to 150 bar) and relatively low temperature (250 to 450°C).

F. Hydrogenation

Hydrogenation mainly involves **olefins**, notably polyolefins, and **aromatics**:

$$R-CH=CH_2+H_2 \rightarrow R-CH_2-CH_3$$

$$C_6H_5R + 3H_2 \rightarrow C_6H_{11}R$$

Refining seeks to selectively hydrogenate dienes, or even acetylenes, into simple olefins:

$$R-CH=CH-CH=CH-R' + H_2 \rightarrow R-CH_2-CH_2-CH=CH-R'$$

Hydrogenation works best at low temperature and high pressure. The reaction is evidently more difficult with aromatics than with olefins.

G. Dehydrogenation

The dehydrogenation reaction is part of a very–important refining process: **catalytic reforming**. This reaction usually produces aromatic structures, hence it is called **aromatization**.

Reactions involving naphthenes are the first ones involved,

$$C_6H_{11}R \rightarrow C_6H_5R + 3H_2$$

but dehydrocyclisation of paraffins can also occur:

$$R-CH_2-CH_2-CH_2-CH_2-CH_2-CH_3 \rightarrow C_6H_5R + 4\,H_2$$

Unlike hydrogenation, aromatization works best at high temperature and low pressure. The process is exclusively catalytic and it takes place between 5 and 40 bar at around 500°C.

This brief review of the chemistry of hydrocarbons has shown the diversity of the structures and the properties, both physical and chemical, of hydrocarbon chains. These characteristics are referred to often in the sections on refining and combustion.

1.2 Other chemical families

The structure of each of these chemical families typically contains a particular combination of atoms referred to as a **functional group**. Table 1.7 indicates the nomenclature of the main functional groups found in organic chemistry. An understanding of these formulas is important since they are mentioned frequently throughout the book, especially with regard to fuel additives such as colorants, anti-oxidants, detergents, etc.

Table 1.7 *Major functional groups found in organic chemistry (except hydrocarbons).*

Chemical group	Structure
Primary alcohol	$R-CH_2-OH$
Secondary alcohol	$R(R')CH-OH$
Tertiary alcohol	$R-C(R')(R'')-OH$
Phenol	C_6H_5-OH
Ether	$R-O-R'$
Aldehyde	$R(H)C=O$
Ketone	$R(R')C=O$
Carboxylic acid	$R(OH)C=O$
Ester	$R(OR')C=O$
Anhydride	$R-C(=O)-O-C(=O)-R'$
Acid halogenide	$R(X)C=O$
Peroxyacid	$R(OOH)C=O$
Peroxide	$R-O-O-R'$
Hydroperoxide	$R-O-OH$
Primary amine	$R-NH_2$
Secondary amine	$R-NH-R'$
Tertiary amine	$R-N(R'')-R'$
Nitrile	$R-C\equiv N$
Azo compound	$R-N=N-R'$
Amide	$R-C(=O)-NH_2$
Imine	$R-C(R')=NH$
Nitroso compound	$C_6H_5-N=O$
Isocyanate	$R-N=C=O$
Nitroparaffin	$R-NO_2$
Nitroaromatic	$C_6H_5-NO_2$
Alkyl nitrate	$R-O-NO_2$
Thiol or mercaptan	$R-S-H$
Sulfide or thioether	$R-S-R'$
Disulfide	$R-S-S-R'$
Sulfoxide	$R(R')S=O$
Sulfone	$R-S(=O)_2-R'$
Sulfonic acid	$R-S(=O)_2-OH$
Alkyl sulfate	$SO_4(R)_2$

The following section contains some supplementary information on **simple oxygenated organic compounds** (alcohols, ethers, aldehydes, and ketones) and on **sulfur-containing products**.

1.2.1 Oxygenated organic products

In the order of increasing oxidation, these products are alcohols, ethers, aldehydes, and ketones.

1.2.1.1 Alcohols

Alcohols can be found in weak concentrations—usually less than 5%—in commercial fuels and they are generally derived from sources other than standard refining (see Chapter 6).

The general formula for alcohols is $R-OH$, but three distinct classifications exist within this group:

- **primary alcohols** $R-CH_2-OH$
- **secondary alcohols** $R-\underset{\displaystyle R'}{\underset{|}{CH}}-OH$
- **tertiary alcohols** $R-\underset{\displaystyle R''}{\underset{|}{\overset{\displaystyle R'}{\overset{|}{C}}}}-OH$

The classifications are arranged according to the radical's relationship to OH, which is either primary, secondary, or tertiary.

Alcohols are designated in the nomenclature by an "ol" ending. Table 1.8 provides a list of the first items in the alcohol series, which can be effectively detected in certain fuels.

1.2.1.2 Ether-oxides (or ethers)

The general formula is $R-O-R'$. The simplest items in the alcohol group are

- dimethylether CH_3-O-CH_3 or dimethyl oxide
- diethylether $C_2H_5-O-C_2H_5$ or diethyl oxide, commonly known as "ether"

Fuels can contain significant concentrations of organic ethers—up to 10 or 15%—such as methyl tertiary butyl ether (MTBE), tertiary amyl methyl ether (TAME), or ethyl tertiary butyl ether (ETBE), which are obtained from the reaction of methanol or ethanol with light olefins.

For example:

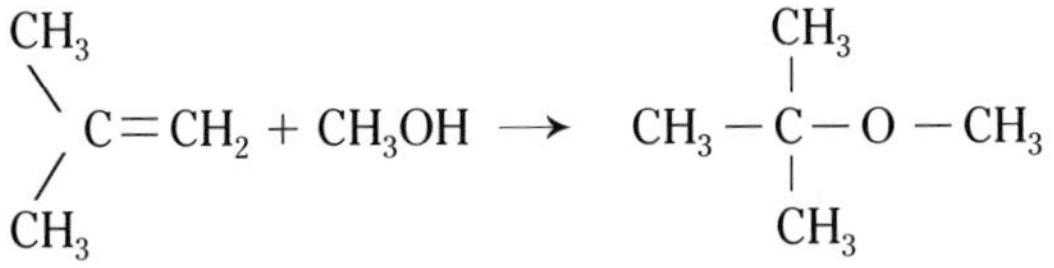

$$(CH_3)_2C=CH_2 + CH_3OH \longrightarrow CH_3-\underset{\displaystyle CH_3}{\underset{|}{\overset{\displaystyle CH_3}{\overset{|}{C}}}}-O-CH_3$$

isobutene + methanol → methyl tertiary butyl ether

Table 1.8 *The nomenclature of common alcohols.*

Name	Formula
Methanol	CH_3-OH
Ethanol	CH_3-CH_2-OH
n-Propanol (Propan-1-ol)	$CH_3-CH_2-CH_2-OH$
Isopropanol	$CH_3-CHOH-CH_3$
n-Butanol (Butan-1-ol)	$CH_3-CH_2-CH_2-CH_2OH$
Isobutanol	$(CH_3)_2CH-CH_2OH$
sec-Butanol (Butan-2-ol)	$CH_3-CH_2-CHOH-CH_3$
tert-Butanol	$CH_3-C(CH_3)_2-OH$
n-Pentanol (Pentan-1-ol)	$CH_3-CH_2-CH_2-CH_2-CH_2OH$
2-Ethylhexan-1-ol	$CH_3-CH_2-CH_2-CH_2-CH(CH_2-CH_3)-CH_2OH$
Ethylene glycol	CH_2OH-CH_2OH
Glycerol (glycerin)	$CH_2OH-CHOH-CH_2OH$

1.2.1.3 Aldehydes and ketones

The respective formulas for aldehydes and ketones are

$$\underset{\text{aldehydes}}{R(H)C=O} \qquad \underset{\text{ketones}}{R(R')C=O}$$

Aldehydes and ketones are not used in fuels but they can be detected in trace amounts in exhaust gases and they are considered to be pollutants. According to the rules of official nomenclature, aldehydes and ketones are designated by the suffixes "al" and "one" respectively. However, as listed in Table 1.9, the primary groups have retained older names.

Table 1.9 *Nomenclature of the main aldehydes found in exhaust gas.*

Name	Formula	Name	Formula
Formaldehyde (or methanal)	$H-CHO$	Benzaldehyde	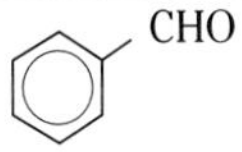 C_6H_5-CHO
Acetaldehyde (or ethanal)	CH_3-CHO	*o*-Tolualdehyde	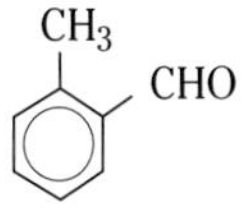 $CH_3-C_6H_4-CHO$ (ortho)
Propionaldehyde (or propanal)	CH_3-CH_2-CHO	*m*-Tolualdehyde	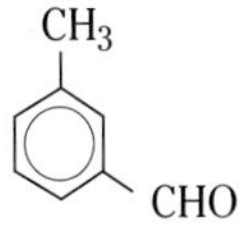 $CH_3-C_6H_4-CHO$ (meta)
Acrolein (or prop-3-ene-1-al)	$CH_2=CH-CHO$	*p*-Tolualdehyde	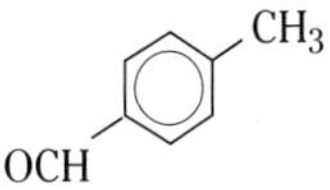 $OCH-C_6H_4-CH_3$ (para)
Butyraldehyde (or butanal)	$CH_3-CH_2-CH_2-CHO$		
Crotonaldehyde (or but-2-ene-1-al)	$CH_3-CH=CH-CHO$		
Valeraldehyde (or pentanal)	$CH_3-CH_2-CH_2-CH_2-CHO$		

1.2.1.4 Acids and esters

Acids

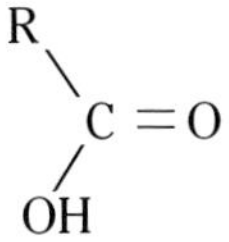

$$R-C(=O)-OH$$

are not found in fuels but they are found in very weak concentrations in exhaust gases. Certain acids, notably

formic acid $H-\underset{\underset{O}{\|}}{C}-OH$ and **acetic acid** $CH_3-\underset{\underset{O}{\|}}{C}-OH$,

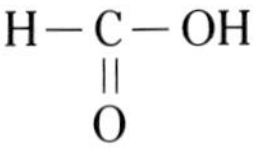

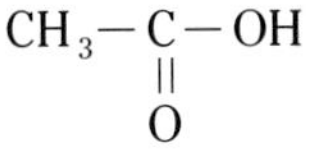

are pollutants like aldehydes.

Esters

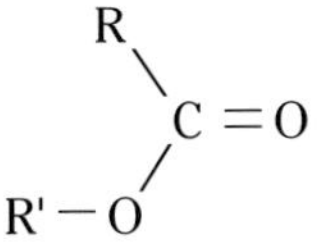

$$R-C(=O)-O-R'$$

are obtained from the action of alcohols on acids

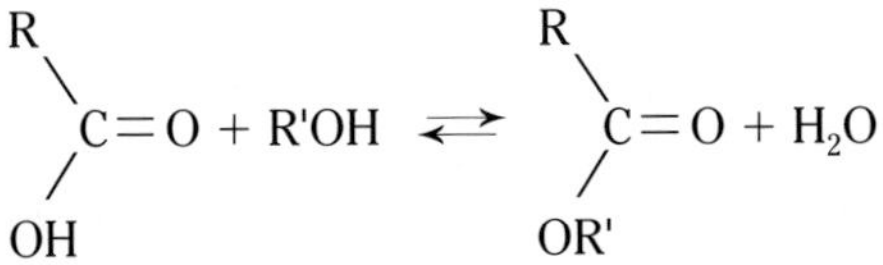

$$R-C(=O)-OH + R'OH \rightleftarrows R-C(=O)-OR' + H_2O$$

Some compounds of this type, such as tert-butyl acetate, have been proposed as fuel additives.

1.2.2 Sulfur-containing products

Sulfur is found in almost all petroleum products, either in free form or more frequently in chemical compounds with bonds like S — H, S — C, or S — S. The concentration of these compounds is intimately linked with the type of crude oil, the distillation interval of the cut, and the refining procedures used. In general, sulfur-containing compounds are concentrated in the heaviest fractions. For example, the sulfur content in regular gasoline is always low (100 to 500 ppm), but it can reach 4% in certain heavy fuels.

The sulfur compounds that affect the fuels industry are

- **Hydrogen sulfide** (H_2S), which is a highly volatile compound (boiling point of –61.8°C) that can be found in light petroleum fractions and in some natural gas deposits, notably the Lacq deposit in France. Hydrogen sulfide acts like a weak mineral acid, but it is corrosive. In addition, it is a toxic compound with a distinct odor.
- **Thiols or mercaptans**, which have the general formula R — SH. Thiols appear as counterparts to the R — OH alcohols, but they have a far more acidic characteristics.

  The first compounds in the series—methylmercaptan and ethylmercaptan—are highly volatile (boiling points of 6 and 35°C) and they have a very disagreeable odor.
- **Sulfides or thioethers and disulfides**, with general formulas R — S — R' and R — S — S — R', which are relatively unstable and they are easily converted back to mercaptans by heating them.
- **Sulfide ring compounds** like thiophene

Thiophene and its derivatives have reasonably good thermal stability.

The nomenclature for sulfide compounds usually results in the use of a "thio" prefix, such as thiol, thioether, and thiophene.

Refining techniques are usually applied in ways that reduce the level of sulfide products because of their disagreeable odor, their corrosiveness, and their contribution to atmospheric pollution—because their combustion produces sulfurous and sulfuric anhydrides (SO_2 and SO_3).

Sulfur is eliminated in modern refineries using **hydrodesulfurization**. Hydrogen is the reactant used to transform the sulfur found in complex chemical structures into H_2S. The following types of reactions are used:

$$R-S-H + H_2 \longrightarrow RH + H_2S$$

$$R-S-R' + 2H_2 \longrightarrow RH + R'H + H_2S$$

The acidic gas released from these reactions is treated in a Claus unit and the valuable sulfur is retained. The process consists of transforming H_2S into sulfur by controlled combustion and the Claus catalytic process. The sulfur, which is in a vapor state, is condensed and produced as a liquid or a solid.

The chemical reactions can be summarized with the following equations:

$$H_2S + \frac{3}{2}O_2 \longrightarrow SO_2 + H_2O$$

$$2H_2S + SO_2 \rightleftarrows 3S + 2H_2O \qquad \text{(Claus reaction)}$$

The equations indicate that the Claus reaction can be optimized if only one third of the feedstock is converted by combustion.

1.3 Composition and properties of various types of fuels

Fuels can be classified and differentiated by using either their **general characteristics**, which are based on physical and chemical properties, or by using more precise **analytical techniques**, which are based on the exact structure of the compounds.

Also, since fuels must undergo complete combustion, a review of some definitions and general rules of thermochemistry is appropriate.

1.3.1 General classification

Table 1.10 lists some specific properties of various types of fuels. The densities and boiling points already serve as distinctive criteria. These distinctions are directly related to the number of carbon atoms in each of the hydrocarbon compounds:

- 3 or 4 for liquefied petroleum gas (LPG)
- about 4 to 10 for gasoline
- 9 to 13 for kerosene
- 10 to 20 for diesel fuel
- at least 13 to 40 for heavy fuel oil

More accurate information on the chemical formulation can be obtained by correlation methods or analytical techniques.

1.3.1.1 Characterization factor

A method of representing the characteristics of hydrocarbons is shown in Fig. 1.2, where the abscissa represents the boiling point and the ordinal—top to bottom—represents the density at 15°C. With this representation, the members of the same chemical group are evenly distributed along **characteristic curves**. Also, mixed products containing a ring structure with substituted paraffin chains are distributed along intermediate lines. For long-chain hydrocarbons, the density follows an asymptotic trend (d = 0.856), which represents the density of chemical chains consisting of many CH_2 links. The information in Fig. 1.2 is logical because the density is linked to the H/C ratio, and the boiling point depends primarily on the number of carbon atoms present.

Empirical formulas were developed to relate density and boiling point within the same family group. This led to the idea of the **characterization factor**, proposed in 1937 by the Universal Oil Products (UOP) company.

Table 1.10 *General characteristics and compositions of various types of fuels.*

Type of product	General characteristics								
	Density at 15°C	Distillation range at atmospheric pressure		Number of carbon atoms in the constituents	H/C atomic ratio	Composition by chemical family (volume%)			
		Initial point (°C)	Final point (°C)			Paraffins	Naphthenes	Olefins	Aromatics
LPG fuel	0.51-0.58	< 0	0	3 and 4	2.0-2.67	60-100	0	0-40	0
Gasoline	0.72-0.77	30-35	180-200	4-10	1.7-1.9	40-65	0-5	0-20	15-45
Kerosene	0.77-0.83	140-150	250-280	9-13	1.9-2.1	50-65	20-30	0	10-20
Diesel	0.82-0.86	160-180	340-370	10-20	1.9-2.1	50-65	20-30	0	10-30
Heavy fuel	0.95-1.05	180-200	450*	> 15	0.8-1.7	**	**	0	**

* Temperature at 20 to 30% evaporated. The major fraction (70 to 80%) does not distill at atmospheric pressure.

** Molecules that are purely paraffinic, naphthenic, or aromatic do not exist in heavy fuels. The percentage of carbon atoms that are bound in aromatic structures can vary from 55 to 100.

The classification factor (K) is defined by the following equation:

$$K = \frac{(T/1{,}8)^{1/3}}{d}$$

The temperature (T) is in degrees Kelvin and the density (d) is taken at 60°F.

Figure 1.2 shows that K is about 13 for paraffins, 12 for mixed hydrocarbons (chains bonded to a saturated ring), and 11 and 10 respectively for naphthenes and aromatics.

To represent fuels with this methodology, the **average boiling point** must be defined. An interesting and practical method proposed by Maxwell (1977) consists of using the distillation curve and choosing temperature reference points at T_{10}, T_{50}, and T_{90}, which respectively correspond to 10, 50, and 90% of the distilled product.

$$T_m = \frac{T_{10} + 2T_{50} + T_{90}}{4}$$

where:

T_m = average boiling point temperature

T_i = temperature at which i% of the product is distilled

The characterization factor for a petroleum fraction can then be calculated quickly from readily available data—the density and the distillation curve.

The value of K varies between 10 and 13, which represents the range of pure products, and its value represents the chemical character of a fraction. This value will often be used in refining to classify a feedstock or an effluent.

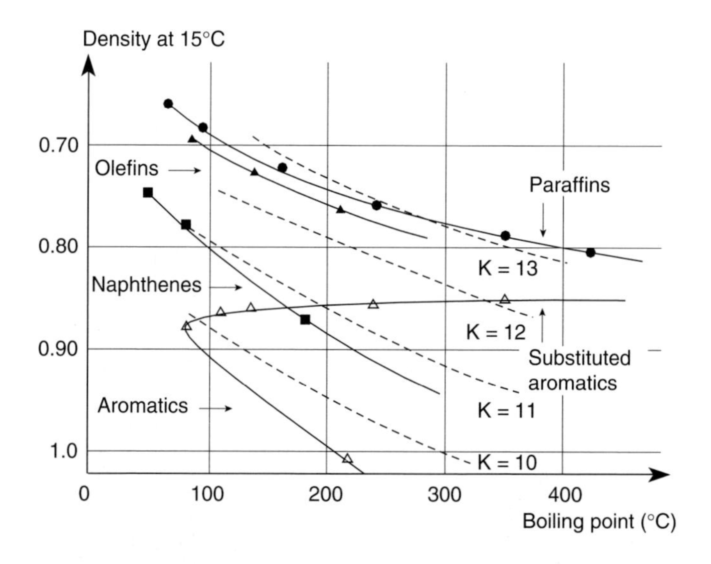

Figure 1.2 *Density to boiling point relationship. Definition of the characterization factor.*

1.3.1.2 Average molecular weight

The average molecular weight of a petroleum fraction is expressed with the following formula:

$$M = \frac{\Sigma\, n_i M_i}{\Sigma\, n_i}$$

where n_i represents the number of molecules in the sample i with a molecular weight of M_i.

The average molecular weight can be estimated by correlating the average distillation temperature and the density (see Fig. 1.3). The estimate is an approximation with about 10% precision. The average molecular weight of a gasoline is close to 0.09 kg/mol and the molecular weight of diesel fuel is about 0.2 kg/mol.

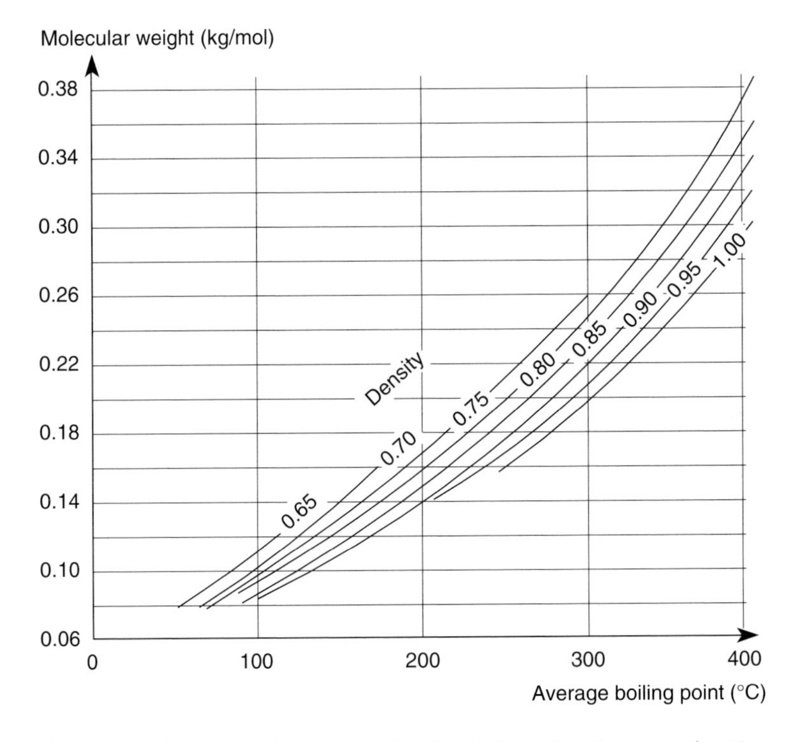

Figure 1.3 *Average molecular weight of petroleum fractions as a function of the density and boiling point.*

1.3.1.3 FIA analysis

Fluorescent Indicator Adsorption (FIA) analysis is a simple laboratory method that can be used to classify fuels based on their overall content of paraffins, olefins, and aromatics. The procedure (ASTM D 1319, IP 156) consists of eluting a fuel sample with isopropanol in an adsorption column lined with active silica gel that is impregnated with yellow and reddish-brown **fluorescent dies**—

the dies react with olefins and aromatics respectively. Different groups of hydrocarbons are selectively retained. Inspecting the column with an ultra-violet light reveals three distinct zones where **paraffins**, **olefins**, and **aromatics** are respectively located. The height of each zone is proportional to the content by volume of each constituent in the sample.

The procedure can only be rigorously applied to fuels that distill at less than 315°C (gasoline and kerosene). The absolute values of the repeatability and reproducibility deviations are 2 and 3% respectively. Diolefins, sulfide compounds, and products consisting of an olefinic chain attached to an aromatic ring, can be confused with aromatics during FIA analysis. The method is sometimes used to analyze diesel fuel, but the results are relatively imprecise.

FIA analysis is highly inaccurate if oxygenated products (alcohols or ethers) are present in gasoline.

1.3.1.4 Bromide number

The bromide number method (ASTM D 1159 and D 2710, NF M 07–017) can be used to estimate the **quantity of olefins** in a fuel. The method uses the double bond $\rangle C{=}C\langle$ capacity to retain two bromide atoms by an addition reaction:

$$\rangle C{=}C\langle \; + Br_2 \longrightarrow -\underset{Br}{\overset{|}{\underset{|}{C}}}-\underset{Br}{\overset{|}{\underset{|}{C}}}-$$

The bromide number is expressed as the mass of bromide that is added to a given mass of analyzed product. If the average molecular weight of the product is known, the olefinic hydrocarbon content can be calculated using the bromide number. The bromine number method is imprecise in the presence of polyolefins or heavily branched molecules.

1.3.1.5 Maleic anhydride value

The maleic anhydride value (MAV) index is based on the potential that conjugated **olefin double bonds** have for adding maleic anhydride according to the following reaction:

O O
+ O → O
O O

The method was developed by *Amoco*. The MAV is expressed as the number of milligrams of anhydride consumed by one gram of the product being tested. It is still in widespread use to detect the presence of conjugate diolefins in fuels. This type of structure is undesirable because it can lead to the formation of gums and sediments by polymerization.

1.3.2 Structural analysis—proportioning of components

Gas chromatography can be used to evaluate the nature and the concentration of almost all the individual hydrocarbons found in light fuels such as LPG or gasoline. Tables 1.11, 1.12a, and 1.12b show examples of the results of an analysis that detected and measured eight constituents in LPG, and over 180 constituents in standard gasoline.

In this last example, the composition can be presented more concisely by forming groups of constituents based on family similarity or on the number of carbon atoms (Table 1.13).

All constituents obviously cannot be detected in medium and heavy products. However, the concentration of various hydrocarbon groups can be measured using a **combination of different methods**, such as gas and liquid chromatography, mass spectrometry, nuclear magnetic resonance (NMR) of hydrogen and carbon, ultra-violet spectrometry, and infrared spectrometry (Wauquier 1995). Tables 1.14 and 1.15 show that paraffins, condensed or uncondensed naphthenes, simple aromatics, and dicyclics or polycyclics can be measured in diesel fuel or jet fuel (Heinze 1994). The methods are very efficient for studying the relationships between fuel constituents and a fuel's behavior in engines—such as octane and cetane ratings and the formation of pollutants.

Finally, there are specific analytical methods, mostly chromatographic, that enable the measurement of the very-complex molecules (heterocyclic compounds) found in fuels; these molecules are often present in trace amounts.

Table 1.11 *Example of the composition of a European LPG fuel.* *

LPG Fuel	
Constituent	**% mass**
Propane	21.3
Propylene	3.7
n-Butane	37.5
Isobutane	22.5
But-1-ene	3.9
Isobutene	3.9
(*E*)-But-2-ene	4.5
(*Z*)-But-2-ene	2.7
Total	**100.0**

* *Gas chromatography analysis.*

Table 1.12a *An example of the composition of a reformulated gasoline from California.* *

Normal paraffins	
Constituent	**% mass**
n-Butane	1.24
n-Pentane	1.44
n-Hexane	0.99
n-Heptane	0.81
n-Octane	0.61
n-Nonane	0.12
n-Decane	0.03
n-Undecane	0.02
n-Dodecane	0.01
n-Tridecane	0.01
Total	**5.28**

* *Analysis by gas chromatography in accordance with standard NF M 07-068.*

Table 1.12a *(continued)*

Naphthenes			
Constituent	**% mass**	**Constituent**	**% mass**
1-Methylcyclopentane	0.50	*trans*-1-Methyl-2-Ethylcyclopentane	0.05
Cyclohexane	0.06	1-Methyl, 1-Ethylcyclopentane	0.02
1,1-Dimethylcyclopentane	0.19	1-Methyl, 4-Ethylcyclopentane	0.08
cis-1,3-Dimethylcyclopentane	0.17	C_8-Naphthenes	0.03
trans-1,3-Dimethylcyclopentane	0.15	1,1,3-Trimethylcyclohexane	0.05
trans-1,2-Dimethylcyclopentane	0.11	1,3,5-Trimethylcyclohexane	0.04
1-Methylcyclohexane	0.30	Dimethylethylcyclopentane	0.38
1-Ethylcyclopentane	0.10	Methylethylcyclohexane	0.06
r-1,*trans*-2,*cis*-4-Trimethylcyclopentane	0.08	C_9-Naphthenes	0.33
r-1,*trans*-2,*cis*-3-Trimethylcyclopentane	0.05	C_{10}-Naphthenes	0.24
r-1, *cis*-2,*trans*-4-Trimethylcyclopentane	0.12	C_{12}-Naphthenes	0.03
cis-1,3-Dimethylcyclohexane	0.04	C_{13}-Naphthenes	0.02
trans-1-Methyl-3-Ethylcyclopentane	0.08		
cis-1-Methyl-3-Ethylcyclopentane	0.06	**Total**	**3.34**

Table 1.12a *(continued)*

Isoparaffins			
Constituent	**% mass**	**Constituent**	**% mass**
Isobutane	0.08	2,4,4-Trimethylhexane	0.10
Neopentane	0.02	2,3,5-Trimethylhexane	0.21
Isopentane	7.38	2,2-Dimethylheptane	0.04
2,2-Dimethylbutane	0.56	2,4-Dimethylheptane	0.13
2,3-Dimethylbutane	1.36	2,6-Dimethylheptane	0.09
2-Methylpentane	2.93	2,5-Dimethylheptane	0.23
3-Methylpentane	1.51	3,4-Dimethylheptane	0.04
2-2-Dimethylpentane	0.08	4-Ethylheptane	0.07
2,4-Dimethylpentane	2.39	4-Methyloctane	0.17
2,2,3-Dimethylbutane	0.03	2-Methyloctane	0.20
3,3-Dimethylpentane	0.10	3-Ethylheptane	0.06
2-Methylhexane	1.15	3-Methyloctane	0.24
2,3-Dimethylpentane	4.35	C_9-Isoparaffins	0.05
3-Methylhexane	1.27	2,2-Dimethyloctane	0.06
3-Ethylpentane	0.16	3,5-Dimethyloctane	0.01
2,2,4-Trimethylpentane	8.30	2,7-Dimethyloctane	0.08
2,2-Dimethylhexane	0.10	2,6-Dimethyloctane	0.04
2,5-Dimethylhexane	0.95	3,6-Dimethyloctane	0.01
2,4-Dimethylhexane	1.11	2-Methyl-3-ethylheptane	0.01
3,3-Dimethylhexane	0.06	4-Ethyloctane	0.01
2,3,4-Trimethylpentane	2.54	5-Methylnonane	0.02
2,3,3-Trimethylpentane	1.78	2-Methylnonane	0.04
2,3-Dimethylhexane	0.89	3-Methylnonane	0.05
2-Methyl-3-ethylpentane	0.06	C_{10}-Isoparaffins	0.05
2-Methylheptane	0.65	C_{11}-Isoparaffins	0.06
4-Methylheptane	0.30	2.5-Dimethyldecane	0.03
3,4-Dimethylhexane	0.10	2.6-Dimethyldecane	0.06
3-Methylheptane	0.76	C_{12}-Isoparaffins	0.09
3-Ethylhexane	0.26	C_{13}-Isoparaffins	0.06
2,2,5-Trimethylhexane	1.17	**Total**	**44.71**

Table 1.12a *(continued)*

Aromatics			
Constituent	**% mass**	**Constituent**	**% mass**
Benzene	1.32	1,4-Dimethyl-2-ethylbenzene	0.15
Toluene	6.87	1,3-Dimethyl-4-ethylbenzene	0.17
Ethylbenzene	3.20	1-Methylindane	0.04
m-Xylene	7.09	1,2-Dimethyl-4-ethylbenzene	0.25
o-Xylene	2.27	1,2-Dimethyl-3-ethylbenzene	0.09
Isopropylbenzene	0.10	1,2,3,5-Tetramethylbenzene	0.13
n-Propylbenzene	0.41	1,2,4,5-Tetramethylbenzene	0.17
1-Methyl-3-Ethylbenzene	1.23	5-Methylindane	0.10
1-Methyl-4-Ethylbenzene	0.54	4-Methylindane	0.19
1,3,5-Trimethylbenzene	0.59	1,2,3,4-Tetramethylbenzene	0.06
1-Methyl-2-Ethylbenzene	0.77	Tetraline	0.02
1-2,4-Trimethylbenzene	1.60	Naphthalene	0.33
1-2,3-Trimethylbenzene	0.29	C_{10}-Aromatics	0.19
1-Methyl,4-Isopropylbenzene	0.04	2-Methylnaphthalene	0.29
Indane	0.17	1-Methylnaphthalene	0.14
Indene + 1-Methyl-2-isopropyl-benzene	0.02	C_{11}-Aromatics	0.50
		C_2-Indanes	0.08
1,3-Diethylbenzene	0.13	C_{12}-Aromatics	0.14
1-Methyl-3-*n*-propylbenzene	0.27	C_{13}-Aromatics	0.08
1-Methyl-4-*n*-propylbenzene	0.23	C_2-Naphthalenes	0.06
1,3-Dimethyl-5-ethylbenzene	0.25		
1-Methyl-2-*n*-propylbenzene	0.08	**Total**	**30.65**

Table 1.12a *(last panel)*

Olefins			
Constituent	**% mass**	**Constituent**	**% mass**
Isobut-1-ene + *n*-but-1-ene	0.02	(*E*)-Hex-2-ene	0.19
(*E*)-But-2-ene	0.04	2-Methylpent-2-ene	0.20
(*Z*)-But-2-ene	0.05	(*E*)-3-Methylpent-2-ene	0.18
3-Methylbut-1-ene	0.02	3-Methylcyclopent-1-ene	0.02
Pent-1-ene	0.09	(*Z*)-Hex-2-ene	0.11
2-Methylbut-1-ene	0.18	(*Z*)3-Methylpent-2-ene	0.20
(*E*)-Pent-2-ene	0.32	3-Methylhex-1-ene	0.01
(*Z*)-Pent-2-ene	0.16	Cyclohex-1-ene	0.03
2-Methylbut-2-ene	0.43	C_7-Olefins	0.90
Cyclopent-1-ene	0.06	C_8-Olefins	0.79
4-Methylpent-1-ene	0.02	C_9-Olefins	0.47
3-Methylpent-1-ene	0.03	C_{11}-Olefins	0.02
2-Methylpent-1-ene	0.20		
(*E*)-Hex-3-ene	0.09	**Total**	**4.86**
(*Z*)-Hex-3-ene	0.03		

Oxygenated	
Constituent	**% mass**
MTBE	11.13

Table 1.12b *Example of a standard European commercial gasoline. (Analysis by gas chromatography in accordance with standard NF M 07-068).*

Normal paraffins	
Constituent	**% mass**
n-Butane	5.14
n-Pentane	1.26
n-Hexane	0.64
n-Heptane	0.65
n-Octane	0.48
n-Nonane	0.11
n-Decane	0.01
Total	**8.29**

Table 1.12b *(continued)*

Naphthenes			
Constituent	**% mass**	**Constituent**	**% mass**
1-Methylcyclopentane	1.09	*cis*-1-Methyl-3-Ethylcyclopentane	0.04
Cyclohexane	0.10	*trans*-1-Methyl-2-Ethylcyclopentane	0.03
1,1-Dimethylcyclopentane	0.01	1-Methyl, 4-Ethylcyclopentane	0.04
cis-1,3-Dimethylcyclopentane	0.24	C_8-Naphthenes	0.01
trans-1,3-Dimethylcyclopentane	0.20	1,1,3-Trimethylcyclohexane	0.02
trans-1,2-Dimethylcyclopentane	0.15	1,3,5-Trimethylcyclohexane	0.02
1-Methylcyclohexane	0.34	Dimethylethylcyclopentane	0.04
1-Ethylcyclopentane	0.10	C_9-Naphthenes	0.09
r-1,*trans*-2,*cis*-4-Trimethylcyclopentane	0.06	C_{10}-Naphthenes	0.09
r-1,*trans*-2,*cis*-3-Trimethylcyclopentane	0.06	C_{12}-Naphthenes	0.09
r-1, *cis*-2,*trans*-4-Trimethylcyclopentane	0.11	*cis*-Decaline	0.01
cis-1,3-Dimethylcyclohexane	0.02	**Total**	**2.92**
trans-1-Methyl-3-Ethylcyclopentane	0.05		

Table 1.12b *(continued)*

Isoparaffins			
Constituent	**% mass**	**Constituent**	**% mass**
Isobutane	0.30	3-Methylheptane	0.55
Isopentane	7.84	3-Ethylhexane	0.17
2,2-Dimethylbutane	0.05	2,2,5-Trimethylhexane	0.28
2,3-Dimethylbutane	1.46	2,4,4-Trimethylhexane	0.04
2-Methylpentane	3.12	2,3,5-Trimethylhexane	0.05
3-Methylpentane	1.71	2,2-Dimethylheptane	0.02
2,2-Dimethylpentane	0.02	2,4-Dimethylheptane	0.05
2,4-Dimethylpentane	0.63	2,6-Dimethylheptane	0.04
2,2-3-Dimethylbutane	0.05	2,5-Dimethylheptane	0.11
3,3-Dimethylpentane	0.08	3,4-Dimethylheptane	0.02
2-Methylhexane	0.92	4-Methyloctane	0.10
2,3-Dimethylpentane	0.46	2-Methyloctane	0.13
3-Methylhexane	0.90	3-Methyloctane	0.26
3-Ethylpentane	0.16	C_9-Isoparaffins	0.02
2,2,4-Trimethylpentane	4.11	2,2-Dimethyloctane	0.01
2,2-Dimethylhexane	0.08	2,7-Dimethyloctane	0.02
2,5-Dimethylhexane	0.82	2-Methylnonane	0.02
2,4-Dimethylhexane	0.56	3-Methylnonane	0.02
3,3-Dimethylhexane	0.06	C_{10}-Isoparaffins	0.02
2,3,4-Trimethylpentane	2.22	C_{11}-Isoparaffins	0.10
2,3,3-Trimethylpentane	1.55	C_{12}-Isoparaffins	0.61
2,3-Dimethylhexane	0.57	C_{13}-Isoparaffins	0.01
2-Methyl-3-ethylpentane	0.03		
2-Methylheptane	0.47	**Total**	**31.10**
3,4-Dimethylhexane	0.28		

Table 1.12b *(continued)*

Aromatics			
Constituent	**% mass**	**Constituent**	**% mass**
Benzene	1.23	1-Methyl-4-*n*-propylbenzene	0.17
Toluene	8.11	1,3-Dimethyl-5-propylbenzene	0.22
Ethylbenzene	2.89	1-Methyl-2-*n*-propylbenzene	0.07
m-Xylene	5.70	1,4-Dimethyl-2-ethylbenzene	0.16
p-Xylene	1.96	1,3-Dimethyl-4-ethylbenzene	0.15
o-Xylene	3.06	1-Methylindane	0.03
Isopropylbenzene	0.19	1,2-Dimethyl-4-ethylbenzene	0.24
n-Propylbenzene	0.64	1,2,3,5-Tetramethylbenzene	0.18
1-Methyl-3-Ethylbenzene	2.05	1,2,4,5-Tetramethylbenzene	0.22
1-Methyl-4-Ethylbenzene	0.91	5-Methylindane	0.07
1,3,5-Trimethylbenzene	0.88	4-Methylindane	0.09
1-Methyl-2-Ethylbenzene	0.92	1,2,3,4-Tetramethylbenzene	0.07
1,2,4-Trimethylbenzene	2.94	Tetraline	0.03
1,2,3-Trimethylbenzene	0.61	Naphtalene	0.30
1-Methyl,4-Isopropylbenzene	0.10	C_{10}-Aromatics	0.13
Indane	0.36	2-Methylnaphtalene	0.06
Indene + 1-Methyl-2-Isopropylbenzene	0.13	1-Methylnaphatalene	0.03
1,3-Diethylbenzene	0.11	C_{11}-Aromatics	0.17
1-Methyl-3-*n*-propylbenzene	0.20		
2-Methylindane	0.11	**Total**	**35.49**

Table 1.12b *(last panel)*

Olefins			
Constituent	**% mass**	**Constituent**	**% mass**
Isobut-1-ene + *n*-but-1-ene	0.28	(*E*)-Hex-3-ene	0.33
(*E*)-But-2-ene	0.56	(*Z*)-Hex-3-ene	0.12
(*z*)-But-2-ene	0.60	(*E*)-Hex-2-ene	0.65
C_4-Olefins	0.05	2-Methylpent-2-ene	0.70
3-Methylbut-1-ene	0.29	4-Methylcyclopent-1-ene	0.15
Pent-1-ene	0.89	(*E*)-3-Methylpent-2-ene	0.46
2-Methylbut-1-ene	1.72	3-Methylcyclopent-1-ene	0.08
Isoprene	0.06	(*Z*)-Hex-2-ene	0.36
(*E*)-Pent-2-ene	2.25	(*Z*)-3-Methylpent-2-ene	0.64
(*Z*)-Pent-2-ene	1.27	2,4-Dimethylpent-1-ene	0.02
2-Methylbut-2-ene	3.13	3-Methylhex-1-ene	0.05
(*E*)-Penta-1,3-diene	0.06	Cyclohex-1-ene	0.06
(*Z*)-Penta-1,3-diene	0.06	C_6-Olefins	0.28
Cyclopent-1-ene	0.38	C_7-Olefins	1.49
4-Methylpent-1-ene	0.13	C_8-Olefins	0.34
3-Methylpent-1-ene	0.19	C_9-Olefins	0.07
2-Methylpent-1-ene	0.53	C_{10}-Olefins	0.02
Hex-1-ene	0.24		
2-Ethylbut-1-ene	0.15	**Total**	**18.66**

Oxygenated	
Constituent	**% mass**
MTBE	0.50
ETBE	3.00
Total	**3.50**

See Tables 1.13 to 1.15 on pages 54 to 56.

1.3.3 Combustion parameters

This section covers some of the fundamentals of thermo-chemistry by describing how these fundamentals apply to the combustion of fuels in engines.

1.3.3.1 Stoichiometric combustion equation and stoichiometric ratio

The chemical structure of a fuel or a combustible (number of carbon atoms in the chain, number of bonds, ...) does not have a direct effect on the overall chemical balance of combustion; only the **general composition**, which is the total mass content of carbon, hydrogen, and oxygen (if alcohols or ethers are used) has an effect.

Therefore, the elementary quantitative analysis of a fuel produces the general formula $(CH_yO_z)_x$, where the coefficient x, which is related to the average molecular weight, has no effect on the air-fuel ratio.

Table 1.13 *An example of a simplified analysis of a standard commercial gasoline. Constituents are distributed by the number of carbon atoms and by chemical family (content by % mass).*

Number of carbon atoms	Chemical groups						
	n-Paraffins	**Isoparaffins**	**Naphthenes**	**Aromatics**	**Olefins**	**Oxygenated**	**Total**
4	5.14	0.30			1.49		**6.93**
5	1.26	7.84			10.11	0.50	**19.71**
6	0.64	6.34	0.19	1.23	5.07	3.00	**17.47**
7	0.65	3.22	1.05	8.11	1.56		**14.59**
8	0.48	11.47	0.43	13.61	0.34		**26.33**
9	0.11	1.12	0.16	9.49	0.07		**10.95**
10	0.01	0.09	0.09	2.80	0.02		**3.01**
11		0.10		0.25			**0.35**
12		0.61					**0.61**
13		0.01					**0.01**
Total	**8.29**	**31.10**	**2.92**	**35.49**	**18.66**	**3.50**	**99.96**

Table 1.14 *An example of the chemical family composition of a commercial diesel fuel (mass spectrometry analysis).*

Chemical family	Structural symbol	% mass
Paraffins	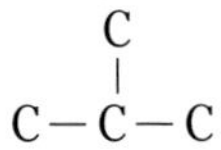	30.90
Non-condensed naphthenes	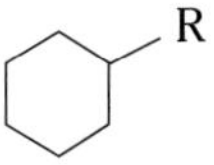	23.70
Condensed naphthenes	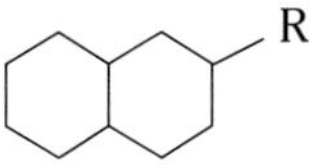	15.10
Alkylbenzenes	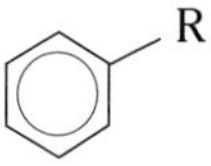	9.20
Indanes and tetralins	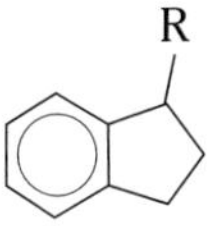 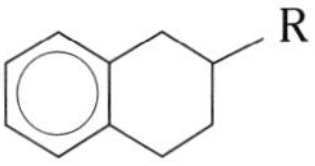	6.40
Indenes	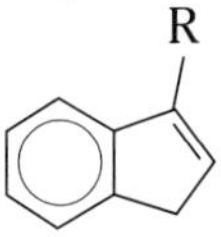	1.80
Naphthalenes	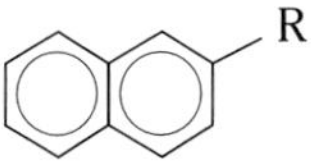	5.50
Acenaphthenes and diphenyls	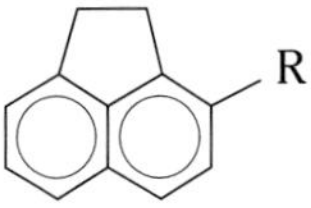 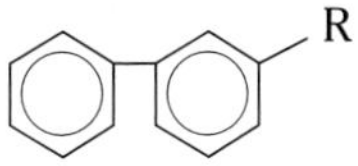	2.75
Fluorenes and acenaphthylenes	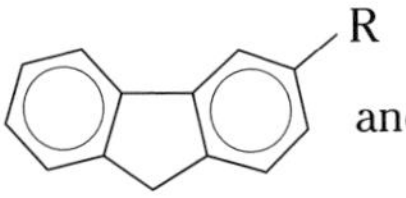 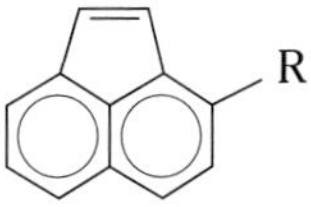	1.50
Anthracenes and phenanthrenes	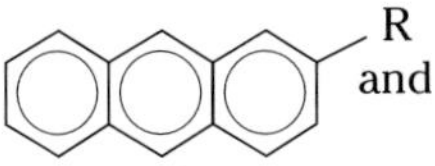 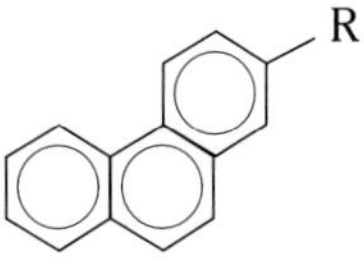	1.30
Benzothiophenes	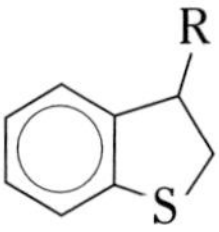	1.60
Dibenzothiophenes	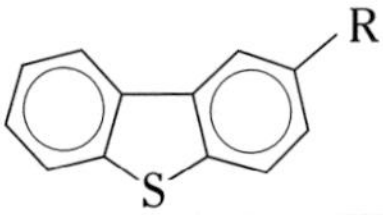	0.25
Total		**100.00**

Table 1.15 *An example of the chemical family composition of a commercial jet fuel. Analysis prepared by mass spectrometry.*

Chemical family	Structure symbol	% mass
Paraffins	C–C(–C)–C	58.30
Non-condensed naphthenes	R	23.85
Dicyclic naphthenes	R	2.40
Simple aromatics	R	13.40
Indanes and tetralins	R and R	1.70
Naphthalenes	R	0.35
Total		**100.00**

The following equation describes the chemical combustion:

$$CH_yO_z + \left(1 + \frac{y}{4} - \frac{z}{2}\right)(O_2 + 3.78N_2) \longrightarrow CO_2 + \frac{y}{2}H_2O + 3.78\left(1 + \frac{y}{4} - \frac{z}{2}\right)N_2$$

Air contains 20.9% O_2 and 79.1% N_2 by volume. It also contains argon at 0.93% by volume and other rare gases, which are present in trace amounts and usually associated with nitrogen.

Stoichiometry is the combination of fuel and air required to obtain complete combustion according to the previous equation. The **stoichiometric ratio** (r) is the quotient of the masses of air and fuel (m_a and m_c respectively), when they are combined under stoichiometric conditions:

$$r = \frac{m_a}{m_c}$$

The range for r is between 13 and 15 for hydrocarbons in general and between 14 and 14.5 for standard liquid fuels such as gasoline, diesel fuel, and jet fuel. The value of r increases with the H/C ratio, varying from 11.49 for carbon to 34.46 for pure hydrogen.

1.3.3.2 Equivalence ratio: representation, calculation, and importance in various types of combustion processes

The combustion conditions in engines and turbines, in practice, do not correspond to stoichiometry and they feature either an **excess** or a **lack** of fuel with respect to oxygen.

The composition of the elements involved in the reaction is expressed by the equivalence ratio (φ), which is defined by the following equation

$$\varphi = \frac{\dfrac{m_{ce}}{m_{ae}}}{\dfrac{m_c}{m_a}}$$

where m_{ce} and m_{ae} are the masses of fuel and air respectively. Therefore, the equivalence ratio is expressed as a function of the stoichiometric ratio in the following way:

$$\varphi = \frac{m_{ce}}{m_{ae}} r$$

The equivalence ratio is expressed with respect to the noblest reactant, which is the fuel, and the mixture is referred to as rich or lean, depending on the quantity of fuel present with respect to stoichiometry.

In some countries such as Germany, the composition of the combustion mixture is referred to by the **coefficient of excess air** (λ), where λ is the inverse of the equivalence ratio φ

$$\lambda = \frac{1}{\varphi}$$

In general, diesel engines and aircraft turbines run on a **lean mixture** and the equivalence ratio has little effect on their performance.

However, gasoline engines, depending on their technology, run with rich, stoichiometric, or lean mixtures. With older vehicles, the equivalence ratio can vary from about 0.85 to 1.20. Operation at lean mixtures (0.85 to 0.95) is preferred for cruising speeds, since it provides better engine efficiency; however, rich mixtures (1.05 to 1.15) provide additional power for acceleration while at the same time having a detrimental effect on emissions.

The equivalence ratio for vehicles with 3-way catalytic converters (see Chapter 5) is controlled within a narrow stoichiometric range (0.98 to 1.02).

1.3.3.3 Heating value

The heating value does not usually appear in the specifications of fuels, but it is essential

- **for the mechanical engineer** who is trying to establish thermal efficiency
- **for vehicle operators** who wish to both minimize their expenses on a particular trip and to expand their operating range

A. The definition of heating value

The heating value, in units of mass or volume, represents the amount of energy released by a unit mass or a unit volume of fuel, when burned completely in a chemical reaction that forms CO_2 and H_2O. Unless otherwise indicated, the fuel is used as a liquid at a standard temperature (usually 25°C). The air and combustion products are also measured at the same temperature.

The difference between the **gross heating value** (GHV) and the **net heating value** (NHV) is the amount of the heat released by the condensation of the water in the exhaust. The only really useful measure is the NHV, because water vapor is usually released in the combustion products of engines and burners.

B. Methods of measuring heating value

The gross heat of combustion is measured in a laboratory using the procedures in ASTM D240, which consist of burning a fuel sample in an oxygen atmosphere inside a bomb calorimeter that contains water. The thermal effect is calculated from the rise in temperature of the medium and from the thermal characteristics of the apparatus.

Calorimeter techniques provide the mass GHV (GHV_m). To determine the mass NHV (NHV_m), the **mass content of the hydrogen** (W_H) in the fuel must also be known. When expressed as a percent, the mass of the water (M) released by the combustion of one kilogram of fuel is

$$M = \frac{W_H}{100} \frac{18}{2}$$

Considering the mass enthalpy of the vaporization of the water at 25°C (ΔH_v = 2 358 kJ/kg), the following equation is the result,

$$NHV_m = GHV_m - 212.2\, W_H$$

where NHV_m and GHV_m are in kJ/kg.

The precision of this method (calorimeter measurement, determination of hydrogen content, and final calculation) is satisfactory with an NHV_m repeatability of 0.3% and a reproducibility of 0.4%.

The heat of combustion of pure organic products can be calculated from their **heat of formation**, which is determined from thermodynamic tables. The NHV_m is obtained by using the thermochemical relationships established at standard temperature and pressure (10^5 Pa and 25°C):

$$\underset{\text{combustion}}{\Delta H^\circ} = \underset{\text{products}}{\sum (\Delta H^\circ_{f,T})} - \underset{\text{reactants}}{\sum (\Delta H^\circ_{f,T})}$$

where:

$\underset{\text{combustion}}{\Delta H^\circ}$ = mass enthalpy variation or the thermal effect of the combustion, equal in absolute value but opposite in sign to NHV_m

$\underset{\text{products}}{\sum (\Delta H^\circ_{f,T})}$ = sum of the mass enthalpy variations from the formation reactions of the products

$\underset{\text{reactants}}{\sum (\Delta H^\circ_{f,T})}$ = sum of the mass enthalpy variations from the formation reactions of the reactants.

Table 1.16 *An example of the calculation of the NHV of toluene using thermodynamic data (Source: "Thermodynamic tables - Hydrocarbons" TRC, TX, USA).*

$$*\Delta H^{\circ}_{v,\,298}\left(C_6H_5\ CH_3\right) = 38\ 003\ \text{J/mol}$$
$$\Delta H^{\circ}_{f,\,298}\left(C_6H_5\ CH_3\right) = 49\ 999\ \text{J/mol}$$
$$\Delta H^{\circ}_{f,\,298}\left(H_2O\right) = -\ 241\ 818\ \text{J/mol}$$
$$\Delta H^{\circ}_{f,\,298}\left(CO_2\right) = -\ 393\ 510\ \text{J/mol}$$

$$C_6H_5\ CH_{3l} + 9\ O_{2g} + n\ N_{2g} \longrightarrow 7\ CO_{2g} + 4\ H_2O_g$$
$$\Delta H = 7\ (-393\ 510) + 4\ (-241\ 818) - 49\ 999 + 38\ 003 = -3\ 733\ 838\ \text{J}$$
$$M\left(C_6H_5\ CH_3\right) = 0.092\ \text{kg/mol}$$
$$\text{NHV}_m = \frac{3\ 733\ 838}{0.092} = 40\ 585\ \text{kJ/kg}$$

$*\Delta H^{\circ}_{v,\,298}$ *is the heat of vaporization of toluene at 298 K*

Table 1.16 includes an example of the NHV_m calculations for toluene, which are based on the mole enthalpies of the formation reactions of the products and the reactants; the reaction also includes any change-of-state enthalpies.

This calculation method is of limited interest and it is useful only when applied to certain pure organic products for which the direct measurement of the NHV_m is difficult to achieve for various reasons (cost, availability, etc.).

Relatively simple **correlation formulas** are available for relating a fuel's heat of combustion to the volumetric mass and to the composition of the chemical groups provided by FIA.

The following two examples were provided by Sirtori *et al* (1974):

$$\text{NHV}_m = 4.18\ (106.38\ \text{PAR} + 105.76\ \text{OL} + 95.55\ \text{ARO})$$

and

$$\text{NHV}_v = 4.18\ (10672.05\ \rho - 8.003\ \text{ARO})$$

where NHV_m and NHV_v are the mass and volumetric NHV respectively in kJ/kg and kJ/liter; PAR, OL, and ARO are the respective paraffin, olefin, and aromatic contents from the FIA; ρ is the density (kg/liter) at 15°C.

C. Numeric heating values

Table 1.17 lists the numeric values of NHV_m for a number of pure organic compounds. Table 1.18 lists the values of NHV_m and NHV_v for common fuels and combustibles.

The NHV_m for a large number of hydrocarbons and petroleum fractions appears to be between 40 000 and 45 000 kJ/kg. In general, NHV_m increases with the H/C ratio. For instance, the heats of combustion of hydrogen and carbon are 120 970 and 32 800 kJ/kg respectively. Otherwise, the presence of oxygen in the molecule obviously tends to reduce the NHV. Also, the NHV_m of methane

Table 1.17 *Mass net heating values for pure organic compounds at 25°C.*

Compound	NHV_m (kJ/kg)	Compound	NHV_m (kJ/kg)
Paraffins		**Acetylenics**	
Methane	50 009	Acetylene	48 241
Ethane	47 794	Methylacetylene	46 194
Propane	46 357	But-1-yne	45 590
Butane	45 752	Pent-1-yne	45 217
Pentane	45 357		
Hexane	44 752	**Aromatics**	
Heptane	44 566	Benzene	40 170
Octane	44 427	Toluene	40 589
Nonane	44 311	*o*-Xylene	40 961
Decane	44 240	*m*-Xylene	40 961
Undecane	44 194	*p*-Xylene	40 798
Dodecane	44 147	Ethylbenzene	40 938
		1,2,4-Trimethylbenzene	40 984
Isoparaffins		Propylbenzene	41 193
Isobutane	45 613	Cumene	41 217
Isopentane	45 241		
2-Methylpentane	44 682	**Alcohols**	
2,3-Dimethylbutane	44 659	Methanol	19 937
2,3-Dimethylpentane	44 496	Ethanol	28 865
2,2,4-Trimethylpentane	44 310	*n*-Propanol	30 680
		Isopropanol	30 447
Naphthenes		*n*-Butanol	33 075
Cyclopentane	43 636	Isobutanol	32 959
Methylcyclopentane	44 636	Tertiobutanol	32 587
Cyclohexane	43 450	*n*-Pentanol	34 727
Methylcyclohexane	43 380		
		Ethers	
Olefins		Methoxymethane	28 703
Ethylene	47 195	Ethoxyethane	33 867
Propylene	45 799	Propoxypropane	36 355
But-1-ene	44 334	Butoxybutane	37 798
(*Z*)-But-2-ene	45 194		
(*E*)-But-2-ene	44 124	**Aldehydes and ketones**	
Isobutene	45 055	Methanal (formaldehyde)	17 259
Pent-1-ene	45 031	Ethanal (acetaldehyde)	24 156
2-Methylpent-1-ene	44 799	Propanal (propionaldehyde)	28 889
Hex-1-ene	44 426	Butanal (butyraldehyde)	31 610
		Acetone	28 548
Diolefins			
Buta-1,3-diene	44 613	**Other types of chemicals**	
Isoprene	44 078	Carbon (graphite)	32 808
		Hydrogen	120 971
Nitrogen derivatives		Carbon monoxide	10 112
Nitromethane	10 513	Ammonia	18 646
Nitropropane	20 693	Sulfur	4 639

is about 50 009 kJ/kg and that of methanol is 19 937 kJ/kg. Lastly, the NHV_m does not depend only on compositional weighting, but it also varies with the energy of formation of each fuel under consideration. Note that cycloparaffins have a lower NHV_m than olefins, which have the same number of atoms and the same general formula (H/C = 2).

The volumetric heat of combustion (NHV_v) can be obtained from the density ρ at 25°C and the NHV_m by the following relationship:

$$NHV_v = \rho \, NHV_m$$

For all hydrocarbons, NHV_v is usually between 27 000 and 36 000 kJ/liter. NHV_v is usually much less for alcohols (for example, 15 870 kJ/liter for methanol).

The NHV_v measurement is of real economic importance, because fuel consumption and the price of fuels are usually measured in liters/100 km or dollars/liter respectively. From a technical standpoint, the NHV_v determines the operating range of a transportation system for a given volume of fuel, which is critically important in some applications, such as aviation.

It is interesting to compare the variations between the mass and volumetric NHVs of hydrocarbons with respect to their chemical structure.

Figure 1.4 shows a clear distinction between the aliphatic compounds (paraffins and olefins) on the one part and aromatic compounds on the other.

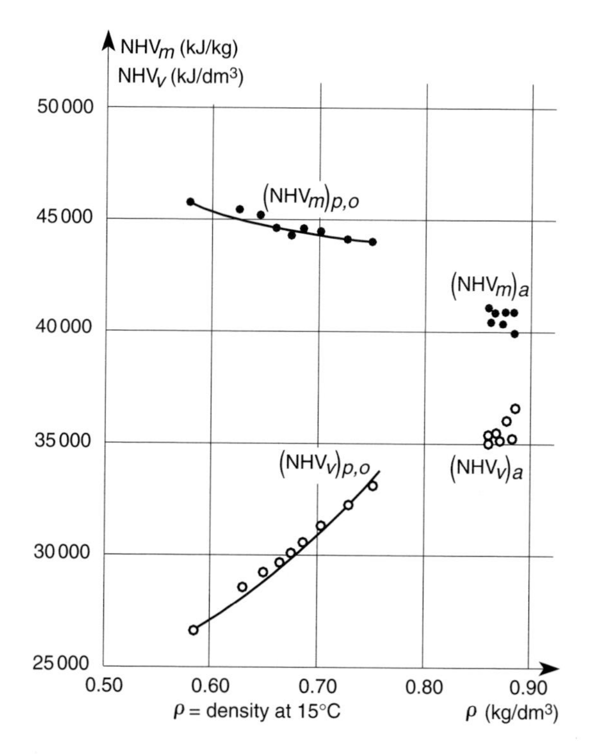

Figure 1.4 *Mass net heating value NHV_m and volume net heating value NHV_v of hydrocarbons. Where ρ, o, a = paraffins, olefins, and aromatics respectively.*

The aliphatic compounds have a greater NHV_m but a lesser NHV_v. In general, the **NHV_v increases with the density**, which, for aliphatic hydrocarbons, is correlated with the number of carbon atoms in the molecule. With regard to aromatics, the simplest product, benzene, has the greatest density. However, all the representatives of this group have high densities (greater that 0.850 kg/liter), which is consistent with a greater NHV_v.

In Table 1.18, the data for gasoline show that the NHV_v is narrowly correlated with density, which itself depends on the aromatic content. NHV_v increases from regular through to premium gasoline; in most refining schemas, obtaining higher octane indexes implies the use of significant quantities of high quality reformates that are rich in aromatics.

The data in Table 1.18 also show that diesel fuel has a mass NHV_m slightly less than that of gasoline; however, considering its much greater density (approximately 0.840 kg/dm³ instead of the average of 0.750 kg/dm³) the NHV_v is in all cases **markedly greater** than that of gasoline; this difference of almost 10% must always be considered when comparing the specific fuel consumption in liters/100 km of gasoline- and diesel-powered vehicles.

Table 1.18 *Mass and volume heats of combustion of commercial fuels (average values).*

Type of fuel	Mass NHV (kJ/kg at 25°C)	Density (kg/dm³ at 25°C)	Volume NHV (kJ/dm³ at 25°C)
LPG fuel	46 000	0.550	25 300
Regular gasoline	43 325	0.735	31 845
Premium gasoline	42 900	0.755	32 390
Jet fuel	42 850	0.795	34 065
Diesel fuel	42 600	0.840	35 785

The physical state of the reactant is an important consideration in the reporting of heats of combustion; for common fuels (gasoline, diesel, jet fuel; the liquid state is most often chosen as a reference point. However, if the fuel is vaporized by a gratuitous mechanical activity (atomization) or thermal effect (heat from exhaust gases) before entering the combustion chamber, the result is an increase in available energy that was not accounted for initially.

This gain is a modest one for hydrocarbons, because their enthalpy, which is between 300 and 500 kJ/kg, represents only 0.8 to 1% of the NHV. Therefore, whether the fuel is introduced as a liquid or as a gas is not particularly useful information with respect to hydrocarbons. However, that is not the case with other fuels such as **alcohols**. The mass enthalpy of vaporization of methanol at 25°C (1100 kJ/kg) represents 5.5% of its NHV in the liquid state. A gratuitous vaporization of methanol at the engine inlet (for example, from the heat of the exhaust gases) would be an elegant and efficient way to increase the performance of this type of fuel.

D. Specific energy

When the **compactness** of an energy conversion system is important, as it is with engines, knowing the quantity of energy contained in the **specific volume**

of fuel-air mixture undergoing combustion is important. This information can be used to predict the relationship between a type of fuel and the power developed by an engine; this is essential in the formulation of racing fuel.

If it is assumed that the fuel is delivered as liquid droplets (fuel injection), the specific energy (SE) of the fuel is given in kilojoules per kilogram of air used under defined mixture conditions. At stoichiometric conditions for example, the SE is NHV_m/r, where r is the stoichiometric ratio defined earlier.

Table 1.19 lists the specific energy of some liquid organic compounds. The variation between products is low (1 to 5%) with respect to isooctane, which is given as the reference base. Some chemical structures are the exception however, such as the **short chained nitroparaffins** (nitromethane, nitroethane, and nitropropane) that are very "energetic".

Table 1.19 *The specific energy of some organic compounds.*

Product	Specific energy	
	kJ/kg of air	Relative value
Isooctane	2 932	1.000
Decane	2 940	1.003
Hexane	2 938	1.002
Butane	2 961	1.010
Cyclohexane	2 942	1.003
Hex-1-ene	3 008	1.026
Benzene	3 032	1.034
Toluene	3 011	1.027
o and *m*-Xylene	3 000	1.023
Methanol	3 086	1.054
Ethanol	2 982	1.019
Isopropanol	2 945	1.004
Tertiobutanol	2 915	0.994
Nitromethane	6 221	2.122
Nitropropane	5 010	1.709

This is why nitromethane, for example, is the preferred fuel for the engines used in scale models; it has also been used in the past as a racing fuel (Formula 1), before the regulations were changed to prohibit its use for safety reasons.

For fuels that are gases in their natural state (LPG or methane), or that are used in this form (total vaporization upstream of the engine combustion chamber), the energetic content of the carburetted mixture (ECCM) per unit volume of the carburetted mixture not only depends on the NHV_m/r ratio, but it also depends on the molecular mass of the fuel.

Table 1.20 *The energy content of carburetted mixtures (ECCM) of various types of fuels. The fuel is assumed to be completely vaporized. Experimental conditions: equivalence ratio 1.00, temperature 25°C, and pressure 1 bar.*

Fuel	ECCM	
	kJ/liter of the carburetted mixture	Relative value
Isooctane (reference)	3.42	1.000
Methane	3.10	0.906
Ethane	3.29	0.962
Propane	3.35	0.980
Butane	3.38	0.988
Decane	3.44	1.006
But-1-ene	3.50	1.023
Acetylene	3.96	1.158
Cyclohexane	3.42	1.000
Hex-1-ene	3.47	1.015
Methylcyclohexane	3.42	1.000
Benzene	3.515	1.028
Toluene	3.505	1.025
o-Xylene	3.50	1.023
1,2,4-Trimethylbenzene	3.48	1.018
Methanol	3.38	0.988
Ethanol	3.41	0.997
Isopropanol	3.43	1.003
n-Butanol	3.44	1.006
Nitromethane	6.03	1.763
Nitropropane	4.06	1.187
Hydrogen	2.92	0.854
Carbon dioxide	3.42	1.000
Ammonia	2.83	0.827
Commercial gasoline	3.46	1.012

Table 1.20 lists the calculated ECCM values for various fuels. Compared to the isooctane reference base, gaseous fuels in their natural state have noticeably lower ECCM values (9.4% for methane, 3.8% for ethane, and 2% for propane).

The reduction is even more acute with hydrogen (15%). Alcohols also demonstrate a slight reduction in ECCM (1.2% for methanol), whereas the nitroparaffins demonstrate considerable gains (76% for nitromethane).

1.3.3.4 Exhaust gas composition

Combustion research with engines and vehicles often involves the relationship between the composition of the air-fuel mixture and the exhaust gases (Raynal 1975).

A. *Calculating exhaust gas composition*

After the complete combustion of a fuel, the carbon is normally found in the form of **carbon dioxide** and the hydrogen is found in the form of **water**. In reality, the final composition of the mixture is always far more complex. With rich mixtures, for example, the oxygen available is insufficient for complete oxidation. In addition to the combustion products mentioned above, **carbon monoxide** and **hydrogen** also occur. In any case, the resulting products can eventually undergo a number of additional reactions (dissociations and combinations) to various extents, depending the temperature and pressure. If it is assumed that **thermal equilibrium is attained**, the final composition can be calculated. Thus, for a fuel with the general formula C_nH_{2n}, Table 1.21 shows the molar fraction of the 13 major combustion products, in equilibrium at 1000 K and under a pressure of 1 bar. The calculations represent 3 levels of equivalence ratio: 0.80, 1.00, and 1.20. In fact, the real state of exhaust gases is rarely an equilibrium situation because many of the chemical reactions that lead to stability are slow and become "**quenched**" below a certain temperature threshold. As well, the presense of small quantities of unburned products is not taken into account in these estimates.

Table 1.21 *The composition of combustion products under conditions of thermal equilibrium (hydrocarbon C_nH_{2n}, temperature 1000 K, and pressure 1 bar).*

Products of combustion	**Molar fraction**		
	φ = **0.80**	φ = **1.00**	φ = **1.20**
CO_2	1.06×10^{-1}	1.31×10^{-1}	1.17×10^{-1}
H_2O	1.06×10^{-1}	1.31×10^{-1}	1.07×10^{-1}
O_2	3.97×10^{-2}	–	–
H_2	–	5.74×10^{-7}	4.23×10^{-2}
N_2	7.39×10^{-1}	7.30×10^{-1}	6.94×10^{-1}
NO	1.48×10^{-5}	1.47×10^{-9}	–
$OH^\bullet$	7.67×10^{-8}	–	–
NH_3	–	–	4.33×10^{-6}
CH_4	–	–	8.67×10^{-7}
CO	–	3.99×10^{-7}	3.20×10^{-2}
NO_2	3.39×10^{-7}	–	–
Ar (argon)	8.85×10^{-3}	8.74×10^{-3}	8.30×10^{-3}
Ne (neon)	2.84×10^{-5}	2.80×10^{-5}	2.66×10^{-5}

In practice, these complete calculations are rarely used and they are unrealistic, considering the quenching of reactions. Using only the major combustion products (CO_2, CO, H_2O, H_2, O_2, and N_2), it is possible to determine an **approximate composition** of the exhaust gases.

The data to consider then, are the elementary composition of the fuel and the mixture characteristics for lean, stoichiometric, and rich conditions.

For a **stoichiometric mixture** (φ=1) the normal products of combustion are CO_2 and H_2O only; their composition is determined by the following equation:

$$CH_yO_z + \left(1 + \frac{y}{4} - \frac{z}{2}\right)(O_2 + 3.78\,N_2) \longrightarrow CO_2 + \frac{y}{2}H_2O + 3.78\left(1 + \frac{y}{4} - \frac{z}{2}\right)N_2 \qquad (1.1)$$

For a **lean mixture** (φ < 1) the exhaust gases also contain residual oxygen according to the following equation:

$$CH_yO_z + n(O_2 + 3.78\,N_2) \longrightarrow CO_2 + \frac{y}{2}H_2O + \left(n + \frac{z}{2} - 1 - \frac{y}{4}\right)O_2 + 3.78\,nN_2 \qquad (1.2)$$

with

$$n = \left(1 + \frac{y}{4} - \frac{z}{2}\right)/\varphi \qquad (1.3)$$

For a **rich mixture** (φ > 1), CO_2, CO, H_2O, and H_2 occur simultaneously according to the following formula:

$$CH_yO_z + n(O_2 + 3.78\,N_2) \longrightarrow aCO_2 + (1 - a)\,CO + bH_2O$$

$$+ \left(\frac{y}{2} - b\right)H_2 + 3.78\,nN_2 \qquad (1.4)$$

The relative concentrations of the combustion products are linked by the following equilibrium reaction:

$$CO_2 + H_2 \rightleftarrows CO + H_2O$$

In reality, experience has shown that the composition of exhaust gases corresponds in all cases to a **quenching** of the previous reaction at a temperature of 1700 K, which leads to an equilibrium constant of about 3.8.

$$K = \frac{[CO][H_2O]}{[CO_2][H_2]} = \frac{(1 - a)(b)}{a\left(\frac{y}{2} - b\right)} = 3{,}8 \qquad (1.5)$$

By involving the oxygen balance

$$2n + z = 2a + (1 - a) + b = a + b + 1, \qquad (1.6)$$

the numeric values of a and b and the relative concentration of the effluents can be determined.

Thus, Figure 1.5 shows an example calculation of the theoretical composition of exhaust gases as a function of the equivalence ratio, based on a fuel with the general formula $CH_{1,75}$, which is close to the composition of commercial gasoline.

B. *Equivalence ratio calculations based on an analysis of the exhaust gases*

Considering some legitimate hypotheses and approximations, and with a reduced number of exhaust gas constituents, it is possible to calculate the equivalence ratio of the initial mixture. Several calculation methods have been

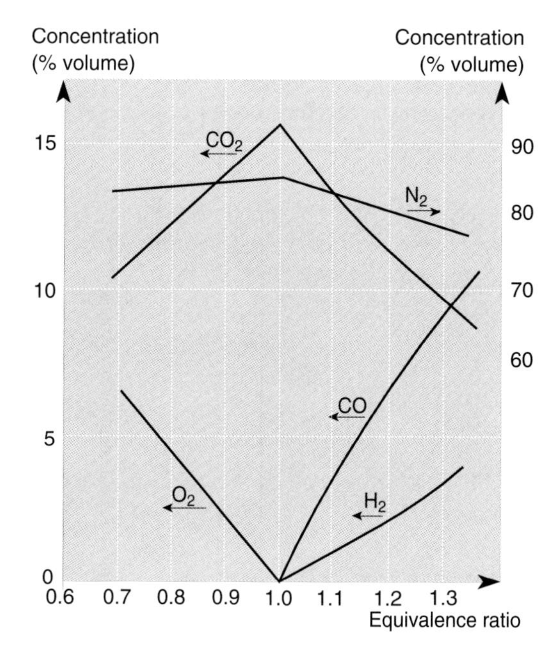

Figure 1.5
Theoretical composition of exhaust gases by volume (dry gas). The fuel composition is $CH_{1.75}$.

proposed (Spindt 1965; Eltinge 1968) and one of the most often-used methods is called "total CO_2", which consists of determining a **carbon balance**.

The balance is based on the CO_2, CO, and HC content (% volume) of the carbon products in the dry exhaust gas.

The "total CO_2" is designated by T and it consists of the following sum,

$$T = (CO_2) + (CO) + (HC)$$

where (HC) represents the methane content (1 carbon atom) equivalent to the total unburned hydrocarbons. Using the preceding equations (1.2) and (1.3), a lean mixture results in the following equation

$$T = \frac{100}{1 + \left(n + \frac{z}{2} - 1 - \frac{y}{4}\right) + 3.78n} = \frac{100}{4.78n + \left(\frac{z}{2} - \frac{y}{4}\right)} \quad (1.7)$$

and a rich mixture results in the following equation

$$T = \frac{100}{1 + \left(\frac{y}{2} - b\right) + 3.78n} \quad (1.8)$$

In the first case—a lean mixture—φ is derived directly from Eqs (1.3) and (1.7) as follows:

$$\varphi = \frac{4.78T\left(1 + \frac{y}{4} - \frac{z}{2}\right)}{100 + T\left(\frac{y}{4} - \frac{z}{2}\right)} \quad (1.9)$$

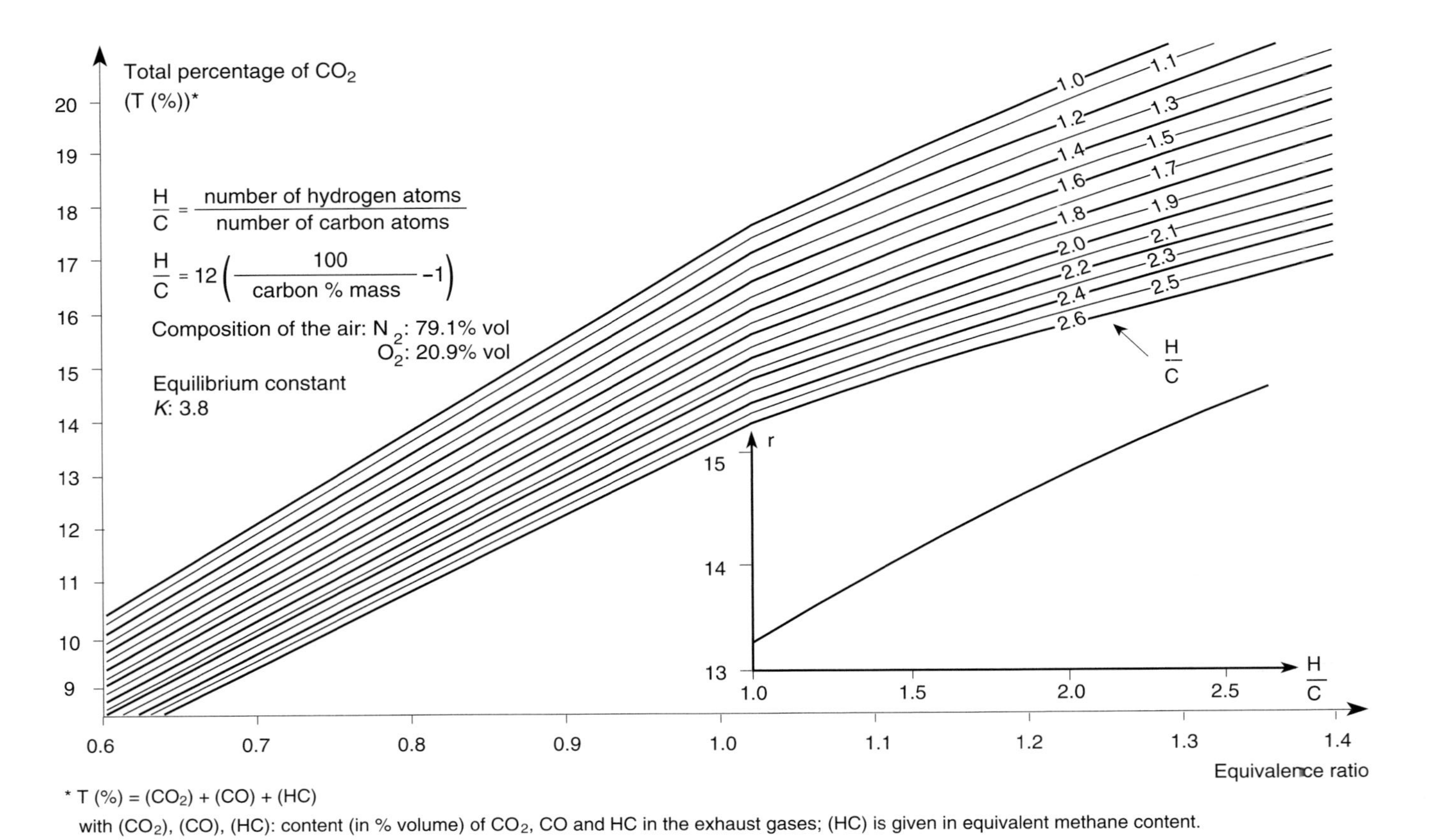

* T (%) = (CO_2) + (CO) + (HC)

with (CO_2), (CO), (HC): content (in % volume) of CO_2, CO and HC in the exhaust gases; (HC) is given in equivalent methane content.

Figure 1.6 *Equivalence ratio calculation based on the composition of exhaust gases. The method is called "CO_2 total"*

With a rich mixture, it can be seen from equations 1.5 and 1.6 that b can be expressed as a function of n using a second degree equation. Next, using Eqs. (1.4) and (1.8), φ can be calculated as a function of T.

Figure 1.6 shows the relationship between "total CO_2" and the equivalence ratio for various CH_y fuels, where the range is between 1 and 2.6.

Sometimes, apparently inconsistent results occur with multicylinder engines, which are exemplified by the presence of CO with lean mixtures and oxygen with rich mixtures. These results reflect the reality of **equivalence ratio variations from cylinder to cylinder** due to imperfect fuel delivery techniques. As a result, the exhaust manifold can simultaneously receive exhaust gases from lean, stoichiometric, and rich combustion processes.

Chapter 2

Refining techniques

The refining industry uses a large number of **chemical separations and transformations**, which enables it to obtain a full range of finished products from crude oil that have characteristics tailored to each type of application (Wauquier 1994).

This chapter describes only those aspects of refining that relate directly to the production of fuels. Following a review of the composition and characteristics of crude oil, this chapter briefly describes the various processes and devices leading to the production of the basestocks used for fuels. The specific problems involved in the formulation and fabrication of each type of product are examined in subsequent chapters (Chapters 3, 4, 7).

The chapter concludes with some information about the current status of global refining and a forecast of possible changes that may occur after 2000.

2.1 Principal characteristics of crude oils

The physical and chemical characteristics of crude oil can vary widely depending on the producing region, and the characteristics can vary widely from one petroleum deposit to another within the same **production zone**.

Table 2.1 lists the major characteristics of some crude oils (crudes) from different regions of the world. An understanding of these specifics is important because they affect transportation and storage, and the ease of obtaining finished products, and therefore the cost of refining.

Some finished products—like gasoline—only slightly resemble the crude oil from which they were produced, due to the numerous and complex transformations required to produce them. On the other hand, middle and heavy petroleum fractions (diesel, heating oil, heavy fuel oil), depend on the nature of the crude oil used in their production for some characteristics (for example, aromatic content).

Table 2.1 *General characteristics of various crude oils.*

Characteristics	Name of the crude							
	Arabian light (Saudi Arabia)	**Salaniya (Saudi Arabia)**	**Zakum (Abu Dhabi)**	**Koweït (Kuwait)**	**Kirkuk (Iraq)**	**Iran light (Iran)**	**Iran heavy (Iran)**	**Brent (UK)**
Density d_4^{15}	0.858	0.888	0.822	0.869	0.849	0.856	0.871	0.834
Viscosity (mm^2/s)	10 (at 21°C)	37 (at 21°C)	4,3 (at 20°C)	10 (at 38°C)	13 (at 10°C)	11 (at 21°C)	17 (at 21°C)	31 (at 5°C)
Pour point (°C)	–35	–29	–21	–15	–22	–29	–21	–6
Sulfur content (%)	1.79	2.85	1.05	2.52	1.97	1.35	1.65	0.26
Yields (% vol)								
gas and gasoline	20	18	28	20	23	20	18	27
kerosene and diesel fuel	35	27	37	33	32	35	34	36
atmospheric residue	45	55	35	47	45	45	48	37
Sulfur content of the atmospheric residue (350^+) (% mass)	3.1	4.3	2.2	4.1	3.8	2.4	2.55	2.1

Table 2.1 *General characteristics of various crude oils.* (continued)

Characteristics	Name of the crude							
	Statjord (Norway, UK)	**Zarzaitine (Algeria)**	**Es Sider (Libya)**	**Brega (Libya)**	**Tia Juana Light (Venezuela)**	**Bachaquero (Venezuela)**	**Rainbow (Canada)**	**North Slope (U.S., Alaska)**
Density d_4^{15}	0.833	0.811	0.841	0.823	0.865	0.954	0.822	0.896
Viscosity (mm^2/s)	7.3 (at 15°C)	6.9 (at 10°C)	5 (at 38°C)	5.6 (at 21°C)	11 (at 38°C)	300 (at 38°C)	3.8 (at 40°C)	42 (at 15°C)
Pour point (°C)	+4	−12	+7	−1	−43	−23	+2.5	−18
Sulfur content (%)	0.27	0.07	0.37	0.21	1.1	2.4	0.5	1.06
Yields (% vol)								
gas and gasoline	32	28	25	27	21	7	40	19
kerosene and diesel fuel	33	40	37	34	30	20	22	26
atmospheric residue	35	32	38	39	49	73	38	55
Sulfur content of the atmospheric residue (350^+) (% mass)	0.51	0.15	0.78	0.41	1.87	3.0	0.8	1.63

Table 2.1 *General characteristics of various crude oils. (end)*

Characteristics	Name of the crude							
	Maya (Mexique)	**Bonny Medium (Nigeria)**	**Lucina Marine (Gabon)**	**Handil (Indonesia)**	**Bombay Light (India)**	**Shengli (China)**	**Export Blend (Russia)**	**Chateau-Renard (France)**
Density d_4^{15}	0.922	0.903	0.827	0.861	0.829	0.909	0.863	0.892
Viscosity (mm^2/s)	70 (at 38°C)	12 (at 38°C)	16 (at 20°C)	4 (at 38°C)	3 (at 38°C)	100 (at 50°C)	11,5 (at 20°C)	62 (at 20°C)
Pour point (°C)	–18	–27	+15	–27	+7	+21	–23	–39
Sulfur content (%)	3.32	0.23	0.05	0.08	0.15	1.0	1.38	0.35
Yields (% vol)								
gas and gasoline	20	5	21	18	32	8	19	5
kerosene and diesel fuel	22	49	38	42	34	19	24	35
atmospheric residue	58	46	41	40	34	73	57	60
Sulfur content of the atmospheric residue (350^+) (% mass)	5.0	0.37	0.06	0.13	0.23	1.23	2.6	0.5

2.1.1 Crude oil density

The density of crude oil is a very important characteristic, mainly because it determines the price. Density is expressed in API degrees (°) related to the density at 60°F, by the following formula

$$API° = 141.5/d - 131.5$$

where d is the density (60°F/60°F).

Crude oil density generally runs from 0.800 to 1.000, although some extreme examples can occur outside this range.

Crude oil density is a general reflection of the relative proportions of the constituents: gas, gasoline, and medium and heavy fractions (see Table 2.1). Crude can be classified into the following types:

- **light crudes** with densities between 0.800 to 0.836, which come from the North Sea, the Sahara, or Libya for example, and which provide high yields of gasoline and medium fractions directly from distillation
- **intermediate crudes** with densities between 0.825 and 0.875, which usually come from the Middle East (Arabian Light)
- **heavy crudes** with densities greater than 0.890, which come from Canada, Venezuela, and Iran, and which can yield up to 80% heavy fuel oils by direct distillation

2.1.2 Other physical properties of crude oils

Refiners are also interested in other physical properties of fuels such as pour point, viscosity, and vapor pressure.

Viscosity at 20°C can vary over a wide range, for example, from 5 mm^2/s for Algerian crude (Zarzaitine) to 5500 mm^2/s for Venezuelan crude (Bachaquero).

Pour point is also an important characteristic because it affects the "pumpability" of crude oil. The pour point range can be very wide—from –60°C to +30°C.

Vapor pressure is directly related to the light-hydrocarbon content. At the well head, vapor pressure can be very high (about 20 bar) due to the presence of very-light constituents. These constituents are eliminated by a series of releases in special "separators" that enable a liquid-vapor equilibrium to be attained.

The vapor phase that results from this separation is termed "**associated gas**" and it is not very highly valued, even though its production can be significant: for example, about 0.15 tonnes per tonne of crude from some Middle-Eastern wells. The liquid element is stored and then transported to refineries. The vapor pressure at 37.8°C (50°F) is between 0.10 and 0.75 bar, depending on the origin of the crude. It is worth mentioning that some crude oils (Algeria's, for instance) have a vapor pressure that is very close to that of gasoline.

2.1.3 Overall chemical composition of crude oils

As previously indicated, the chemical composition of crude oil, other than light and medium fractions, cannot be precisely known. The constituents are essentially paraffinic, naphthenic, or aromatic hydrocarbons. Based on the proportions of these hydrocarbon groups, oil is referred to as **paraffinic** (Algeria), **naphthenic** (Nigeria), or **aromatic crude** (Venezuela).

Due to their instability, unsaturated aliphatic compounds (olefins and diolefins) are not found in crude oil. These products are produced during refining, particularly in thermal and catalytic cracking processes.

Within the heaviest petroleum fractions, the constituents are not classified by exact structure, but on purely operational definitions. Thus, the fractions that precipitate in the presence of paraffinic solvents, especially heptane (NF T 60-115, IP 143) or pentane, are called **asphaltenes**. The soluble portions are referred to as **maltenes**.

The asphaltenes consist of condensed polyaromatic rings that are linked by saturated chains whose structural organization resembles sheets of graphite. Asphaltenes appear as bright black solids with a molecular mass usually between 1 and 100 kg/mol.

Ashphaltenes in heavy fuels cause numerous problems (storage instability, incomplete combustion, deposit formation, and the emission of pollutants) in all types of uses, from industrial burners to stationary engines and marine diesels (see Chapter 7).

To prepare lubricant basestocks, asphaltenes are eliminated by using a light paraffinic solvent such as propane or butane. The industrial operation is called **deasphalting**, which results in an oil referred to as deasphalted oil (DAO) and an asphalt precipitate.

The maltenes, which are defined as that part of crude oil that is soluble in *n*-heptane, also contain heavy fractions; these fractions can be separated by liquid chromatography; the products that can be diluted by most polar solvents are called **resins**. Resins are usually molecules with a very marked aromatic character whose molecular mass can vary from 0.5 to 1 kg/mol.

2.1.4 The presence of impurities and hetero-elements in crude oils

Crude oils contain small quantities—a few dozen ppm—of **mineral salts** such as chlorides of sodium, magnesium, or calcium, gypsum, calcium carbonate, ... These salts must be eliminated before any treatment can begin. A desalting operation (water wash and caustic wash) precedes atmospheric distillation.

Among other impurities in crude oil, water is found in variable quantities (from a few ppm to more than 1%) as well as various solids (sand, drilling mud, rock debris, metals from pipelines and storage tanks). These undesirable impurities are most often eliminated by decantation.

Lastly, in the fine structure of crude oil, there are molecules, or atoms of carbon and hydrogen, that are linked to **hetero-elements** (sulfur, nitrogen, oxygen) or to **metals** (nickel, vanadium).

Sulfur is the most frequent hetero-element found in crude oils. Its content can vary from 0.1% to more than 8% (by mass), depending on the production zone (refer to Table 2.1). The density and the sulfur concentration of crude oil are the two criteria most often used to evaluate quality, and hence price. The desulfurization of feedstocks and finished products constitutes one of the major constraints of modern refining.

Within a given crude oil, the concentration of sulfur constituents increases progressively from the lightest to the heaviest fractions. This situation is shown in Table 2.2, which lists the distribution of sulfur compounds by distillation product: gasoline, kerosene, diesel oil, residual .... This comparison is based on Arab light, whose light gasoline product contains sulfur at a concentration of 1 molecule in 2000, which increases to 1 molecule in 2 for residuals!

Also, crude oils usually contain small quantities of **nitrogen** (between 0.1 and 0.5%) and **oxygen** (between 0.2 and 2%). These elements are usually bound into complex chemical structures (phenols, carboxylic acids, furans and benzofurans, amines, amides, carbazoles, pyridines, ...) and they are also concentrated in the heavy fractions like asphaltenes and maltenes. This portion of crude oil also contains organometalic compounds, which include nickel and vanadium in concentrations of a few dozen parts per million.

Subsequent sections (Chapter 7) describe usage problems that are created by the presence of nitrogen and metals in heavy fuel. Metals in particular can cause severe corrosion in large diesel engines.

Table 2.2 *An example of the distribution of sulfur products in the various fractions of a crude oil (Arabian light).*

Characteristics	Fractions					
	Light gasoline	Heavy gasoline	Kerosene	Diesel oil	Residual	Crude
Distillation range (°C)	20-70	70-180	180-260	260-370	370 $^+$	
Density d_4^{15}	0.648	0.741	0.801	0.856	0.957	
Average molecular mass (g/mol)	75	117	175	255	400	
Sulfur content (% mass)	0.02	0.03	0.20	1.44	3.17	1.80
$\frac{\text{Sulfurized mols}}{\text{Total mols}}$	$\frac{1}{1800}$	$\frac{1}{855}$	$\frac{1}{90}$	$\frac{1}{9}$	$\frac{1}{2.5}$	

2.2 The various refining operations

Refining consists of a large number of processes that can be combined under the following two major categories:

- **separation processes**, which divide feedstocks into narrow fractions according to given criteria (temperature, solubility) without using chemical reactions
- **transformation or purification processes**, which create new molecules or eliminate undesirable compounds—all these transformations involve many chemical reactions

A third process for obtaining finished products also exists, which is not based on crude oil, but based instead on natural gas.

2.2.1 Separation processes

The two major separation techniques are **distillation at atmospheric pressure**, and **vacuum distillation** with solvent extraction (or **deasphalting**) of the residual. Fig. 2.1 shows the general schema of these operations and the various types of products that are obtained.

2.2.1.1 Primary distillation (at atmospheric pressure) of crude oil

The first separation of crude oil takes place in a **distillation column** with multiple plates, which operates not quite at atmospheric pressure, but under low pressure (1.5 to 3 bar).

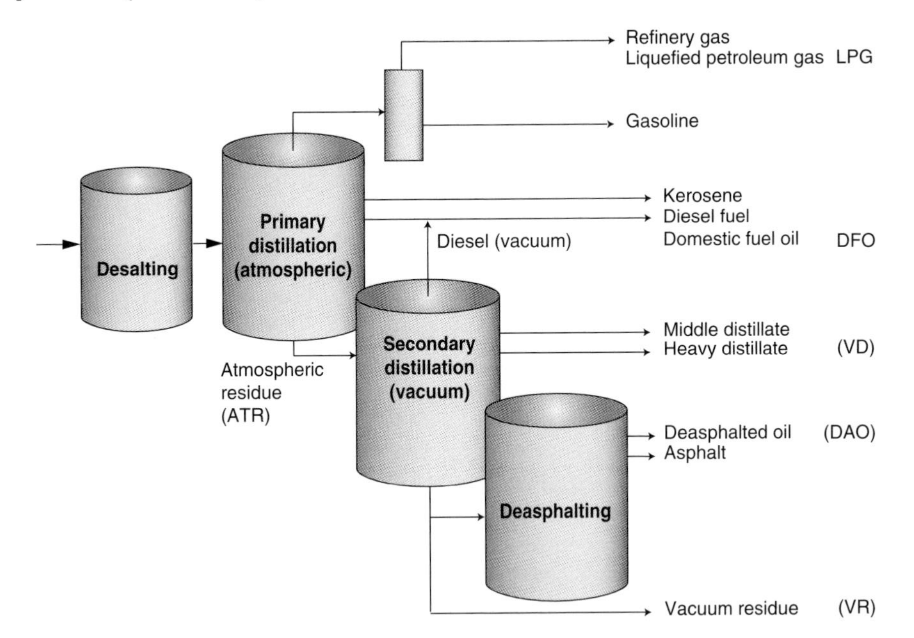

Figure 2.1 *General schema for the separation operations (distillation, deasphalting) in a refinery.*

Partially vaporized crude oil is introduced into the lower part of the column at a temperature of about 350°C. The heating is carried out in a furnace that consumes between 8 and 15 kg of fuel per tonne of crude, which is between 0.8 and 1.5% of the feedstock.

The column has 40 to 50 plates and it can be up to 50 meters tall. The products drawn off on the sidestreams can contain volatile constituents that require elimination. This is accomplished by a partial revaporization or "**stripping**" in subsidiary columns with 4 to 10 outlets.

The following major products are derived from this type of distillation:

- **refinery gas** (fuel gas), which contains light hydrocarbons like ethane and propane
- **liquefied petroleum gas** (LPG), which is made up of propane and butane
- **light and heavy gasoline**
- the **kerosene fraction** (jet fuel, lamp oil)
- straight run **diesel fuel and domestic fuel oil**
- **atmospheric residue** (ATR), which is used as a base for heavy fuel oil or a feed for vacuum distillation

Table 2.3 lists the yields and the major characteristics of the various fractions of a typical crude oil.

Table 2.3 *An example of the properties of various fractions obtained from atmospheric and vacuum distillation.*
Crude oil: 50/50 (% vol) mixture of Arabian light and heavy; d_4^{15} = 0.875.

Characteristics	Fractions								
	Fuel gas	GPL	Light gazoline	Heavy gasoline	Kerosene	Diesel	ATR	VD	VR
Yield (% mass)	0.28	1.09	3.87	13.85	6.74	24.37	49.80	23.50	26.30
Distillation range (°C)			30-80	80-180	180-225	225-375	375$^+$	375-550	550$^+$
d_4^{15}			0.654	0.742	0.793	0.851	0.986	0.935	1.037
Sulfur (% mass)			0.003	0.035	0.15	1.4	3.95	2.8	5.0
Nitrogen (% mass)								1000	3500
Viscosity 50°C (mm²/s)						2.4			
Viscosity 100°C (mm²/s)						1.1	85	9	3300
Ni (ppm)							25		47
V (ppm)							73		138

2.2.1.2 Secondary distillation (under vacuum) of the atmospheric residue

The vacuum distillation process is fed with the residue from the atmospheric distillation process, which has been partially vaporized by being heated to 390 to 430°C in a furnace. The pressure in the column varies from top to bottom from about 60 to 90 mbar. These operating conditions reduce the boiling points, maintaining them under 400°C, which prevents any thermal decomposition of the effluents.

Two types of products are extracted:

- one or several **vacuum distillates** (VD), which are used as feedstocks for transformation process (catalytic cracking) or as basestocks for formulating lubricants
- **vacuum residues** (VR), which are used to make bitumen, heavy fuels, or feedstocks for conversion processes

The energy consumption of the distillation processes (atmospheric and vacuum), when expressed as a ratio of the treated crude oil, is around 2.5%.

2.2.1.3 Deasphalting of the vacuum residue

Deasphalting is a **solvent extraction** process that separates the lighter hydrocarbons from the vacuum residue. The process yields deasphalted oil (DAO) and asphalt (Fig. 2.1). The process uses light paraffins such as propane, butane, and pentane as solvents. The yield of DAO increases with the molecular weight of the solvent, but the quality declines. DAO is used as a feedstock for conversion processes (see 2.2.2.2); most of the impurities in crude oil (metals, sediments, asphaltenes, and mineral salts) are concentrated in the asphalt, which means that the only practical use for it is as an industrial fuel for power plants.

2.2.2 Conversion processes

Conversion processes can be classified under three very general types of operations:

- techniques used to obtain **specific basestocks**
- thermal or catalytic **conversion processes**
- **finishing processes** such as hydrotreating, hydrogenation, and sweetening

2.2.2.1 Techniques for obtaining specific basestocks

These techniques consist of chemical transformations that mainly involve the various groups of hydrocarbons, but it can also involve alcohols such as methanol, ethanol, ...

A. *Catalytic reforming*

Catalytic reforming is actually a basic process in the production of gasoline (Antos et al. 1995) and it represents a major source of **hydrogen** for refineries.

The reactions involved in catalytic reforming are the dehydrogenation of naphthenes and the dehydrocyclization of paraffins (see 1.1.4.2) to obtain structures that are rich in **aromatics** (aromatization).

The major feedstock for catalytic reforming is heavy gasoline (80 to 180°C) from the primary distillation process. If necessary, catalytic reforming can also use the gasoline produced by other processes such visbreaking, coking, hydrocracking, as well as the middle gasoline fractions from catalytic cracking. Feedstocks must receive a high level of hydrotreatment to remove impurities (sulfur, nitrogen, and metals) or undesirable constituents (olefins), which can act alone or in combination to poison the catalysts.

Modern reforming processes operate at low pressure (2 to 5 bar) and at high temperature (510 to 530°C) by **continuous regeneration** of the catalysts.

The catalysts usually consist of aluminum chloride impregnated with platinum. Aluminum chloride's acidity helps to change molecular structures. Platinum aids dehydrogenation, which yields aromatic compounds; it also aids hydrogenation, which limits the formation of coke.

The products are essentially hydrogen, C_3 and C_4 hydrocarbons, and a liquid fraction called **reformate**, which is rich in aromatics. Table 2.4 provides an example of an operational summary of a modern catalytic reforming operation, which includes the characteristics of the feedstocks and the products, as well as the yields obtained. The process is very flexible, meaning that the severity of treatment, which affects the proportion and composition of the products, can be easily adjusted by changing the operating conditions.

Table 2.4 *A sample material balance for a catalytic reforming process.*

Feedstock		Products		
Distillation range (°C)	80-120	**Yields** (% mass)	H_2	3.00
			$C_1 + C_2$	3.75
Volumetric mass (kg/dm³)	0.742		C_3	3.50
			iC_4	1.75
Chemical composition (% vol)			nC_4	2.50
n-paraffins + isoparaffins	73		C_{5+}	85.50
naphthenes	15		**Total**	**100.00**
aromatics	12	**Characteristics of the reformate** C_{5+}:		
Sulfur (ppm)	< 0.5			
Nitrogen (ppm)	< 0.5	**Volumetric mass** (kg/dm³)		0.810
Water (ppm)	< 4	RON		102
Olefins (ppm)	< 1	MON		92
		Aromatic content (% vol)		60
		Vapor pressure (bar)		0.3

B. Isomerization

The isomerization process consists of transforming normal paraffins into **isoparaffins**, either to obtain products for other transformation reactions (production of isobutane from *n*-butane to feed an alkylation process) or to increase the octane rating of the light gasoline derived from direct distillation, which cannot be treated by catalytic reforming.

In the latter case, the feed is primarily paraffins that have little C_5 and C_6 branching. The process operates at a temperature of 130 to 160°C and at a pressure of 20 to 40 bar in the presence of hydrogen, to limit parasitic coking reactions and dismutation ($C_5 + C_5 \rightarrow C_4 + C_6$).

The catalysts consist of aluminum chloride or zeolite acid with small quantities of platinum to limit the formation of coke. The feed must be free of impurities such as sulfur, nitrogen, and water.

The products are rich in C_5 isoparaffins (isopentane) and C_6 isoparaffins (mostly 2,2- and 2,3-dimethylbutane). The severity of the process can be increased by separating the unreacted pentane and hexane from the primary outflow and recycling them through the reactor.

Table 2.5 provides an example of the feedstock characteristics and the products involved in an isomerization process.

C. Alkylation

The alkylation process consists of the **addition** of isobutane to light olefins—preferably butenes, but also propenes or pentenes—to obtain C_7 to C_9 isoparaffins. The theoretical reaction schema is

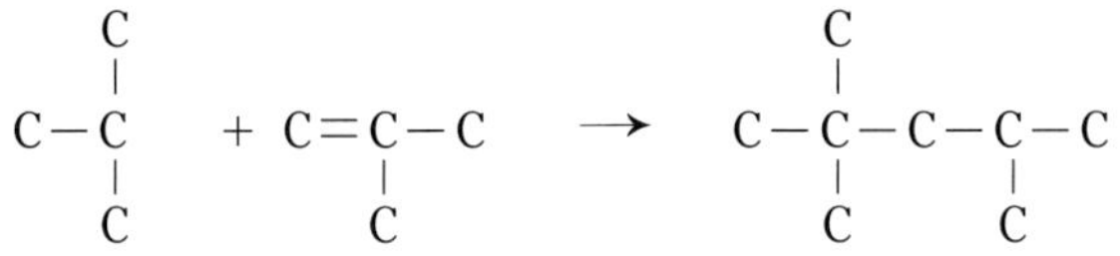

$$\begin{array}{ccccc} \text{C} & & & & \text{C} \\ | & & & & | \\ \text{C}-\text{C} & + & \text{C}=\text{C}-\text{C} & \rightarrow & \text{C}-\text{C}-\text{C}-\text{C}-\text{C} \\ | & & | & & |\qquad\quad | \\ \text{C} & & \text{C} & & \text{C}\qquad\quad \text{C} \\ \text{isobutane} & + & \text{isobutene} & \rightarrow & \text{isooctane} \end{array}$$

This reaction is very exothermic and it takes place in the liquid phase at low temperature (30°C) under a pressure of about 12 bar. The catalysts are **strong acids**, either sulfuric acid ($H_2\,SO_4$) or hydrofluoric acid (HF).

Each catalyst presents an environmental problem, which may consist of the transport and treatment of sulfuric sludge or the risks of an accidental leak of hydrofluoric acid. To solve this problem, research is currently under way to find new alkylation techniques that use solid catalysts.

The olefins used in the alkylation process are derived mainly from catalytic cracking, whereas the isobutane comes from the direct distillation of C_4 crude oil cuts and catalytic reforming.

Table 2.6 provides a sample of feedstock and product characteristics for an alkylation process.

Table 2.5 *A sample of the characteristics of the feedstocks and products of the isomerization of a light gasoline.*

Feedstock		Products		
Distillation range (°C)	30-80	**Yields** (% mass)	H_2	−0.2
			Gas $C_1 + C_4$	2.2
Volumetric mass (kg/dm³)	0.65		Liquid C_{5+}	98.0
			Total	**100.00**
Chemical composition (% vol)		**Characteristics of the isomerate** C_{5+}:		
n-paraffins	50			
isoparaffins	39			
naphthenes	9	**Density** (kg/dm³)		0.647
aromatics	2	RON		88
RON	69	MON		86
MON	66			
Sulfur (ppm)	< 0.5	**Isoparaffin content** (% vol)		90
Nitrogen (ppm)	< 0.5			
Water (ppm)	< 1			

Table 2.6 *A sample material balance for an alkylation process.*

Type de constituent	Feedstock (% mass)	Products* (% mass) Gas	Products* (% mass) Alkylate (raw)
Propene	0.80	–	
Isobutane	39.98	0.40	0.04
Butane	11.14	8.92	2.23
Butenes	47.58	8.13	1.93
Pentane	–	0.45	1.66
Pentenes	0.50		0.16
C_{5+}	–	0.38	75.40
	100.00	18.28	81.42

Alkylate characteristics

		ASTM D86 Distillation	
d_4^{15}	0.710		
RVP (bar)	0.4	IP (°C)	32
RON	97.6	50%	106
MON	94.4	FP (°C)	198

* The estimated losses are 0.3% by mass.

D. Ether synthesis

These processes involve adding alcohols—methanol or ethanol—to C_4 olefins (isobutene) or to C_5 olefins (pentenes and amylenes).

The following reaction schemas apply:

$$CH_3-OH + C{=}\underset{\displaystyle C}{\underset{|}{C}}-C \rightleftarrows C-\overset{\displaystyle C}{\overset{|}{\underset{\displaystyle C}{\underset{|}{C}}}}-O-CH_3$$

methanol + isobutene ⇄ methyl tertiary butyl ether (MTBE)

$$CH_3-CH_2OH + C{=}\underset{\displaystyle C}{\underset{|}{C}}-C \rightleftarrows C-\overset{\displaystyle C}{\overset{|}{\underset{\displaystyle C}{\underset{|}{C}}}}-O-CH_2-CH_3$$

ethanol + isobutene ⇄ ethyl tertiary butyl ether (ETBE)

$$CH_3-OH + C{=}\underset{\displaystyle C}{\underset{|}{C}}-C-C \rightleftarrows C-C-\overset{\displaystyle C}{\overset{|}{\underset{\displaystyle C}{\underset{|}{C}}}}-O-CH_3$$

methanol + isopentene ⇄ tertiary amyl methyl ether (TAME)

The **equilibrium** of these reactions require an adjustment of operating conditions (pressure, temperature, relative concentration of the reactants) to optimize the ether yield.

MTBE is readily synthesized using a sulfonic resin catalyst at a temperature between 50 and 80°C and at a pressure of 15 to 20 bar, which is required to keep the C_4 liquid. The conversion yield is between 95 and 98%.

The preparation of **ETBE** is a more delicate operation because the thermodynamic conditions are less favorable. The catalyst is the same type used to synthesize MTBE and the same installation can be used alternatively to produce either MTBE or ETBE (Chatin 1992).

The reaction to produce **TAME** takes place in the liquid phase at a pressure of 4 or 5 bar. The transformation remains incomplete and the TAME is used in a mixture along with the excess olefins that have not reacted.

The specific properties of the oxygenated basestocks produced by ether synthesis are provided in Chapter 6.

E. Dimerization or oligomerization of olefins

This process consists of producing **dimers** or **trimers** of C_3 light olefins and C_4 light olefins (propene, butenes), which most often result from catalytic cracking.

Propene can be readily dimerized into isohexane at a conversion rate of better than 95% by using complex transition metal (nickel) catalysts in solution. The reaction takes place at a temperature of around 50°C and at low pressure (15 bar). The product is an interesting basestock for the production of gasoline (RON of nearly 100), but its olefinic structure results in some problems (MON of only 80 and a suspicion of its involvement in tropospheric pollution [see Chapter 5]).

Currently, work on oligomerization is directed towards the synthesis of diesel oil fractions (C_9 to C_{12}) by the trimerization or tetratrimerization of propene, followed by hydrogenation (O'Connor et al. 1990).

2.2.2.2 Conversion processes

The term conversion refers to the group of refinery operations that consist of the relatively deep transformation of heavy fractions into a host of lighter products such as gas, gasoline, kerosene, and diesel oil. The conversion can result in a residue that is heavier than the initial feedstock.

The conversion processes are used with high-boiling-point petroleum cuts, which usually exceed a temperature of 380°C. Fig. 2.2 shows the origin of these cuts in a schematic view. The following cuts are the main ones used in the conversion:

- **atmospheric residue**
- **vacuum distillates**, which are usually referred to as Vacuum Gas Oil (VGO)
- **vacuum residue** (deasphalted or non-deasphalted)

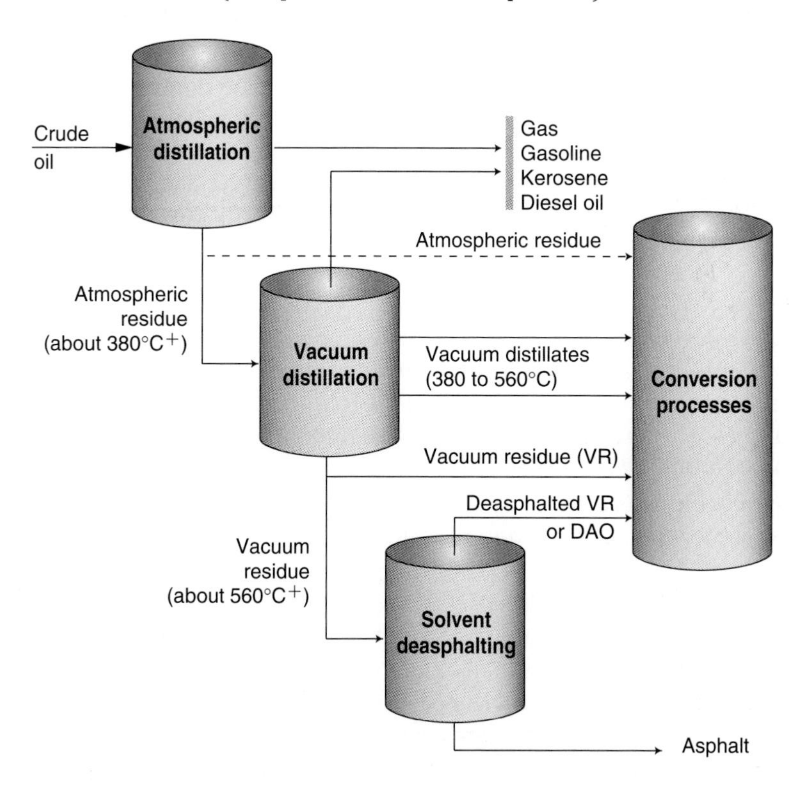

Figure 2.2 *The origin and characteristics of the feedstocks used in conversion processes.*

Table 2.7 describes some examples of the characteristics of feedstocks that could be used for conversion processes. All conversion processes involve chemical cracking reactions that are either thermal or catalytic, and which occur in either the presence or the absence of hydrogen. The various processes are classified according to reaction schemas or operating conditions.

A. Thermal conversion

Thermal cracking begins as soon as a petroleum feedstock reaches about 400 to 420°C, which is the threshold for the decomposition of a large number of hydrocarbons. The severity of the process depends on the temperature and the residence time.

This section describes three thermal cracking processes: visbreaking, coking, and steam cracking.

a. Visbreaking

Visbreaking (or viscosity breaking) is a **low severity** thermal cracking process that reduces the viscosity of atmospheric or vacuum residues without the formation of coke. The feed is heated in a furnace to 450°C and it is kept in a soaking chamber where the cracking reactions take place at constant temperature under a pressure of 10 bar. The products are separated and distilled. Table 2.8 provides a sample operating summary for a visbreaking process. The gas, gasoline, and diesel oil, which are rich in sulfur, nitrogen, and olefins, must be hydrotreated before being mixed into the fuel stream.

This process enables a refinery to **reduce the production of heavy fuel oils** while maintaining the viscosity and thermal stability specifications required by the consumer's fuel systems (marine diesel engines, industrial burners). In addition, the diesel fuel output from visbreaking can be used as a diluent for heavy fuels, thereby saving the higher-quality straight-run distillate for the refinery diesel fuel pool.

b. Coking

Coking is a **severe** cracking process that results in the total conversion of some residues into gas, gasoline, diesel oil, distillates, and coke. The process takes place at temperatures exceeding 500°C. The following two methods can be used:

- **Delayed coking** is a fixed charge discontinuous process with a long residence time (24 h cycles).
- **Fluid coking** is a continuous process that uses a fluidized bed technique.

Delayed coking can be used to produce very pure coke that can be used in the production of electrodes. The coke obtained from fluid coking is used as fuel. A variation of the process (*Flexicoker*) developed by *Exxon* uses an air-blown coke gasifier that produces a low-heating value gaseous fuel.

The coking process produces significant quantities of gas and light to medium fractions (Fig. 2.3). All the products are unstable due to the presence of olefinic or diolefinic constituents, and sulfur and nitrogen compounds. The products must be hydrotreated before being returned to the gasoline or diesel fuel pool.

Table 2.7 *Sample feedstock characteristics for a conversion process.*

Characteristics	Feedstock					
	Arabian light				Arabian heavy	Tia Juana
	Atmospheric residue	Vacuum distillate 1	Vacuum distillate 2	DAO	Atmospheric residue	Atmospheric residue
Crude based yield (% vol)	44.3	23.0	2.75	14.85	54.8	80.2
Density d_4^{15}	0.962	0.912	0.943	0.973	0.995	1.017
Viscosity at 100°C (mm^2/s)	36.9	6.0	24.5	120.0	180.0	810.0
Sulfur content (% mass)	3.10	2.20	2.80	3.65	4.40	3.18
Asphaltenes content (% mass)	2.60	–	< 0.05	< 0.05	7.0	7.20
Metal content Ni + V (ppm)	42	< 1	< 2	21	125	550
Conradson carbon (% mass)	10.8	< 0.1	2.1	8.5	15.3	15.5

Table 2.8 *A sample material balance for a visbreaking process. (VR feedstock—Arabian heavy).*

Characteristics	Feedstock	Products			
		Light gasoline	Heavy gasoline	Diesel oil	Residue
Cut (°C)	538 +	C_5*-80	80-150	150-350	350 +
Yield (% mass **)	100	1.20	2.40	12.45	82.00
d_4^{15}	1.048	0.680	0.745	0.865	1.065
Sulfur (% mass)	5.78	0.5	1.4	3.35	6.1
Viscosity at 100°C (mm^2/s)	4500			1.1	2600
Nitrogen (ppm)	3500	5	25	300	4100
Bromine number (g/100 g)		85	70	20	
MAV (mg/g)		20	12		
Ni + V (ppm)		290			350

* The lightest constituents of the cut are C_5 hydrocarbons (pentane, isopentane) with boiling points between 28 and 35°C.

** The other products obtained are H_2S (0.35%) and C_1-C_4 (1.60%).

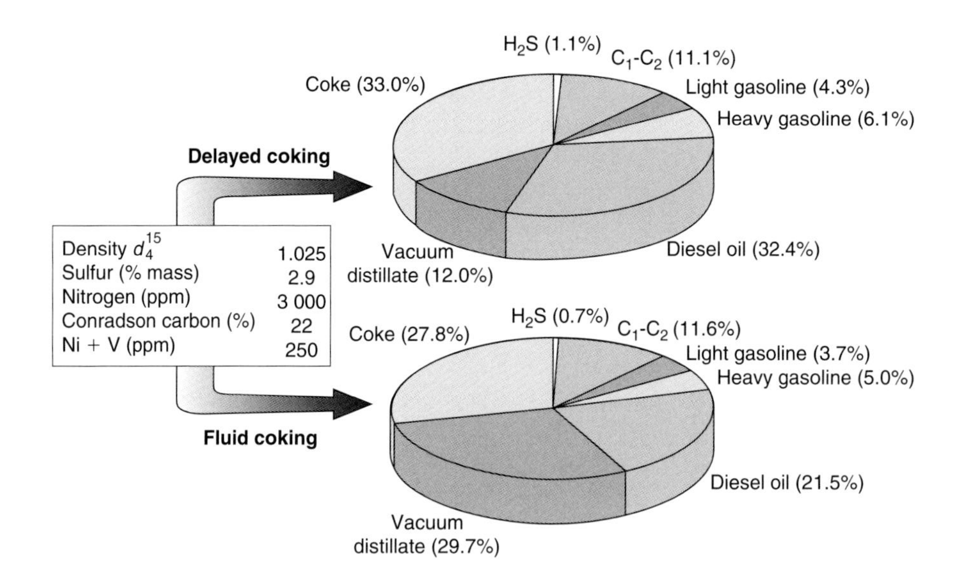

Figure 2.3 *Delayed and fluid coking processes. Sample material balance for a VR feedstock.*

c. Steam cracking

Steam cracking should not really be classified as a refining process. It is essentially a basic **petrochemical** process, which produces light olefins (mostly ethylene) and aromatics (benzene, toluene, xylene) from light crude fractions (LPG, naphtha) and also from some heavy fractions.

Steam cracking is a thermal cracking process that operates at **high temperature** (800 to 850°C) with a very-short residence time (about one hundredth of a second) in the presence of steam.

Some liquid fractions (C_5 cuts) that result from steam cracking can be added to the gasoline pool after hydrogenation.

B. Fluidized catalytic cracking

Fluidized catalytic cracking (FCC) has been understood for some time, since the first industrial unit was built in 1936 at Marcus-Hook in California. French engineer Eugène Houdry was one of the pioneers of this refining technology.

Catalytic cracking has progressed a long way since then, due to the development of new catalysts and other technological improvements. This process is now used in almost all industrialized countries with developed markets for fuels.

Modern catalytic cracking operates in the gaseous phase at low pressures of 2 to 3 bar. The catalyst is used as a solid heat transfer medium. The temperature of the reaction varies from 500 to 540°C and the residence time is about one second. The products are essentially of an **olefinic** nature for the light fractions and of an **aromatic** nature for the heavy fractions.

The catalysts currently in use are an acidic mixture of amorphous aluminum silicate and **zeolite** (10 to 40%). In general, zeolites are three dimensional crystalline aluminosilicate structures with evenly spaced interconnected pores. During a reaction, molecules penetrate the structure of the zeolite and adsorb on the active areas to react. A catalyst of this type has the advantage of presenting a large reactive surface, which is larger than an amorphous catalyst, and it also provides better reaction selectivity. A large pore (100 nm diameter) zeolite catalyst is used for catalytic cracking, which allows the largest molecules access to the active areas. Its acidic characteristics contribute to its cracking activity, which is sometimes improved by the presence of rare earths like lanthanum, cerium, ...

The catalysts used in catalytic cracking are not poisoned by impurities like sulfur and nitrogen, however, they can be poisoned by metals, even in trace amounts. Therefore, feedstocks must be entirely free of metals.

A typical catalytic cracker consists of two major sections (see Fig. 2.4):

- the **reactor** (or riser) in which the cracking reactions take place with a coating of coke on the catalyst
- the **regenerator**, where air injection enables the burning of coke at 650 to 720°C

This general schema is open to appreciable technological improvements. Thus the R2R (1 riser-2 regenerator) process developed by *Total* and *IFP*

(Mauléon et al. 1994) uses two-stage regeneration and it optimizes the operation of the riser, which results in important advantages: reduced coke formation, easier maintenance, possible use with non-demetalized feedstocks, … .

Catalytic cracking delivers a wide-ranging group of products that are subjected to subsequent separations and other treatments.

The following products are obtained from catalytic cracking:

- **very light hydrocarbons** (C_1 or C_2), which are called refinery gas
- **LPG** (propane, propene, butanes, butenes), which can be used as a gasoline component and as feedstock for the synthesis of basestocks for gasoline (alkylation, oligomerization, etherification, …)
- a **gasoline fraction** that is relatively rich in aromatics and olefins, which can be subdivided into the following three sub-products:
  - a **light cut** (C_5), which is highly olefinic and which can be used to make TAME

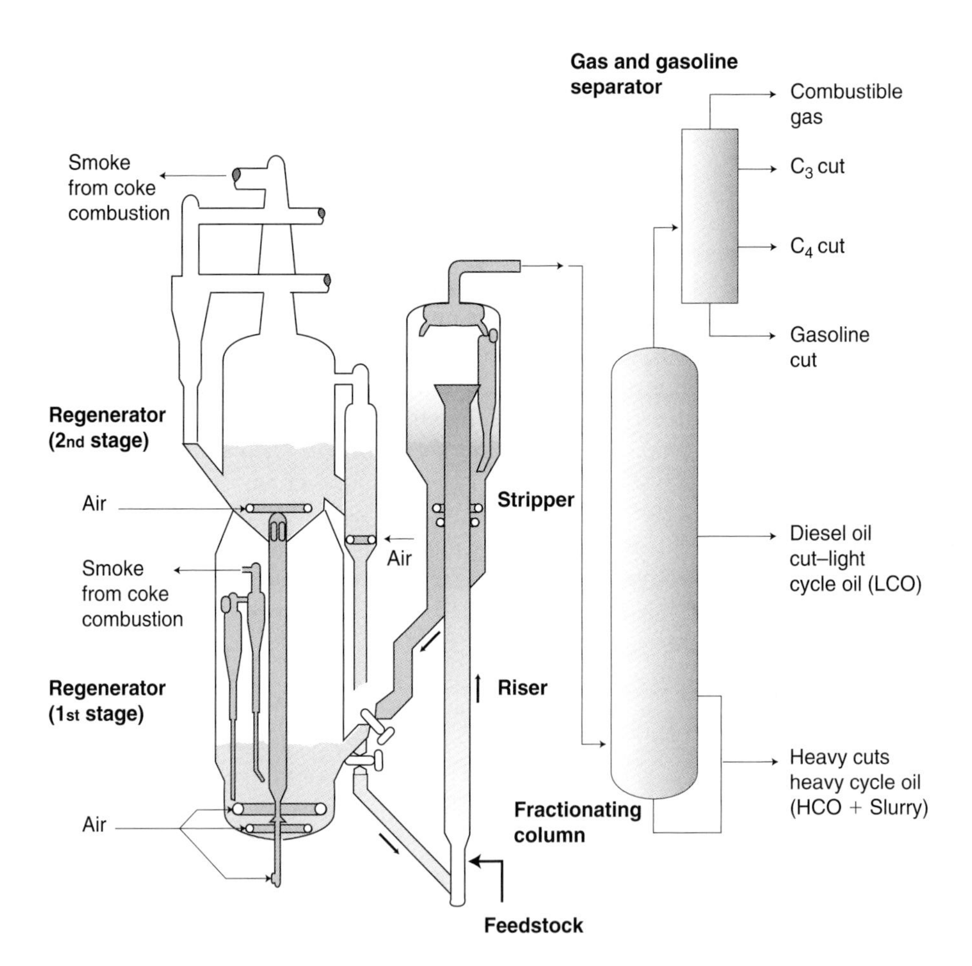

Figure 2.4 *General schema of a catalytic cracking unit.*

  - a **central cut** (75 to 125°C) with a generally weak octane rating that could, for example, be sent to catalytic reforming
  - a **heavy cut** (125 to 210°C), which is highly aromatic and most often used directly in the gasoline pool
- a **light distillate**, referred to as *Light Cycle Oil* (LCO), which is close to being diesel fuel, but it is too aromatic to be sent in large amounts to the diesel pool
- **heavy fractions** that consist of a dense ($d_4^{15} \simeq 1.000$) aromatic liquid that is referred to as *Heavy Cycle Oil* (HCO), which is an almost-solid slurry that contains numerous impurities (coke, catalyst particles). The HCO and the slurry are not separated in some cases

The **coke**, which is deposited on the catalyst and burned in the regenerator, must be accounted for in the final balance. The combustion of the coke and the slurry produces energy (electricity, steam) that is used by other processes in the catalytic cracking unit.

Table 2.9 provides some examples of data related to the operation of an FCC unit for three possible types of feedstock, which are hydrotreated and unhydrotreated. The precise characteristics of the products are covered in more detail in Chapters 3 and 4.

C. Hydrocracking

Hydrocracking is a cracking process carried out under **high hydrogen pressure** (Maier 1988). This technique combines cracking reactions and hydrogenation and it leads primarily to the formation of saturated paraffinic and naphthenic hydrocarbons; it does not produce coke. As opposed to catalytic cracking, which is oriented to the production of gasoline, the hydrocracking process offers great **flexibility** in the range of products that can be obtained. Its operation can be directed toward maximizing diesel fuel, kerosene (Bigeard et al. 1994), or gasoline, depending on the needs of the refinery. This provides an important level of flexibility to meet market demands.

The feedstocks are distillates from the vacuum distillation of crude oil or from other conversion processes (catalytic cracking, visbreaking, coking, deep conversion. Feedstocks are severely hydrotreated before hydrocracking.

This step is essential, because it eliminates the nitrogen atoms and the organic sulfur that is present in heavy hydrocarbons, which can poison the hydrocracking catalysts.

The hydrocracking process takes place at high temperatures of 350 to 400°C and under heavy hydrogen pressures of 150 to 200 bar. The process uses a fixed bed, because the production of coke is limited and continuous regeneration is unnecessary. The hydrocracking complex must include an adjacent **hydrogen production unit**, because hydrogen is consumed at the rate of 15 to 25 kg per tonne of feedstock.

Hydrocracking catalysts have two functions, one for **acid** reactions and the other for **hydrogenation**. The acidity is usually supplied with the aid of the zeolite base and the hydrogenation is supplied with the aid of one or more

Table 2.9 *An example of the feedstocks and products from an FCC unit.*

Feedstock	VD*	Hydrotreated VD*	Hydrotreated ATR*
Distillation interval (°C)	375-550	375-550	375^+
d_4^{15}	0.935	0.923	0.933
Sulfur (%)	2.80	0.15	0.35
Nitrogen (ppm)	1000	300	1000
Viscosity at 100°C (mm^2/s)	9	8.5	23
Ni (ppm)	< 1	< 0.5	2
V (ppm)	< 1	< 0.5	4
Products		**(% mass)**	
H_2S	1.35	0.09	0.18
Refinery gas (C_1-C_2)	3.50	2.80	3.90
Propane	1.16	1.32	1.25
Butane	1.77	3.40	3.08
Propene	4.62	5.28	5.00
Butenes	5.95	6.63	6.27
Gasoline (C_5-220°C)	42.70	49.00	45.80
LCO (220-360°C)	20.95	17.55	17.55
Slurry ($360°C^+$) **	12.80	9.33	9.37
Coke	5.20	4.60	7.60
Total	**100.00**	**100.00**	**100.00**

* Origin of the crude : Arabian light and heavy 50/50.
** This fraction includes the HCO.

metal compounds deposited on the catalyst base. The type of catalyst chosen depends on the type of product being sought and it also determines the operating schema of the process.

A palladium-based zeolite catalyst requires a completely sulfur-free feedstock because sulfur poisons the palladium. The products that flow from previous hydrotreatment processes are free from NH_3 and H_2S gases; these products are then distilled and only the residue is used for hydrocracking. This **two-stage** configuration is usually used when the demand for gasoline is strong because high yields of light and heavy gasoline can be obtained (see Table 2.10), which are excellent feedstocks for catalytic reforming.

Hydrocracking catalysts can also consist of zeolite with deposits of nickel oxide and molybdenum. In this case, all the products from previous hydrotreatments are hydrocracked in one or several reactors in series. The NH_3 and H_2S

Table 2.10 *Flexibility of the hydrocracking process.*

Products	Yields (% mass)		
	Maximum diesel oil	**Maximum kerosene**	**Maximum gasoline**
Gas ($H_2S + NH_3$)	3.7	3.7	3.8
Propane cut	0.6	1.6	2.2
Butane cut	2.0	4.5	11.2
Light gasoline	8.5	12.0	25.3
Heavy gasoline	16.5	24.0	61.0
Kerosene	23.0	57.0	
Diesel fuel	48.2		
Total	**102.5**	**102.8**	**103.5**
Hydrogen consumption (% mass)	2.5	2.8	3.5

gases released in this phase of purification are not separated from the other products, because the presence of sulfurized gas is essential to retain, and sometimes to initiate, the reactivity of the hydrogenation sites on the catalyst. All the products are distilled and any unconverted residue from the bottom of the column that is not used directly (as additional feed for catalytic cracking, basestock for lubricants, ...) is recycled for additional hydrocracking. This schema, which consists of only **one stage** (no intermediate separation) favors the production of kerosene and diesel fuel (see Table 2.10). The process can be directed toward the production of naphthas by using appropriate sulfur-resistant catalysts that are combined with efficient recycling of the residue.

This remarkable flexibility and the quality of the products (see 4.11.1 and 7.4.1) make hydrocracking an attractive and effective conversion technique. Its growth has been limited by its high cost (investment and operating costs) and its high energy consumption.

Mild hydrocracking (Hennico et al. 1993), which is a less severe version of the technique described above, has recently been proposed. It operates at lower pressure (40 to 60 bar) with conversion rates of 25 to 35%. The objective of the process is to prepare—from heavy vacuum distillates—feedstocks for catalytic cracking and lubricant basestocks.

D. The hydroconversion of residues

This technique can be considered as a **deep conversion process** that is intended to minimize the production of heavy fractions from crude oil.

The intended objective of hydroconversion (Billon et al. 1988) is the partial conversion of all heavy products—both atmospheric and vacuum residues—that contain large amounts of impurities (sulfur, nitrogen, metals, asphaltenes).

The process operates at **high pressure** (150 to 200 bar) in the presence of **hydrogen**. The consumption of catalysts is significant because they are rapidly covered over with the metals contained in the feedstock (nickel, vanadium).

The rate of conversion is also limited by the presence of asphaltenes, which tend to concentrate and precipitate in the heavy products, thus rendering them unsuitable for consumption.

Table 2.11 shows two sample material balances for the hydroconversion for an atmospheric residue and a vacuum residue respectively. Other than small quantities of saturated gas (C_1 to C_4) and gasoline, the process provides a diesel fuel cut, a distillate, and a residue.

With subsequent treatment, the gasoline and diesel fractions can be mixed to the fuel pool; the heavy fractions, which are almost completely free of metals, can be used as feedstocks for other processes (catalytic cracking, hydrocracking).

Table 2.11 *Sample material balances for the hydroconversion of residues.*

Characteristics	Type of feedstock	
	Atmospheric residue	Vacuum residue
Distillation range (°C)	375^+	550^+
d_4^{15}	0.986	1.037
Sulfur (% mass)	3.95	5,0
Nitrogen (ppm)	2300	3500
Viscosity at 100°C (mm²/s)	85	3300
Asphaltenes (% mass)	5.6	10.6
Conradson carbon (% mass)	12.5	22.6
Ni (ppm)	25	47
V (ppm)	73	138
Types of product	**Yields (% mass)**	
$H_2S + NH_3$	3.88	4.84
$C_1 + C_2$	0.86	1.00
C_3 (propane)	0.53	0.66
C_4 (saturated)	0.76	0.90
Gasoline	3.47	4.13
Diesel fuel	21.55	20.47
Vacuum distillate	42.95	34.25
Vacuum residue	27.55	35.50
Total	**101.55**	**101.75**
Hydrogen consumption (% mass)	1.55	1.75

The cost and complexity of the hydroconversion of residues limits its actual use to very specific situations where there is a weak demand for heavy products. However, in the medium and long term, such a scenario could occur with a change in the demand for finished products.

2.2.2.3 Hydrotreatment or hydrorefining

Hydrotreatment and hydrorefining are the terms that refer to all of the processes that use **hydrogen**, in the presence of a **catalyst**, to treat various petroleum fractions. The essential objective of these processes is the elimination of impurities from feedstocks. Impurities include sulfurized, oxygenated, or nitrogenated compounds and metals. These impurities affect the various processes that use catalysts that are subject to poisoning (reforming or cracking) and the finished products that have very-stringent quality specifications (sulfur content, storage stability ...). Hydrorefining also includes hydrocarbon transformation operations, which are especially useful in improving some product characteristics (hydrogenation of polyolefins into monoolefins, and aromatics into naphthenes or paraffins ...).

The primary objective of the processes is the **elimination of sulfur compounds** at numerous points in the production process, such as the treatment of light cuts (LPG, light gasoline, reforming feeds), medium distillates (diesel and domestic fuel oil), and heavy cuts (fuels).

Sulfur-related problems are well known; the most important problem is the risk of poisoning the precious metal catalysts used in the reforming processes. The sulfur in finished products is transformed by combustion into the sulfur anhydrides SO_2 and SO_3, which cause corrosion and reduce the effectiveness of the catalytic converters installed in vehicle exhaust systems. On the other hand, sulfur compounds, even in very weak concentrations, and especially in gasoline, produce a very disagreeable odor.

Hydrotreatments also eliminate some undesirable hydrocarbons (polyunsaturates) and some metallic or nitrogenated heterocyclic compounds that would cause uncontrolled oxygenation of the finished products during storage.

A. Chemical and catalytic processes

The reactions that take place are both **chemical hydrogenations** (aromatics → naphthenes, olefins → paraffins) and **scissions** of the bonds between the carbon atoms and the atoms of sulfur, nitrogen, oxygen, or metals.

The operating processes can be called by different names, depending on the primary chemical reactions involved. The following terms are defined:

- **hydrodesulfurization** (HDS), which has the following reaction schema

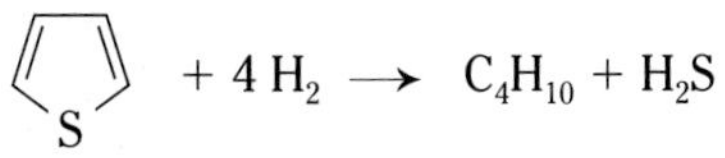

$$\text{(thiophene)} + 4\,H_2 \longrightarrow C_4H_{10} + H_2S$$

- **hydrotreating** of aromatics (HDT)

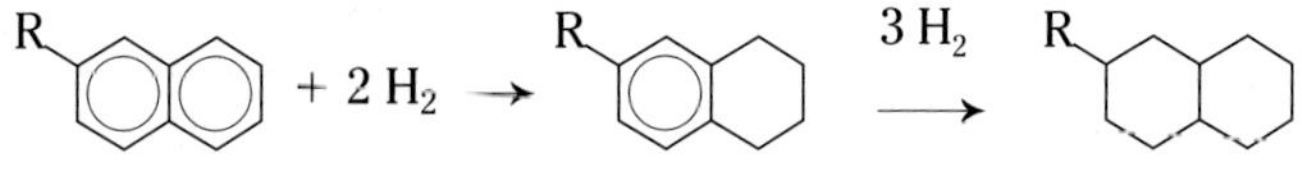

$$\text{R-(naphthalene)} + 2\,H_2 \rightarrow \text{R-(tetralin)} \xrightarrow{3\,H_2} \text{R-(decalin)}$$

- **hydrodenitrification** (HDN)

$$-\overset{|}{\underset{|}{C}}-\overset{|}{\underset{|}{C}}-\overset{|}{\underset{|}{N}} + H_2 \rightarrow -\overset{|}{\underset{|}{C}}-\overset{|}{\underset{|}{C}}- + NH_3$$

- **hydrodeoxygenation**

$$-\overset{|}{\underset{|}{C}}-\overset{|}{\underset{|}{C}}-\overset{|}{O} + H_2 \rightarrow -\overset{|}{\underset{|}{C}}-\overset{|}{\underset{|}{C}}- + H_2O$$

- **hydrodemetalization** (HDM)

organo-metallic compounds + $H_2 \rightarrow$ hydrocarbons + metals

All these reactions are **exothermic** and they take place between 320 and 400°C under pressures from 20 to 150 bar, depending on the severity required. Hydrogen consumption increases with the degree of purification achieved and the characteristics of the feedstock. Thus, the hydrotreatment of a reforming feedstock will only consume 0.6 kg of hydrogen per tonne of product. However, the hydrotreatment of a direct-distillation diesel fuel requires 3 kg of hydrogen and the hydrotreatment of a cracked diesel fuel requires 5 to 10 kg of hydrogen.

Hydrotreatment catalysts can only be used for a single function and they consist of metallic oxides deposited on porous aluminum. When they are in contact with sulfur, the oxides rapidly transform into **metallic sulfides**, which are chemical species that promote the preferred reactions. Combinations of two metals are often used to obtain specific reactions. For example, the following combinations can be chosen for preferred reactions:

- cobalt-molybdenum for desulfurization
- nickel-molybdenum for denitrification
- nickel-tungsten for desulfurization and hydrogenation

Catalysts that get progressively covered in coke must undergo periodic regeneration by a controlled combustion of the hydrocarbon deposits. In HDM's case, the metals in the feedstock get caught on the catalyst surface and they gradually reduce its effectiveness; in practice, when the mass of the deposits reaches 30 to 40% of the mass of the catalyst, it must be changed.

B. Fields of application

This section describes some hydrotreatment processes that are particularly interesting or important.

a. Hydrodesulfurization of middle distillates

This treatment applies to jet fuel, diesel fuel, and domestic fuel oil that must meet very-stringent specifications for sulfur content (see Chapters 3, 4, and 7). The basestocks used to formulate these products can contain over 1% sulfur (and up to 3 to 4% for some LCO type diesels). The European specifications for finished products are 0.05 and 0.1% (in 1999) respectively for diesel fuel and domestic fuel oil.

The **desulfurization rate** (*DR*) is the ratio between the extracted sulfur content and the initial sulfur content. In reality, this value is essentially equal to the change in sulfur content with respect to the initial content,

$$DR = \frac{S_{in} - S_{fin}}{S_{in}} \ (\%)$$

where S_{in} and S_{fin} are the initial and final sulfur contents.

Thus, a reduction from 1.5 to 0.05% in the sulfur content corresponds to a desulfurization rate of 96.7%.

This very-high rate does not cause any major technical problems, but it does require the availability of the necessary equipment to treat medium distillates (diesel fuel and domestic fuel oil), which represent an important portion of the petroleum outlook in certain regions (for example, Europe). An **increase in the hydrodesulfurization capacity** can be foreseen, due to the remodeling of existing facilities and the commissioning of new ones.

Table 2.12 provides an example of the operating conditions (pressure, temperature, catalyst space velocity) and the efficiency rate when desulfurizing a diesel fuel. The hydrogen consumption can vary from 3 to 4 kg per tonne of treated feedstock. The catalyst space velocity represents the compromise between cost and effectiveness, which is highly dependent on the desired desulfurization rate.

With respect to the chemical mechanisms of hydrodesulfurization, the compounds involved in the desulfurization of diesels are essentially derivatives of dibenzothiophene. For some of these compounds, opening the sulfur heterocyclic rings using hydrogen treatment is very difficult. This is the case with 4,6-dimethyldibenzothiophene, which has the following formula:

S
H_3C CH_3

In fact, some diesel-fuel hydrotreatment processes go beyond simple hydrodesulfurization and achieve a **partial hydrogenation** of aromatics and naphthenes (Kasztelan et al. 1994). This process results in a higher-quality product with respect to the cetane rating and reduced emissions when it is burned in engines (see Chapter 5).

b. Hydrotreatment of light cuts

The petroleum fractions addressed by this process are the gasolines from direct distillation, catalytic cracking, or other conversion processes (coking, visbreaking), before they are mixed into the pool of finished products.

The feedstock for catalytic reforming must also undergo a careful hydrotreatment to almost totally remove sulfur and nitrogen.

The hydrotreatment of light cuts is a relatively **low-severity** process. The temperatures vary from 325 to 350°C and the pressures vary from 20 to 30 bar. Hydrogen consumption is about 0.5 to 1.5 kg per tonne of feedstock.

Table 2.12 *An example of diesel-fuel desulfurization, which includes operating conditions and effectiveness.*

Characteristics	Feedstock	Product
Density d_4^{15}	0.836	0.827
Viscosity (mm²/s)	4.59	4.36
Distillation range (°C)		
IP	189	188
FP	367	362
Sulfur content (ppm)	9420	445
Desulfurization rate (%)		95.3
Nitrogen content (ppm)	97	40

Operating conditions	
Pressure (bar)	27
Temperature (°C)	350
GHSV	2
Hydrogen consumption (kg/t)	3

* GHSV = Gas Hourly Space Velocity

c. Hydrotreatment of heavy products

This process applies primarily to the constituents of heavy fuels and to feedstocks—atmospheric residues, distillates, and vacuum residues—that are destined for conversion operations. The objective is to reduce sulfur, nitrogen, and metal content. Table 2.13 provides an example of the operating conditions and the effectiveness of this type of process when it is used for hydrodemetallization. When compared to previous techniques, the operating conditions are clearly more severe: high pressure, low GHSV, and high hydrogen consumption (5 to 10 kg per tonne of feedstock). The catalyst, which gets contaminated by the metals in the treated product, must be used in larger quantities and it's lifetime is reduced. Catalysts are sometimes added in and drawn off continuously in some reactors, which are fluidized-bed units instead of fixed-bed units.

Table 2.13 *Hydrotreatment of heavy feedstocks. Operating conditions and sample results.*

Characteristics	Feedstock	Product at 370°C $^+$
Density d_4^{15}	0.992	0.962
Sulfur content (% mass)	5.15	2.63
Nitrogen content (% mass)	0.40	0.33
Nickel content (mass ppm)	50	11
Vanadium content (mass ppm)	350	40

Operating variables (operating range)	
Temperature (°C)	380-420
Pressure (bar)	80-150
GHSV	1-3
Hydrogen consumption (kg/t)	5-10

d. Other types of hydrotreatment

Among other types of hydrotreatment, an interesting process is the one used to hydrogenate the **polyolefins** in some of the liquid products (C_5 cuts) from steam crackers. The resulting products contain nothing more than monoolefins and they can be included in the gasoline pool in some cases. The treatment takes place at low temperature using a palladium catalyst.

2.2.2.4 Finish treatments

This section covers the two most common finish treatments, which are amine washing and Merox sweetening.

A. Amine washing

The objective of the amine gas washing process is to eliminate the **hydrogen sulfide** in refinery gases (C_1-C_2 fractions and LPG). The treatment applies primarily to the cuts derived from cracking and hydrotreatment units. The process consists of the chemical absorption of H_2S in an aqueous solution of **monoethanolamine** ($HO—CH_2—CH_2—NH_2$) or **diethanolamine** ($HO—CH_2—CH_2—NH—CH_2—CH_2—OH$).

The reaction results in a mixed compound according to one or the other of the following schemas

$$RHNH + H_2S \longrightarrow RHNH_2HS$$

$$(R)_2NH + H_2S \longrightarrow (R)_2NH_2HS$$

where R represents the following arrangement:

$$HO - CH_2 - CH_2^{\bullet}$$

The gas to be treated passes into an absorption column at 40°C under a pressure of 5 to 10 bar, where it is washed by a counter flow of amine solution. The purified gaseous outflow contains only about 10 ppm of H_2S. The mixed compound liquid is sent to a regenerator under low pressure (1.8 bar) where the H_2S is separated by heating to 100°C. The amine is recycled to the absorber while the H_2S is converted to sulfur using the Claus process (see 1.2.2).

B. Merox process

The Merox process transforms corrosive and malodorous **mercaptans** into neutral and stable **disulfides**.

The reaction occurs according to the following formula:

$$2\,RSH + \frac{1}{2}\,O_2 \longrightarrow R-S-S-R + H_2O$$

The process does not, in principle, reduce the sulfur content, but it enables fuels to meet quality criteria or specifications requiring precise mercaptan content. This is a requirement for LPG and jet fuel (see Chapters 6 and 7). This process can be used instead of more expensive hydrotreatment processes when the total sulfur content does not justify their use.

The reactant used in the Merox process is the oxygen contained in air. The reaction takes place at 50°C under a pressure of 5 to 6 bar in the presence of sodium hydroxide and a cobalt catalyst. There are two process variants referred to as **Merox Sweetening** and **Merox Extraction**.

The **sweetening** variant, which is intended mainly for jet fuel and gasoline, leaves the disulfide in solution in the cut being treated and the total sulfur content does not change.

The **extraction** variant separates the disulfides using decantation and the total sulfur content is reduced. This variant is applied primarily to LPG.

2.2.3 Obtaining fuels from natural gas

Like crude oil, natural gas can be used as a primary source of liquid fuels. The conversion process, which is referred to as **chemical conversion**, is rarely used industrially, but it is included here because of its medium- and long-term prospects (Rojey 1994).

Fig. 2.5 shows the various routes that can be followed to chemically convert natural gas to liquid fuels. The following two scenarios are possible:

- **the indirect route**, which results in the production of the synthetic gas $CO + H_2$, which then leads to the production of methanol CH_3OH or to hydrocarbon mixtures according to the Fischer-Tropsch process
- **the direct route**, which has two variants: thermal coupling and oxidant coupling

2.2.3.1 Chemical conversion of natural gas by indirect means

This is an expensive route, in terms of both investment and energy consumption, but it has the advantage of being at the industrial development stage, whereas the direct route is still at the proposal stage.

A. Production of synthetic gas

Synthetic gas, so called because it constitutes a step in the synthesis of **methanol**, is a mix of carbon monoxide and hydrogen. It can be obtained from natural gas by the following two processes:

- **partial oxidation** of methane

$$CH_4 + \frac{1}{2} O_2 \rightarrow CO + 2H_2$$

- **steam reforming**

$$CH_4 + H_2O \rightleftarrows CO + 3\,H_2$$

The first reaction is highly **exothermic** and it takes place at high temperature (950 to 1250°C) in a burner without a catalyst. The thermal decomposition of methane that leads to the production of carbon must be avoided with this method (Dumon et al. 1984).

Steam reforming is highly **endothermic** and it takes place in the presence of a nickel catalyst at high temperature (840 to 950°C) under a pressure of about 30 bar.

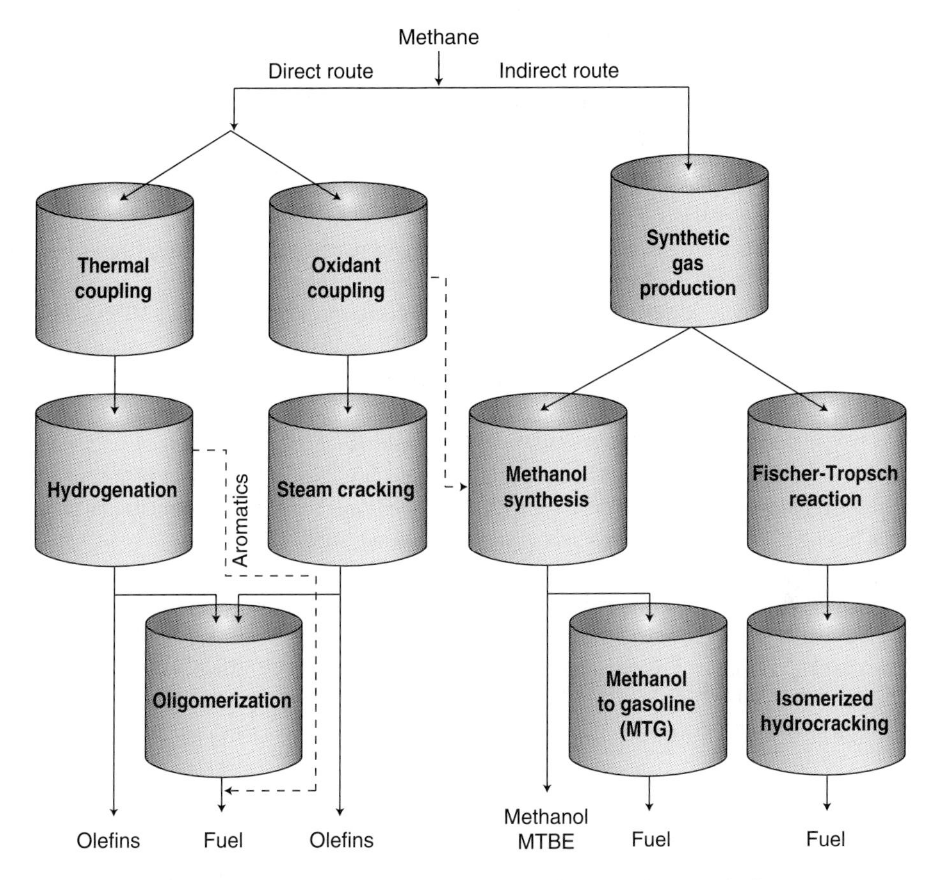

Figure 2.5 *Chemical conversion routes from methane to fuels.*

To complete the subsequent chemical change of the synthetic gas (methanol and Fischer-Tropsch process), the ratio of H_2 to CO must be kept within a range of 2 to 2.5. To stay within this range, steam reforming is used to provide a higher ratio and partial oxidation is used to provide a lower ratio.

This is why modern synthetic gas processes are forced to **combine the two techniques** of partial oxidation and steam reforming. This also results in some compensation between the exothermic and endothermic chemical reactions. However, even with the most optimized processes, producing synthetic gas remains an expensive and energy-inefficient process, because process-consumption can be as much as 10 to 15% of the natural gas feedstock.

B. Methanol production

Methanol by itself is already a possible **alternate fuel** (see Chapter 6), but in this case it is the necessary intermediate step to obtaining **ethers** (MTBE, TAME), which have already been described (see 2.2.2.1).

Methanol is produced from synthetic gas in the presence of small quantities of CO_2 by using the following reactions,

$$CO + H_2O \rightleftarrows CO_2 + H_2$$

$$CO_2 + 3\,H_2 \rightleftarrows CH_3OH + H_2O$$

which result in the following overall reaction:

$$CO + 2\,H_2 \rightleftarrows CH_3OH$$

The two preceding reactions are thermodynamically enhanced by high pressure and low temperature. However, the temperature cannot be too low, because favorable kinetic conditions must be maintained (Chauvel 1985). The best performing copper-based catalysts enable operation at temperatures of 240 to 270°C and under pressures of 5 to 10 bar. A number of methanol production processes exist. The best known and the most widely used process in the world is the one commercialized by *Société ICI*.

The world market for methanol, which is a major intermediary in the chemical industry, was about 28 Mt/yr in 1998. A small part of this production (about 5%) is used as fuel, either as is or in the form of ethers.

C. Producing gasoline from methanol

A process to transform methanol into gasoline, often referred to as MTG (Methanol to Gasoline), was perfected and developed by Mobil Oil (Goulley 1987).

The following equation describes the general chemical reaction:

$$n\,CH_3OH \longrightarrow (-CH_2)_n + n\,H_2O$$

This reaction provides 0.438 tonnes of hydrocarbons for each tonne of methanol treated, which results in an energy efficiency of 95%.

The conversion first involves a balanced reaction to dehydrate methanol into **dimethylether**.

$$n\,CH_3OH \rightleftarrows \frac{n}{2}\,CH_3-O-CH_3 + \frac{n}{2}\,H_2O$$

The ether is then transformed by the following reaction into the biradical $:CH_2$ using a **zeolite** catalyst referred to as ZSM5:

$$\frac{n}{2}\,CH_3-O-CH_3 \longrightarrow n{:}CH_2 + \frac{n}{2}\,H_2O$$

The $:CH_2$ biradical gives rise to olefins and hydrocarbon mixtures of all types:

$$n\,({:}CH_2) \longrightarrow CH_3-(CH_2)_x-CH=CH_2$$

The hydrocarbon products contain 60 to 80% liquids (C_5+ fractions) and 20 to 40% gases. The aromatic content is between 25 and 45%. There are no heavy hydrocarbons greater than C_{10}. In any case, a careful analysis of the products obtained indicates that they contain, in some cases, small quantities (3 to 6%) of 1,2,4,5-tetramethylbenzene, which is also called **durene**. This compound must be eliminated by hydrotreatment because of the risk of it crystallizing into solid deposits due to its high melting point (79°C).

The MTG process was installed commercially at Waitara in New Zealand in 1987. The unit was designed for a production rate of 570 000 t/yr of gasoline from 1.35 billion m^3/yr of natural gas. Production was halted to study the effect of new technological upgrades (replacing the fixed bed with a fluidized bed and integrating the methanol and gasoline production stages).

D. Fischer-Tropsch synthesis

In 1922, two German chemists named Franz Fischer and Hans Tropsch were the first to obtain a hydrocarbon mixture and oxygenated organic compounds by treating a $CO + H_2$ mixture using an iron based catalyst. In 1925, Fischer and Tropsch announced that they had obtained hydrocarbons using catalysts consisting of cobalt and alkalized iron. During the second world war, Fischer-Tropsch synthesis was used extensively to supply the German armed forces with fuels. After 1945, development continued mostly in South Africa where three production complexes with a total capacity of about 4 Mt/yr were completed by SASOL Ltd between 1955 and 1982. In South Africa, the synthetic gas was obtained from coal instead of natural gas. This situation resulted from the availability of cheap and plentiful local supplies of coal as well as political and strategic considerations.

Since 1990, **renewed interest** in the Fischer-Tropsch method of using natural gas as a primary material has resulted in new developments by other companies (Shell, Statoil, Exxon).

The basic reaction for the Fischer-Tropsch process is

$$CO + 2\,H_2 \longrightarrow -CH_2- + H_2O \qquad (1)$$

The $-CH_2-$ bonds produced at each elementary step combine to form hydrocarbon chains.

A competing route to this end is the reaction called methanation:

$$CO + 3\,H_2 \longrightarrow CH_4 + H_2O \qquad (2)$$

To minimize this parasitic reaction, the temperature must be reduced and the pressure must be increased. The Fischer-Tropsch process usually operates at temperatures of 250 to 350°C and at pressures from 20 to 30 bar. Catalysts are usually iron based when gasoline is preferred, and cobalt based when the reaction is directed toward the production of heavier products.

Interest is no longer centered mainly on the production of gasoline, as it was for many years in South Africa, but on the production of **medium and heavy fractions** such as kerosene, diesel fuel, and lubricant basestocks.

As a result, Shell (Eilers et al. 1990) developed a process called SMDS (*Shell Middle Distillate Synthesis Process*). A 470 000 t/yr unit is currently in operation at Bintulu in Malaysia. The crude products obtained from it are mainly paraffins with long straight chains. These products must then undergo **isomerized hydrocracking** (300 to 350°C at 30 to 50 bar) to obtain a sufficiently fluid diesel cut.

The overall energy yield of 60 to 65% is much less than standard refining processes. Furthermore, the investment needed to implement a Fischer-Tropsch unit is considerable (600 million dollars at Bintulu). The existence of mutually favorable circumstances in Malaysia (the ready availability of low-priced natural gas and outlets for byproducts such as paraffins, which are used in the food processing industry and as lubricants) justified the world's first experience in this domain.

However, the process has a large margin for possible improvements based on the choice of technologies, the use of more appropriate catalysts, and improved integration of the synthetic-gas production section. Research continues in this area, by Shell (Senden 1998) and also by other companies like Statoil (Rytler 1990), Exxon (Lahn et al. 1992), Syntroleum (Weick 1998) and IFP (Chaumette 1998).

2.2.3.2 Chemical conversion of natural gas by direct means

This conversion route (Leprince 1988), which is still in the planning stages, consists of directly transforming methane into hydrocarbons that have at least two atoms of carbon (C_{2+}), without passing through the intermediate step of $CO + H_2$ synthetic gas. Two possible paths are envisaged:

- **thermal coupling**, which requires high temperatures
- **oxidant coupling**, which operates at lower temperatures in the presence of oxygen and a catalyst

A. Thermal coupling

The thermal coupling process requires temperatures greater than 1 000°C and very-short residence time. The major difficulty to be resolved is controlling the formation of coke. The products obtained are primarily **ethylene**, **acetylene**, and **light C_{2+} olefins**. This method is expected to appear in the medium term, but it will most likely be applied in the petrochemical industry instead of in refining.

B. Oxidant coupling

The presence of oxygen helps the transformation of methane into C_{2+} hydrocarbons at relatively moderate temperatures (it still requires 700 to 800°C). The use of a catalyst is essential. The processes currently being studied deliver **conversion rates that are too low in C_{2+}**, which is on the order of 25% when 40 to 45% is required. The separation of the products is also very costly. An industrial application of this process appears to be a remote possibility.

2.3 Global refining: some facts and trends

This section includes a discussion of global data on crude oil consumption, finished-product demand, and refining costs.

2.3.1 Fuel consumption and refining capacity

World petroleum consumption reached 3.2 billion tonnes per year in 1996. It could reach 3.7 billion tonnes/yr by the year 2000 and surpass the 4-billion-tonne/yr mark in the 21st century, with most of that increase being attributable to Asia and South America. Petroleum represented about 38% of world energy consumption in 1996 and it will be about 35% by 2000.

In the past, world refining capacity had reached a peak of 4 billion tonnes/yr around 1980; it was 3.7 billion tonnes/yr in 1996. Fig. 2.6 shows the distribution of capacity and the projected increases for various areas of the globe.

North America is the major world refining area with a capacity of 860 Mt/yr. The United States (770 Mt/yr) obviously holds the dominant position.

Western Europe remains an important refining area (705 Mt/yr) despite the large reductions that occurred at the beginning of the 1980s.

The refining capacity of Eastern Europe is about 630 Mt/yr, with 500 Mt of that capacity in the ex-USSR. This data, however, does not represent the real situation in this region because many of the installations are old and underutilized.

In Asia and the Pacific, refining capacity is about 720 Mt/yr, with about 250 Mt/yr belonging to Japan. This is the world region where the greatest growth potential exists in the short and medium term.

The other two major locations with refining capacity are Latin America at 370 Mt/yr, which consists mainly of Venezuela and the Caribbean, and the Middle East, which is at 265 Mt/yr despite the destruction caused during the Iran-Iraq and Iraq-Kuwait conflicts.

The world total is currently over 700 refineries with an average capacity per unit of 5 Mt/yr. However, this figure hides some great disparities because some big refineries (Venezuela, Saudi Arabia, Korea, ...) can process over 30 Mt/yr of crude oil, while some cannot exceed 1 Mt/yr. Most of these smaller refineries are in the United States, where at least forty can be readily accounted for. In effect, there are only about 30 high-capacity refineries in the world with capacities exceeding 15 Mt/yr.

The utilization rate—the difference between the amount of crude oil actually treated and the production capacity—is a world average of 85% (91% in the United States, 87% in Western Europe). This relatively satisfactory situation, from an economic viewpoint, was reached by closing many refineries, particularly in Europe.

2.3.2 Refinery structures. Demand for finished products

A description of worldwide refining that considers only overall refining capacity is a rather imperfect description of the real situation. In reality, distinctions must be made between the following items:

- **simple refineries**, which consist of a single atmospheric distillation column linked with a catalytic reformer and distillate hydrodesulfurization units
- **complex refineries**, which are additionally equipped with standard conversion units (catalytic crackers and hydrocrackers or visbreakers)

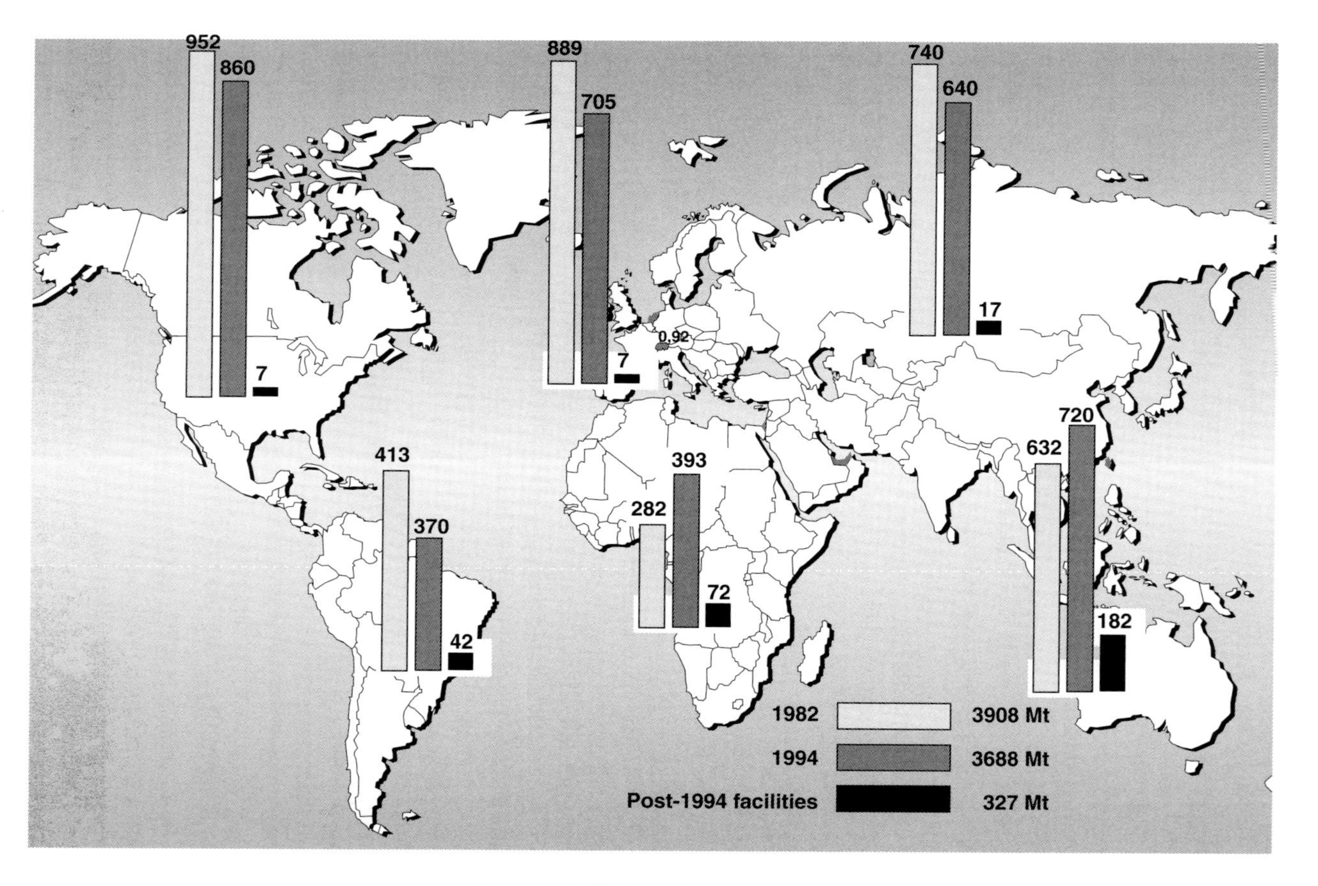

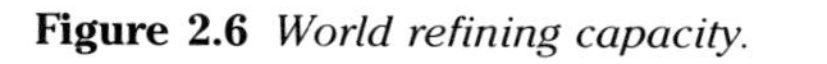

Figure 2.6 *World refining capacity.*

- **deep conversion refineries**, which have a broad range of units for applying sophisticated transformation processes (hydroconversion) to residues to produce light products (Heinrich et al. 1991)

A better view of the global refining situation is provided by the **conversion rate**, which by definition is the difference between the fluid catalytic cracking (FCC) capacity and the atmospheric distillation capacity. The data in Fig. 2.7 show that the conversion rate has increased considerably in all regions in the last 20 years. The rate has reached, for example, 56% in North America, 28% in Western Europe, and 27% in Latin America.

Another way of reflecting the degree of refining sophistication consists of **adding the capacities provided by four key transformation processes** (catalytic cracking + hydrocracking + catalytic reforming + alkylation) and expressing the result as a factor of the distillation capacity. Fig. 2.8 shows that within 10 years—1984 to 1994—refineries have become more complex, especially in the United States, and after some delay, in Europe as well.

This evolution has evidently been accompanied by **increased demand for light products and medium cuts**, at the expense of heavy fractions. Hence, Fig. 2.9 shows the changes in the petroleum balance of the United States and Western Europe in the last 10 years. In Western Europe, the structure of petroleum demand has changed substantially. The consumption of heavy fuels was cut in half (from 39 to 20%) while the demand for distillates increased from 33 to 41%. In France, the extensive development of nuclear power facilities for electrical generation has resulted in a spectacular drop in the market for heavy fuels; demand has dropped from over 35 Mt/yr in 1973 to less than 5 Mt/yr in 1996!

Also worth observing in Fig. 2.9 is the **significant portion** of the demand for petroleum products that is **represented by transportation fuels** (gasoline, diesel fuel, jet fuel). In 1996, the transportation portion was 58% in the United States and 43% in Europe.

2.3.3 Refining costs

The cost of establishing a new refinery depends on its size, its equipment, and its location. The following information establishes an order of magnitude: in 1998, a refinery with a capacity of 160 000 bbl/d (8 Mt/yr) that included several conversion units (catalytic crackers, visbreakers) cost about 1.5 billion dollars. This amount could now be considerably higher due to environmental constraints, which can consist of treating discharges from the refinery itself as well as quality improvements in the finished products.

Refinery costs can be divided into three categories:

- **variable costs** (supplying energy, catalysts, and chemical products) amounting to $3 to $4 per tonne of crude processed
- **fixed costs** (personnel, maintenance, ...), excluding depreciation, of about $10 to $12 per tonne
- **capital cost** of about $25 per tonne

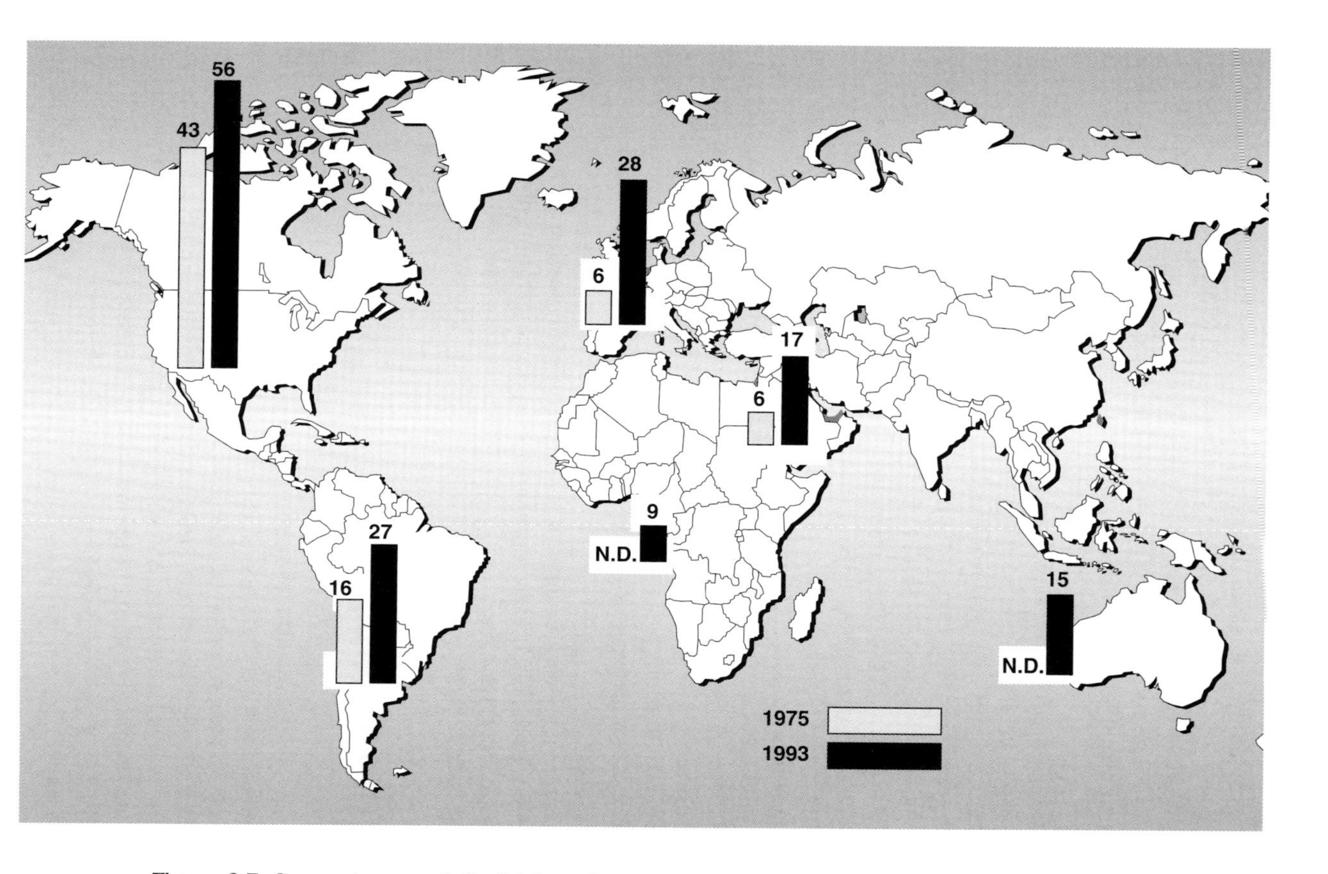

Figure 2.7 *Conversion rates* (in %) for refineries in various world regions in 1975 and in 1993.*
**The ratio of the FCC capacity to the atmospheric distillation capacity.*

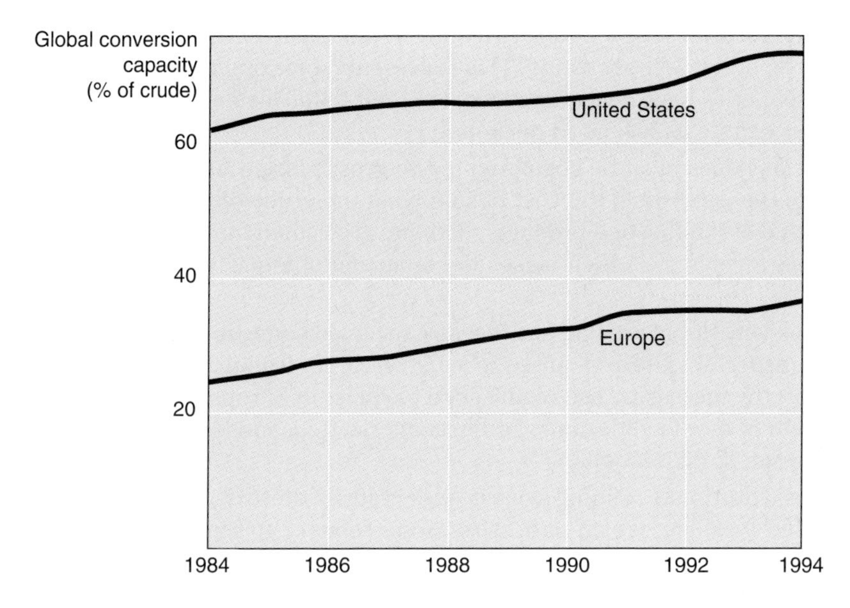

Figure 2.8 *Evolution of the overall conversion capacity* of refineries in the United States and Europe.*

** The total capacity of four refining processes (catalytic cracking + hydrocracking + catalytic reforming + alkylation) compared to the distillation capacity.*

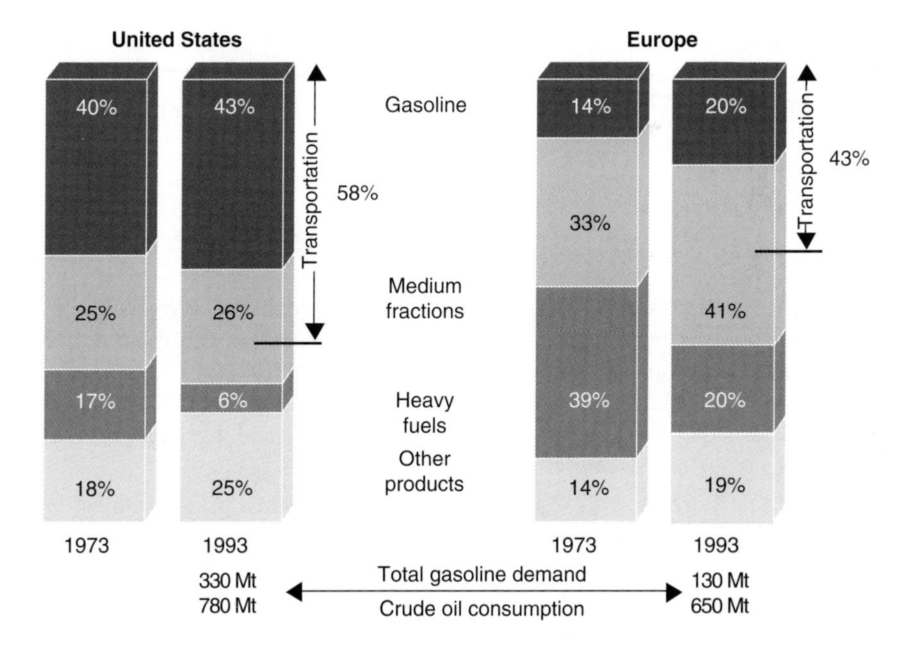

Figure 2.9 *Change in the balance of petroleum products in the United States and in Western Europe between 1973 and 1993.*

Therefore, for a new refinery running at full capacity the total refining cost would be about $40 per tonne. This represents a maximum value since many older refineries have depreciated most of their initial installations and the refining cost can be as low as $3 per tonne.

These values must be compared to the **gross margin for refining**, which is the difference between the cost of supplying the crude oil and the tax-exempt sale price of the finished products. Refining gross margins have been very low and they remain so, which makes the financing of a new installation very difficult.

However, global refining has recently received some **important new investments**, especially those required to satisfy new **environmental regulations**. As a result, the increasing use of unleaded gasoline in Europe has led to the construction of new facilities (alkylation, isomerization, and MTBE synthesis) that cost about $5 billion dollars.

This additional refining cost is reflected in the extra cost of obtaining unleaded gasoline, which varies from one refinery to another and averages about $20 per tonne.

Another challenge faced by the refining industry is the **desulfurization of medium cuts** like diesel fuel and heating oil to a sulfur content of 0.05% and 0.1% respectively. This change, which has already been achieved in the United States and Europe, cost about three billion dollars in each of these regions. This translates into an increase of about $10 to $20 per tonne of product.

Desulfurization will soon be required in other countries and it will require similar increases in refining costs.

Another significant change is the increasing market penetration of reformulated gasoline in the United States, which will involve 30 to 60% of that country's gasoline pool. The costs associated with this change could be between two and four million dollars.

2.3.4 Energy costs associated with refining

Fig. 2.10 shows the order of magnitude of the energy expenditures required to operate various refinery units. In each case, the expense of the various items (cost of supplying thermal energy, steam, and electricity, and the cost of the reactants consumed) has been compiled and this total is referred to as **process-consumption**, which is expressed, in energy equivalent values, as a percent of the calorific value of the feedstock.

The data gathered in this manner are only **estimates**; a precise energy balance depends on the effects of the operating conditions, on the technological details of the process, and on the overall operation of the refinery. However, this method of classifying the data reveals the most energy-consumptive processes: reforming, alkylation, hydrocracking, ...

From an overall refinery perspective, energy process-consumption has gone from 4 to 5% in the 1970s, to 7 to 8% since 1990. This figure will certainly reach

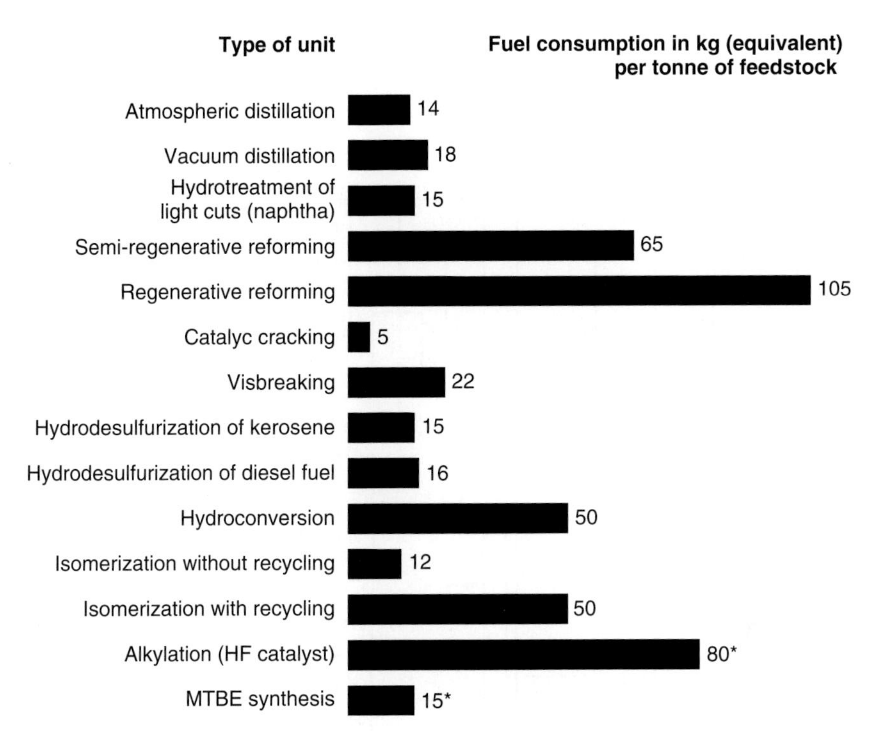

Figure 2.10 *Energy consumption of various refinery processes (order of magnitude).*

* *For alkylate and MTBE, the energy consumption is expressed in kg of fuel per tonne of product obtained.*

11 to 13% in the very-complex refineries of the years 2000 to 2010, which will have schemas similar to the one shown in Fig. 2.11.

One of the major causes of this overall rise in process-consumption will be the heavy **consumption of hydrogen**, which is required for the deep desulfurization of finished products and the coincident reduction in the content of aromatic compounds.

For a long time, the hydrogen balance of a refinery could easily be accommodated by the production output of catalytic reforming and the modest consumption of hydrodesulfurization units. That will no longer be the case with future refineries, like those equipped with a hydrocracker to obtain good-quality medium fractions (diesel fuel or kerosene) and a hydrodesulfurization unit to produce low-sulfur fuels from vacuum residues (Raimbault et al. 1994). Fig. 2.12 shows an estimate of the hydrogen balance that corresponds to such a scenario. The hydrogen needs cannot be met without the installation of a special unit to produce hydrogen from natural gas.

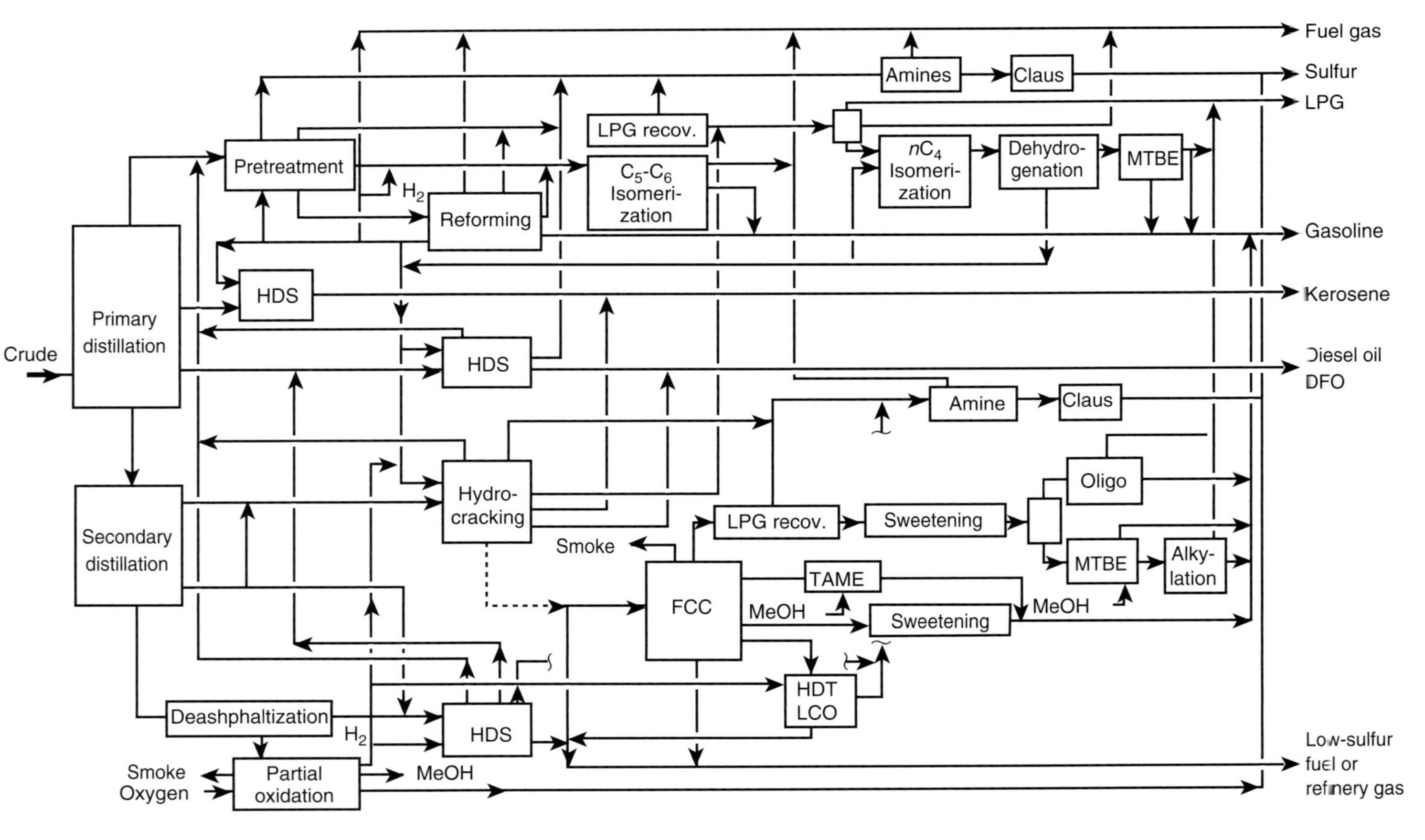

Figure 2.11 *A possible configuration of a future refinery.*

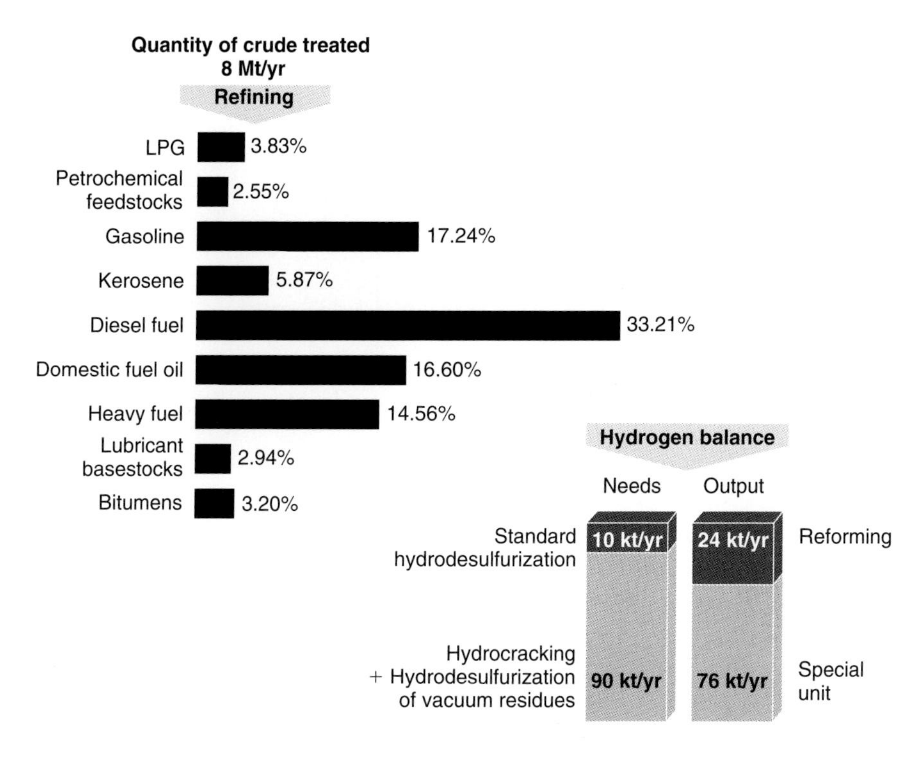

Figure 2.12 *Hydrogen balance for a refinery equipped for hydrocracking and hydrodesulfurization of vacuum residues.*

Chapter 3

Gasoline

The general term "gasoline" refers to all fuels used in spark-ignition engines—engines that were once referred to as "explosion engines". Gasoline is classified into several categories based on octane rating. This distinction is usually expressed by the terms "**regular gasoline**" and "**premium gasoline**". The presence or absence of lead alkyls is another distinguishing characteristic. In Europe, the standard unleaded gasoline is called **Eurosuper**. Therefore, a wide range of gasolines can be offered by the same service station: regular or premium—with and without lead.

The beginning of this chapter presents a review of the major aspects of the operation of automobile engines, such as technology, design and control parameters, performance measurement, and combustion process analysis. These fundamental concepts appear frequently, not only in Chapter 3, but throughout the book. The reader who is unfamiliar with automotive technology can return to this general presentation as needed.

The second part of the chapter describes the characteristics of gasoline and its specifications, which must provide optimized combustion and satisfactory behavior during storage and distribution.

Particular attention is paid to gasoline formulation in the refinery and to the improvement of its behavior by the use of various types of additives. The problems of matching engines and fuels are then covered in detail, with specific emphasis on preventing engine knock.

The behavior of the gasoline engine versus its diesel rival is described in Chapter 5 from the perspective of fuel consumption and environmental impact.

How the spark-ignition engine functions

This section begins with a tribute to some of the pioneers of the engine and automobile industries and then it continues with a description of the functional principles of the spark-ignition engine. This includes a full description of the engine's systems and features that are involved in the use of fuel—the fuel system, the ignition system, and the lubrication system. This section also defines and specifies the major parameters that engine designers use to set the controllability and performance of engines and vehicles.

The section also provides a description of the various possible combustion processes, both normal and abnormal. This description includes an explanation of the fuel's involvement in the combustion process in terms of its physical and chemical characteristics.

3.1 Some milestones in the history of engines and automobiles

The internal combustion engine was born about 1860, almost simultaneously in France and in Germany (Newcomb et al. 1989).

On January 24, 1860, the Frenchman **Étienne Lenoir** (1822-1900) submitted a patent describing an "engine system using air expanded by the combustion of gas ignited by electricity". Lenoir's engine was constructed, but it delivered very-low performance because it did not use pre-compression.

On January 16, 1862, the Frenchman **Alphonse Beau de Rochas**'s (1815-1893) patent described the four-stroke cycle that we recognize today. However, Beau de Rocha's invention remained without a practical application for two years.

The German **Nikolaus Otto** (1832-1891) built the first four-stroke engine in 1864 and put several others on the market beginning in 1876. The competition between the inventors was fierce and at the end of a long and sensational lawsuit (1885-1888), Otto's patents were found to be invalid in France.

Once the engines were conceived, the next step was installing them in automobiles.

The Frenchman **Delamare-Deboutteville** (1856-1901) built the first car in 1883 and it was driven on the roads in the Rouen region. It used lamp gas ($CO + H_2$) as a fuel.

In 1886, the Germans **Gottlieb Daimler** (1834-1900) and **Carl Benz** (1844-1929) started automobile manufacturing in Germany. Carl Benz had already outlined the two-stroke cycle in 1878.

Among the automobile pioneers worthy of mention are the Frenchmen **René Panhard** (1841-1908) and **Émile Levassor** (1843-1897). In 1891, they introduced a car that traveled non-stop across Paris. René Panhard also invented the multi-speed gearbox in 1895.

The period from 1895 to 1905 was the time of the great races and the first rallies (Paris-Bordeaux, Paris-Berlin, Paris-Madrid, ...). Each event was an opportunity to set new records.

This quick tour of automotive history would not be complete without covering some of the engineers and industrialists who have left their names to some of the world's great automobile companies.

The Germans **C. Benz** and **G. Daimler** had collaborated since 1886, but it took 40 years before the Daimler-Benz company was started in 1926.

Armand Peugeot (1849-1915) and his brother **Eugène** (1844-1907) were also among the first auto builders. They showed a three-wheeled model at the World Fair in 1889. In 1897, Armand Peugeot started the *Société des Automobiles Peugeot.*

Henry Ford (1863-1947) was the pioneer of the American auto industry. In 1887, he personally hand-built his first automobile piece-by-piece. He started the Ford Motor Company in 1902, and in 1908 he launched the famous "Model T", of which 15 million were sold. Henry Ford instituted the assembly line, which enabled the price of Model Ts to be dropped from $850.00 in 1908 to $265.00 by 1922. During 1919, the Ford Motor Company succeeded in building over one million vehicles. Henry Ford was a remarkable captain of enterprise. In 1922 and 1926 he published two books (*Today and tomorrow* and *My life and my work*) that described his concept of work, industry, and business.

In 1919, **André Citroën** (1878-1935) opened a large automobile assembly plant on the *quai de Javel* in Paris. He was the first one in the world to conceive of and propose the use of road signs. In 1934, he launched the famous *Traction-avant* (front-wheel drive) and its success was unchallenged for over 20 years.

Louis Renault (1877-1944) also built his first car with his own hands in 1898. Renault's automobiles performed well in the great races of 1900 through 1910, which were unfortunately marred by accidents. **Marcel Renault**, Louis's brother, died during the Paris-Madrid rally. Louis Renault was extremely inventive and he was awarded over 500 patents. His company was nationalized on January 16, 1945.

No doubt some important names have been left out. Automotive history buffs, please forgive us! There are many other publications that cover all the technical details and the anecdotes.

The memory of **Rudolph Diesel** is covered in Chapter 4, which is devoted to diesel fuel.

Henry Ford and André Citroën signed this historic photograph during their meeting in the United States in 1931.
(Reproduced with the kind permission of the French Society of Automotive Engineers.)

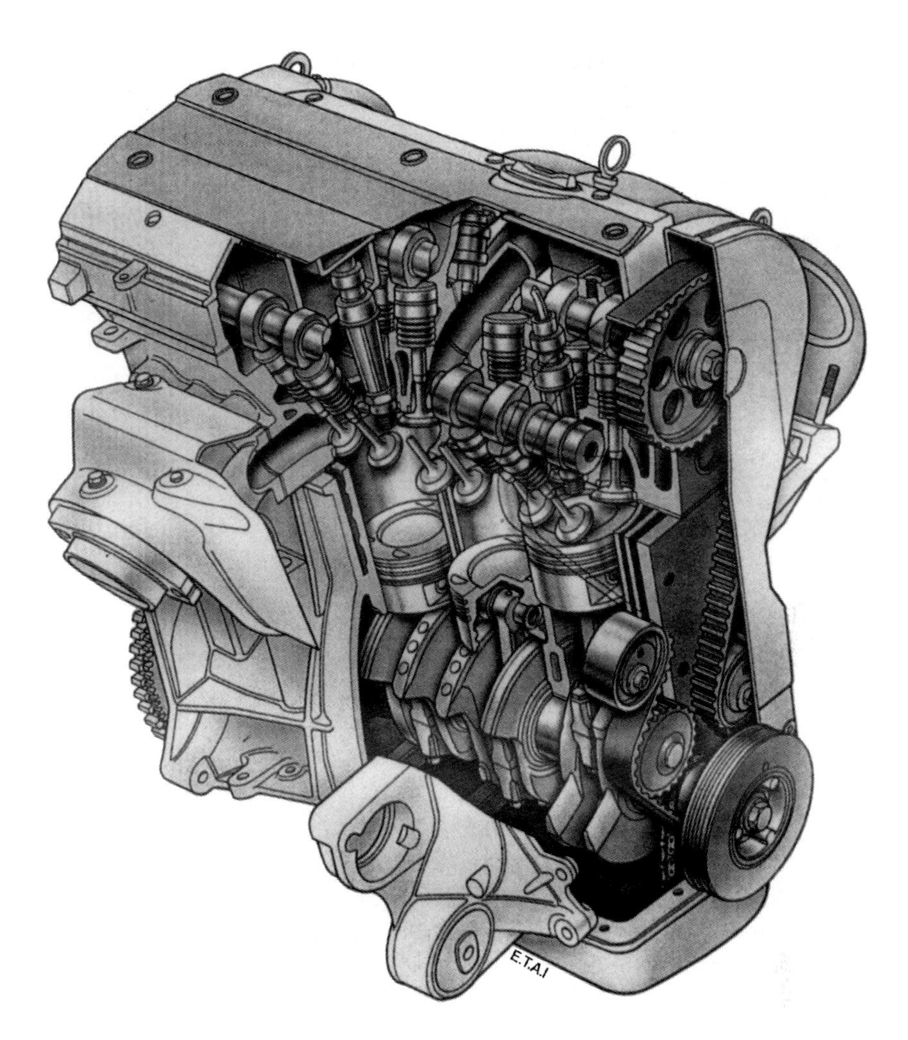

A cut-away view of a modern gasoline engine.
XU 10 J4R (PSA)

- *1998 cm³–97.4 kW*
- *double overhead camshaft*
- *16 valves*

(PSA document)

3.2 Design and production

3.2.1 General description

This section provides a review of the basic construction of a conventional spark-ignition engine. A mixture of air and fuel is ignited inside a cylinder by a **spark**. The cylinder contains a piston that is connected by a rod to a crankshaft. The movement of the piston is constrained between two extreme positions referred to as top dead center (TDC) and bottom dead center (BDC), which correspond respectively to the minimum and maximum volumes of the combustion chamber (see Fig. 3.1). The unit **displacement** (V) is the volume swept by the piston between TDC and BDC. If C represents the piston travel (stroke) and d the cylinder diameter (bore), then

$$V = \frac{C\pi d^2}{4}$$

One end of the piston is open to the combustion chamber volume (v), which is constrained by a cover plate that is referred to as the head. The head contains a spark plug and the intake and exhaust valves.

The **volumetric ratio** (ε), which is referred to as the compression ratio, is determined by the following formula:

$$\varepsilon = \frac{V + v}{v}$$

Compression ratios are usually between 8 and 12 for naturally aspirated engines and between 7 and 9 for supercharged engines.

This ratio will be shown to have a decisive influence on engine performance.

The energy released by the combustion process causes a rectilinear movement of the piston, which is transformed into rotary motion at the crankshaft by the rod and crank system.

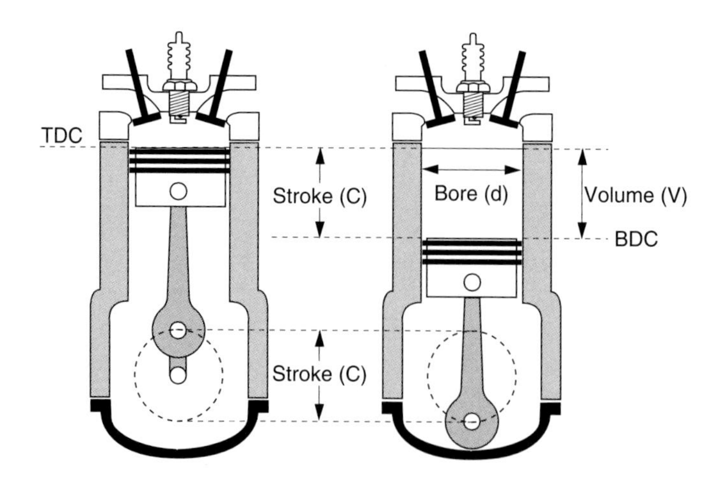

Figure 3.1 *Schematic of a cylinder, piston, and head assembly.*

Almost all automobile engines use the four-stroke cycle (see Fig. 3.2), which was first described by Alphonse Beau de Rochas on January 16, 1862. The four-stroke cycle consists of the following phases:

1. **Intake**: The intake valve opens and the piston, which is moving from TDC to BDC draws in the air-fuel mixture, which progressively fills the available volume. At BDC, the intake valve closes.
2. **Compression-ignition**: The piston moves from BDC toward TDC and compresses the charge to a pressure of about 10 bar. The electric spark provided by the spark plug ignites the charge.
3. **Combustion-expansion**: The energy released by combustion increases the pressure to a range of 50 to 60 bar. This forces the piston toward BDC while the valves remain closed.
4. **Exhaust**: The exhaust valve opens when the piston reaches BDC. The combustion products—usually referred to as burned gases—are expelled from the chamber by the piston as it returns to TDC. At that point, the exhaust valve closes and the cycle begins again.

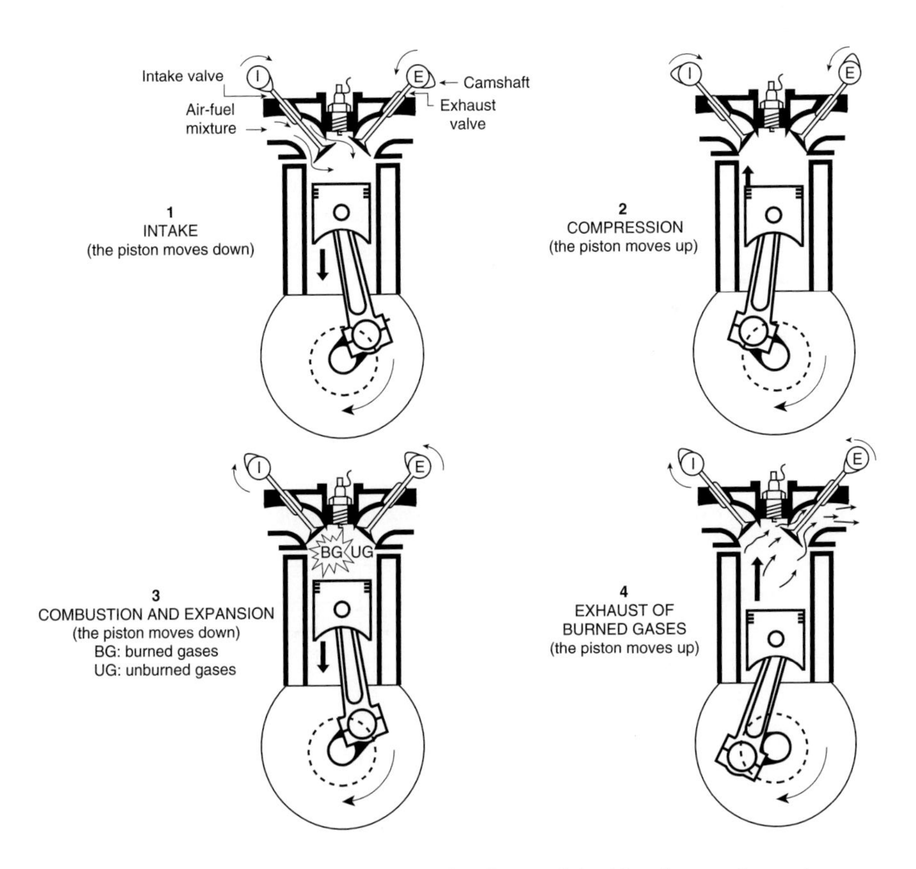

Figure 3.2 *The operating principle of a spark-ignition four-stroke engine.*

Observe that each cycle corresponds to two rotations of the crankshaft. In the study of engines, it is often useful to use the angle of rotation of the crankshaft, or crankangle (°CA), to express time. Thus, at 3000 rpm for example, 1°CA represents 0.056 ms.

The changes in pressure, volume, and temperature that occurs in an engine differ markedly from Beau de Rochas's cycle for the following reasons:

- Combustion is not instantaneous and it does not occur at constant volume. To obtain the best performance, the charge must be ignited before the end of the compression stroke. **Ignition advance** is frequently expressed in°CA with respect to TDC and it can vary from 5 to 50°CA.
- **Heat transferred** from the gases to the engine framework changes the temperature, and consequently the pressure, throughout the cycle.
- The constraints resulting from the inertia of the gases mean increased valve-opening duration (advanced opening and retarded closure). Hence, there is a period of time about TDC—which is referred to as **scavenging** or **overlap**—when the intake and exhaust valves are open simultaneously.
- Within the cylinder, a vacuum occurs during the intake stroke and back pressure occurs during the exhaust stroke. These **pumping losses** can be plotted as a pressure-volume curve that corresponds to the losses from induction and exhaust strokes.

These items appear in Fig. 3.3, which shows a comparison of the shape of a theoretical pressure-volume (PV) diagram and a diagram from an actual engine.

The control of the valve opening points, which are referred to as **valve timing**, can be varied as a function of the desired performance characteristics. For example, a significant increase in valve timing increases volumetric efficiency and power at high speed, instead of at low speed. Table 3.1 lists two examples of the valve timing used with modern engines.

Table 3.1 *Extreme examples of the valve timing used with modern engines.*

Position (°CA)		**Type of engine**	
		1*	**2***
Intake opens	IO	12	30
Intake closes	IC	48	72
Exhaust opens	EO	52	72
Exhaust closes	EC	8	30
Overlap	EC + IO	20	60

* Engine 1: high volumetric efficiency at low rpm;
Engine 2: high volumetric efficiency at high rpm (sports car).

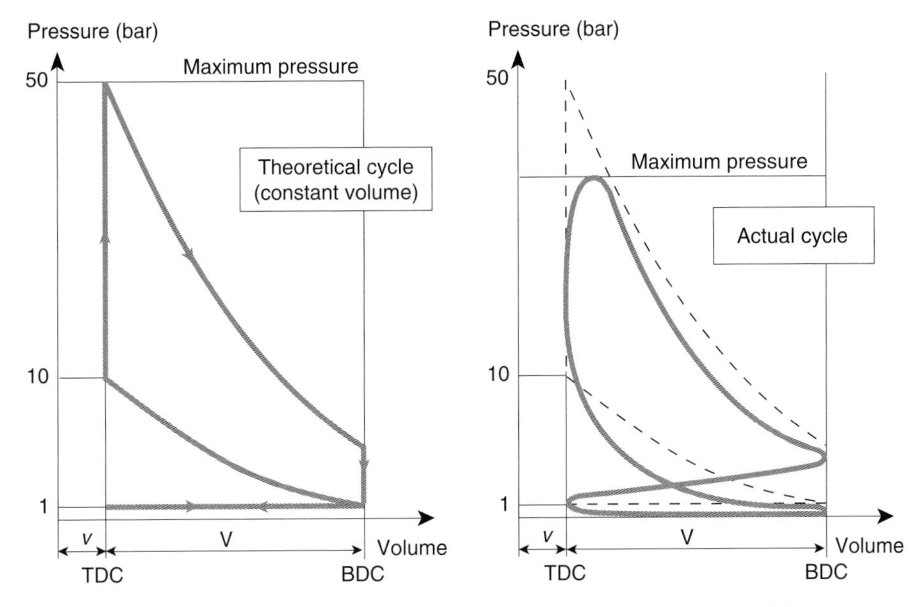

Figure 3.3 *The 4-stroke cycle.*

3.2.2 Technological aspects

3.2.2.1 Cylinder, piston, and head

Automobile engines are usually multicylinder engines: 2, 4, 5, 6, 8, or 12 cylinders, depending on the model. Six- and eight-cylinder engines are widely used in the United States whereas European and Japanese manufacturers prefer four-cylinder engines. The unit volume of cylinders usually varies from 200 to 600 cm^3.

Four-cylinder engines usually have the cylinders arranged in a line and they are referred to as **in-line** engines. Cylinders are usually bored directly into the cylinder block or they consists of "removable liners." The block, which is usually made of cast iron or light alloy, has five bearings in its lower section that support the crankshaft.

Some types of engines have only two cylinders, which are independent of the engine block, and their parallel axes are aligned horizontally. This arrangement, which is referred to as **horizontally opposed**, was used in Europe and Japan for low-power engines, but it is almost completely unutilized today. Conversely, the builders of large 6-, 8-, and 12-cylinder engines have adopted the **oblique** or "**V**" cylinder arrangement, which provides compact size and improved dynamic balance (low vibration).

Pistons are made of light alloy with tops that are flat, dished, or slightly sculptured to accommodate the valves and the spark plug. The piston-to-cylinder seal is maintained by three rings, which are open ended with square cross sections. The rings are located in groves in the piston. The top rings are referred to as "first rings" and bottom one is referred to as a "scraper" ring. The latter ring prevents lubricating oil from getting into the combustion chamber.

The **head**, which is usually made of cast iron or light alloy, can be of many different forms, depending on the engine designer's preferred criteria for the combustion chamber design. Combustion chambers can be flat, hemispheric, pent-roofed, or bowl-shaped recesses in the tops of the pistons. A composite metallic gasket, which is referred to as the head gasket, ensures an effective seal between the combustion chamber and the engine block. Some industrial engines also have separate heads for each cylinder to facilitate maintenance and repair.

3.2.2.2 Mixture distribution

The valves are controlled by a **camshaft** that rotates at one half of the speed of the crankshaft. The crankshaft drives the camshaft by means of a chain or toothed belt. If the camshaft is located in the engine bloc, the valve arrangement is referred to as "overhead valve"; if the camshaft is located in the cylinder head, the valve arrangement is referred to as "overhead cam". The motion of the camshaft in an overhead valve engine is transferred to the valves by a tappet, push rod, and rocker arrangement. In overhead-cam designs, the motion is transmitted directly by a tappet or by a tappet-rod and rocker assembly. Valves are closed by a spring that is compressed when the valves are opened.

3.2.2.3 Cooling

Air cooling, which makes use of fins on the cylinders and the heads, is usually used on motorcycles. This method had been used on two-cylinder horizontally-opposed engines, but these engines have practically disappeared today. Today's engines are almost always equipped with liquid cooling systems.

The coolant is a **mixture of water and ethylene glycol**; water has a high heat capacity and glycol has anti-corrosion and anti-freezing properties. The composition of the mixture is chosen to ensure a low freezing point, good corrosion resistance, and a boiling point greater than 100°C.

The coolant circulates through coolant passages around the cylinder and the combustion chamber. Coolant temperature is controlled by a thermostat. The heat is dissipated to the ambient air by a radiator and a fan. With modern engines, the operation of the fan is controlled by vehicle speed and load conditions.

3.2.2.4 Lubrication

Oil is drawn from a reservoir—the oil pan or sump—located at the bottom of the engine. A pump immersed in the oil sump draws the oil from the sump and directs it under pressure to an oil gallery. Passages from the oil gallery lead to all points needing lubrication: crankshaft bearings, connecting rod bearings, camshaft bearings, ... The cylinder walls and the bottoms of the pistons are lubricated by splash lubrication and oil mist.

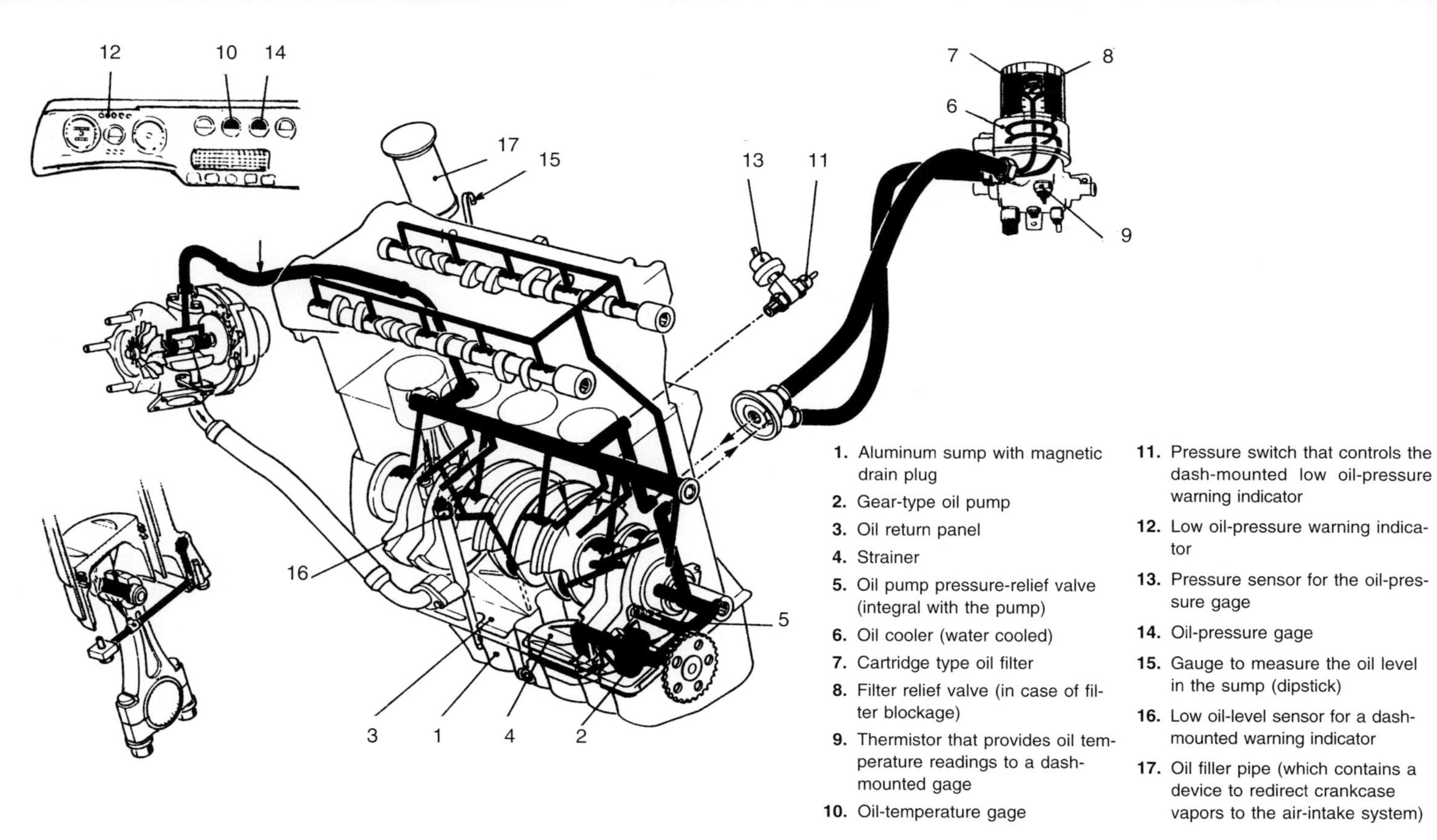

1. Aluminum sump with magnetic drain plug
2. Gear-type oil pump
3. Oil return panel
4. Strainer
5. Oil pump pressure-relief valve (integral with the pump)
6. Oil cooler (water cooled)
7. Cartridge type oil filter
8. Filter relief valve (in case of filter blockage)
9. Thermistor that provides oil temperature readings to a dash-mounted gage
10. Oil-temperature gage
11. Pressure switch that controls the dash-mounted low oil-pressure warning indicator
12. Low oil-pressure warning indicator
13. Pressure sensor for the oil-pressure gage
14. Oil-pressure gage
15. Gauge to measure the oil level in the sump (dipstick)
16. Low oil-level sensor for a dash-mounted warning indicator
17. Oil filler pipe (which contains a device to redirect crankcase vapors to the air-intake system)

Figure 3.4 *Lubrication system for a supercharged automobile engine. (Source: PSA)*

A filter, which is installed in series or in parallel with the oil system, retains the organic and metallic particles released by the combustion process and by engine wear. The oil vapors emitted from the crankcase, which are referred to as "blow-by", are returned to the engine intake to be burned in the combustion chamber. Under heavy load, crankcase gases are usually returned upstream of the throttle. At idle and low loads, the gases are returned downstream of the throttle.

3.2.3 Supercharging–a special technology

This technique is designed to increase engine power without changing engine displacement by compressing the intake charge (1.4 to 3.3 bar) to increase its mass. Supercharging is a well-developed technique that is used with **diesel engines**, which have low specific-power output due to their design (see 4.3.2). With gasoline engines, supercharging was first used in aviation to counter the effects of low air density at high altitude; it was then applied in automobile racing to increase performance for a given size and weight of engine. The technique was applied to luxury gasoline-powered passenger cars in an attempt to combine the advantages of a small displacement and a powerful engine; however, its use is now increasingly rare.

In fact, the very principle of supercharging increases the pressures and temperatures of combustion; this also increases the risks of engine knock. Knock is an unacceptable condition that must be avoided; this is why closed-loop systems that consist of knock sensors and ignition-advance controls are essential for supercharged engines, even if they are expensive. On the other hand, multi-valve technology, which has been developed in the last few years, has enabled improved engine performance (better flow and combustion) without causing knock. Therefore, in the domain of the gasoline-powered passenger automobile, supercharging is being discontinued in favor of systems that use more than two valves per cylinder.

Turbocharger technology and pressure-wave superchargers (Comprex) are discussed in Chapter 4.

3.2.4 The ignition system

Two types of ignition systems can be used on gasoline-powered vehicles. The major difference between them is the energy storage system. This system can consist of a coil (inductive) or a condenser (capacitive).

Coil systems are described here because they are practically the only systems now in use. Coil systems were proposed in 1913 in the United States by Delco Remy,[1] and they were developed and put into general use after the First World War. The elements of the coil system have been consistently modernized and now they are completely controlled by electronics.

1. DELCO: An acronym formed from the company's original name: *Dayton Engineering Laboratories Company* (Ohio).

3.2.4.1 Electromechanical ignition

The schematic circuit shown in Fig. 3.5 consists of the following items:

- The **generator** (not shown), which is also referred to as a dynamo or alternator, is driven by a belt from the engine and it provides the vehicle electrical supply.
- The **battery** provides the low-voltage (12 V) reservoir for primary current.
- The **coil** consists of two windings that are referred to as primary and secondary.
- The **distributor** consists of the following two elements:
  - the contact breaker (or points) that periodically interrupts the primary current and thereby induces a high-tension secondary current;
  - the distributor rotor and the distributor cap, which direct the high-tension current in turn to each spark plug.
- The **condenser**, which is connected to the primary circuit, reduces the breaking time and thereby increases the voltage induced in the secondary circuit.
- The **spark plugs** consist of two electrodes separated by a space (gap) of about 0.5 to 1 mm; the spark jumps the electrode gap between 12 000 and 40 000 V.

A spark must occur in each cylinder on every second rotation of the crankshaft; therefore, the distributor shaft is driven from the camshaft. The cam on the distributor shaft has one lobe for each engine cylinder. The firing order for a four-cylinder engine is 1-3-4-2.

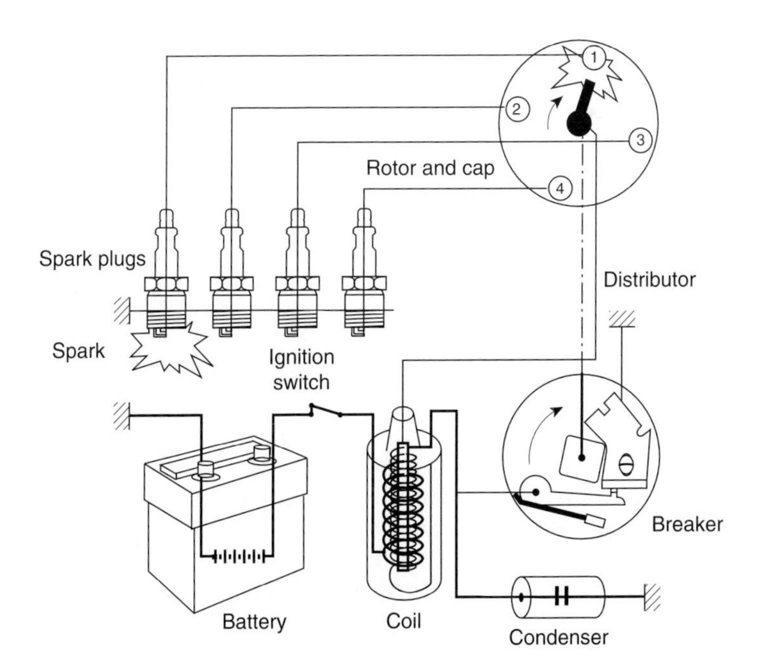

Figure 3.5 *Schematic of an electromechanical ignition system.*

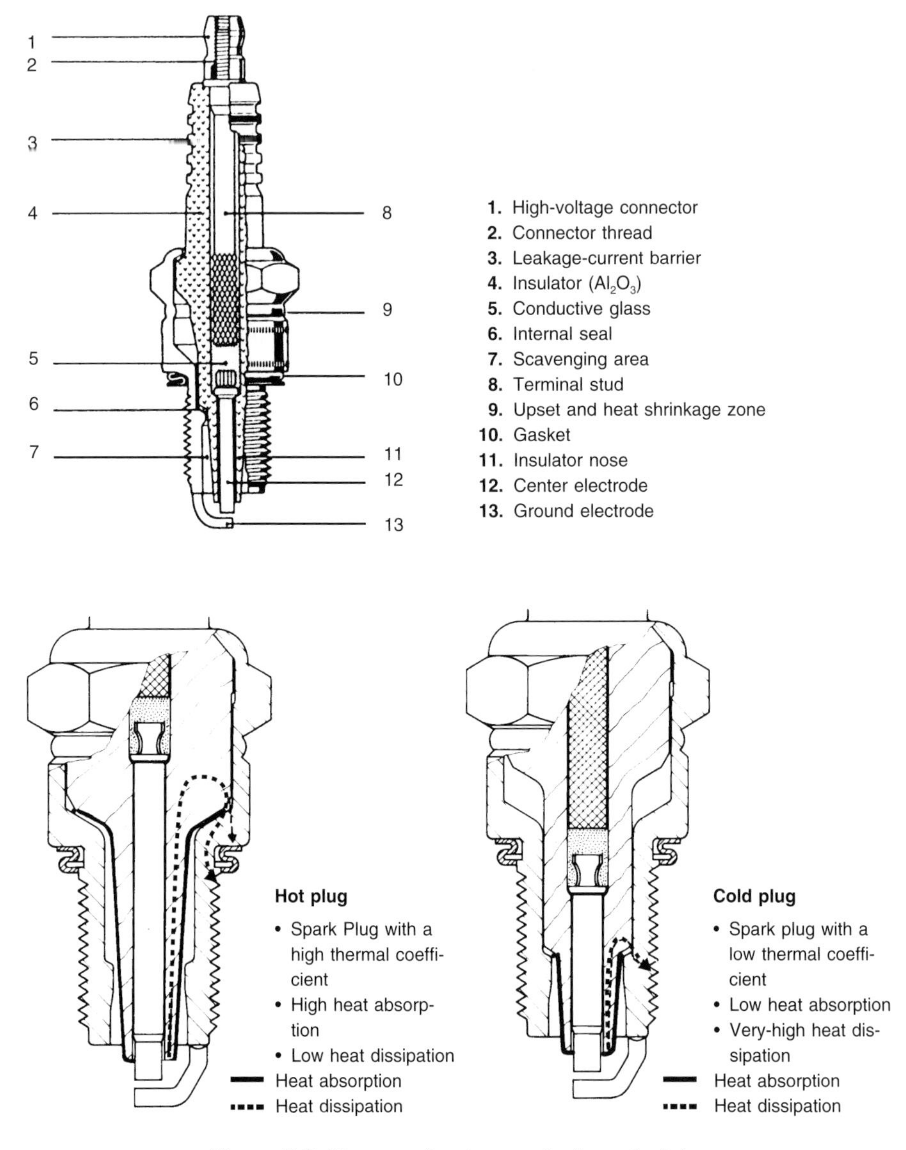

Figure 3.6 *Diagram showing spark plug principles. (Source: Beru)*

The spark plug is always fired just before TDC so that the combustion process, which is always slow when it is compared to engine speed (approx. 60°CA), can be completed early enough in the cycle. The point of ignition is referred to as ignition advance and it is expressed in degrees of distributor rotation (distributor° or °D) or in degrees of crankshaft rotation (°CA). One °D = two °CA.

Electromechanical ignition systems usually have two advance settings, one that depends on speed and one depends on load (Bosch 1988).

Some information about **spark plug technology** is worth mentioning at this point. The center electrode that conducts the high-voltage current is contained in a ceramic insulator and it protrudes near the ground electrode (see Fig. 3.6). The insulator is contained in a nickel-plated shell (or body) that is screwed into the cylinder head. Electrodes typically consist of nickel-based alloys, but they can also be silver or platinum.

Satisfactory spark plug performance requires an adequate temperature level (800 to 950°C) for the electrodes and the insulator nose. If the temperature is too low, deposits accumulate and they interfere with the spark. Conversely, if the temperature is too high, there is a risk of pre-ignition (see 3.3.3).

The **heat range of spark plugs** can be adjusted at the design stage by varying the heat flow through the insulator, the body, and the gasket. A **cold plug** has an electrode-insulator-body path that is short and the center electrode runs cold, whereas a **hot plug** has a longer insulator nose that lengthens the path to the head. Cold plugs are usually used in high-performance engines and those that run at high temperatures. Hot plugs are usually used when deposit accumulation is related to operating conditions (city traffic).

3.2.4.2 Electronic ignition

Electronic ignition is currently the most widely used system. Unlike the previously described arrangement, the ignition point is controlled electronically rather than mechanically.

Speed is measured using a inductive sensor located at the starter ring gear on the flywheel. The load is determined from the absolute pressure in the intake manifold, which is measured using a piezo-electric sensor. These two factors (speed and load) are evaluated by a computer that determines the optimum ignition point based on a **map** stored in its memory (see Fig. 3.7). The computer can also evaluate other factors (air, coolant, and oil temperatures; throttle position, ...) to regulate adjustments and optimize engine performance.

Like the electromechanical ignition system, the high voltage distributor is driven from the camshaft.

3.2.4.3 Fully-electronic ignition

Fully-electronic ignition systems are primarily installed in high-performance automobiles, but they will eventually become widely available (see Fig. 3.8). The ignition point is controlled by a computer that uses a map stored in memory (see 3.2.4.2). The rotating mechanical high-voltage distributor is replaced by an electronic device that is controlled directly by the computer. The single coils used with four-cylinder engines have two twin-outputs and the primary winding has two stages. The computer alternately supplies each of the two stages, which causes the firing of two spark plugs (1 and 4 or 2 and 3) followed by the firing of the other two. There are also dual-coil systems that have two power-output stages and two outputs.

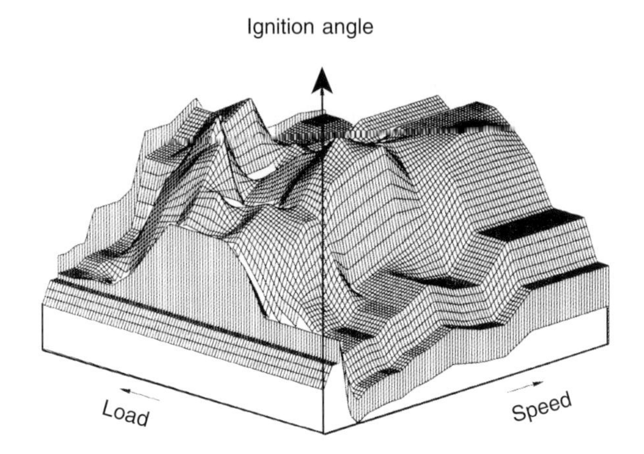

Figure 3.7 *Ignition map (electronic control). (Source: Bosch)*

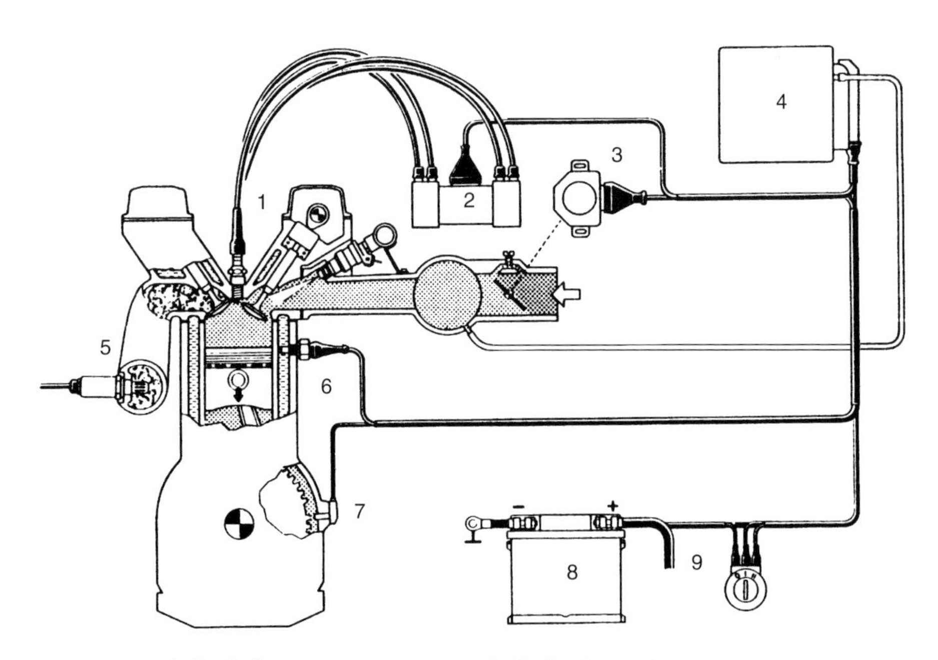

1. Spark plug
2. Dual-spark coils
3. Throttle-valve switch
4. Electronic control unit
5. Lambda sensor
6. Coolant temperature sensor
7. Engine speed and reference-mark sensor
8. Battery
9. Ignition switch

Figure 3.8 *Diagram of a fully-electronic ignition system. (Source: Bosch)*

3.2.5 The fuel system

The triggering and progression of combustion implies the existence of a well-defined fuel-air mixture. The fuel system must accomplish three essential functions:

- precise control of the fuel and air flows
- fuel atomization–to ensure an air-fuel mixture that vaporizes during intake and compression
- even distribution of the mixture to all cylinders

The fuel system consists of various items: air filter, fuel tank and electric supply pump, carburetor or injector, and an intake manifold. There are two major types of fuel systems: carburetion and injection.

For many years–until about 1980 in the United States and about 1990 in Europe–carburation was the most widely-used fuel system for spark-ignition engines. This section includes a review of carburation principles and it explains the imperfections and limitations that have led to its almost complete disappearance. The section continues with a description of modern fuel-injection systems and the *lambda* sensor fuel-mixture control system.

3.2.5.1 Carburation

This technique consists of drawing in and atomizing fuel by using the flow of air in the intake stream. The essential parameter that controls the fuel flow is the air flow through a throttled area (nozzle or venturi) in the intake system. Hence, mixture and atomization occur simultaneously at the venturi point.

Therefore, the simplest version of the carburetor consists of three items:

- a **calibrated orifice**–the jet–through which the fuel is fed into the air stream
- a **venturi** that is located in the air stream, which causes fuel aspiration and vaporization
- a **throttle plate** that is located downstream of the venturi; changing the position of the throttle plate controls the quantity of mixture entering the engine, and hence the engine's power output

In a typical carburetor, air flows from top to bottom (vertical downdraft carburetor) and the venturi has a fixed bore.

Assuming the air flow is continuous and constant, an approximate expression for the air-fuel ratio of the carbureted mixture delivered by the main jet can be established.

The following nomenclature is based on Fig. 3.9:

A_a = area of the venturi throat
A_c = area of the main jet G
C_a = venturi flow coefficient
C_c = main jet flow coefficient
K = numerical coefficient linked to the compressibility of air
P_0, T_0 = air pressure and temperature at the carburetor mouth

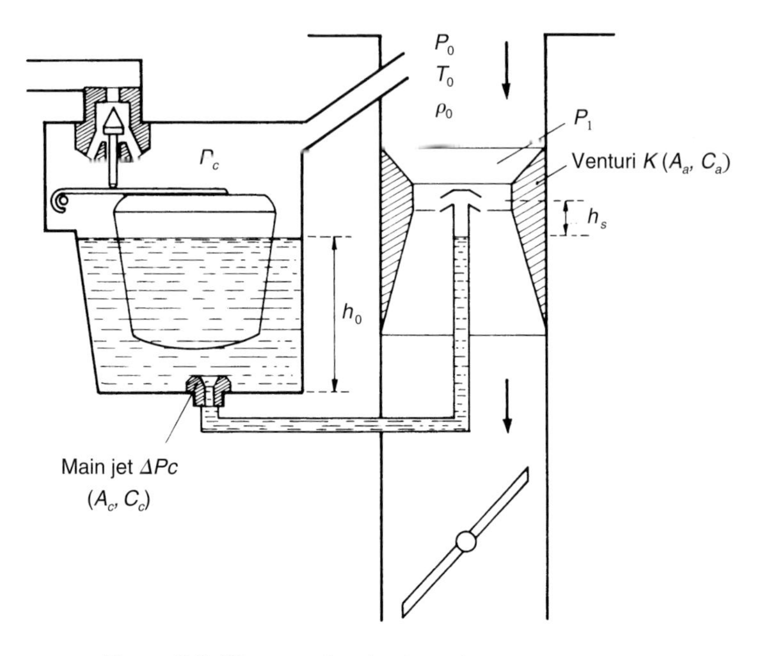

Figure 3.9 *Diagram of a simple carburetor system.*

P_1 = pressure at the throat of the venturi

P_c = pressure in the float bowl at constant level

ρ_0 = air density at P_0, T_0 conditions

ρ_c = fuel density

r = stoichiometric air-fuel ratio
(r was defined in Chapter 1, paragraph 1.3.3.1)

Bernouilli's principle can be applied to get a first approximation of the $\dot{M}_a$ and $\dot{M}_c$ flows of air and fuel.

Thus one obtains

$$\dot{M}_a = KA_aC_a\left[2\rho_0(P_0 - P_1)\right]^{1/2}$$

and

$$\dot{M}_c = A_cC_c(2\rho_c\Delta P_c)^{1/2}$$

where ΔP_c represents the difference in pressure at the jet. This term can be explained by the following definitions and relationships:

ΔP_s = pressure difference required to overcome, on the one hand, the difference in height between the float level and the output tube, and on the other hand the surface tension of the fuel

ΔP_e = pressure drop in the fuel lines

Then:

$$\Delta P_c = P_c - P_1 - \Delta P_s - \Delta P_e$$

In general, the pressure P_c on the float chamber is close to the pressure P_0 at the mouth of the carburetor.

By defining $\Delta P_1 = P_0 - P_1$, we obtain

$$\dot{M}_c = A_c C_c \left[2\rho_c (\Delta P_1 - \Delta P_s - \Delta P_c)\right]^{1/2}$$

The equivalence ratio is expressed by the following relationship:

$$\varphi = r \frac{\dot{M}_c}{\dot{M}_a}$$

or

$$\varphi = r \frac{A_c C_c}{K A_a C_a} \left(\frac{\rho_c}{\rho_0}\right)^{1/2} \left(\frac{\Delta P_1 - \Delta P_s - \Delta P_e}{\Delta P_1}\right)^{1/2}$$

In reality this formula is only approximate because it does not take into account certain **secondary effects**: limitations in the application of Bernouilli's principle, the effect of fuel viscosity, and the pulsing and discontinuous flow of air at low rotation speeds.

The preceding formula also shows that equivalence ratio cannot remain constant when the air flow, which is linked to ΔP_1, varies. A mechanism of correction and adjustment based on engine load must be installed.

In other respects, for fixed aerodynamic conditions and mechanical adjustments, the equivalence ratio is expressed by the following simple relationship

$$\varphi = r\,\mathrm{C} \left(\frac{\rho_c}{\rho_0}\right)^{1/2}$$

where C is a constant.

The mixture equivalence ratio is therefore proportional to the square root of the density of the fuel and inversely proportional to the square root of the density of the air.

Thus, based on reference conditions (20°C, 1 bar), a 30°C increase in temperature or a pressure reduction of 100 mbar leads to an enrichment of 5%.

Lastly, fuel characteristics affect the equivalence ratio through the stoichiometric ratio r and the fuel density ρ_c. Often, r and ρ_c vary inversely in such a way that **a denser fuel results in a decreased mixture equivalence ratio**.

As this analysis shows, for a carburetor to rigorously maintain a constant mixture equivalence ratio under all possible operating conditions, several complex corrective devices must be used—devices that are usually not completely effective. This is why vehicles equipped with "3-way catalysts", which require a perfect stoichiometric mixture, cannot use carburation ant they must resort to more precise technologies.

3.2.5.2 Fuel injection

Fuel injection for gasoline engines, due to its expense, has long been the sole prerogative of sports cars and luxury automobiles. The advent of anti-pollution standards encouraged its further development. In fact, the use of fuel injection, when coupled with the lambda stoichiometry sensor, enables an engine to meet the requirements for 3-way catalytic converters, which would have been impossible with a carburetor due to its imprecise mixture control (especially during transient operation).

Although the carburetor is always located upstream of the throttle valve, the location of fuel injectors varies depending on the device. Fuel can be introduced, for example, directly into the combustion chamber; in this case the injection is referred to as **direct** injection. Gasoline can also be injected outside of the combustion chamber upstream of the intake valve; this is referred to as **indirect** injection. Direct injection is currently the subject of numerous studies, but there are currently only a few individual developments; therefore, the discussion herein is limited to **indirect injection**. These types of systems are essentially divided into two categories consisting of **multipoint** and **single-point** injection systems. The multipoint system provides each cylinder with its own injector, which vaporizes the fuel directly before the intake valve. The single-point systems uses a single injector that is located at a central point in the intake system upstream of the throttle valve.

a. Multipoint injection

Multipoint fuel injection systems are primarily used on luxury automobiles because of their high cost. An arrangement of the devices used in such a system is shown in Fig. 3.10. An electric fuel pump draws fuel from a tank and supplies it through a filter to the distribution manifold. The fuel is then distributed in a uniform manner to the intake tract of each cylinder.

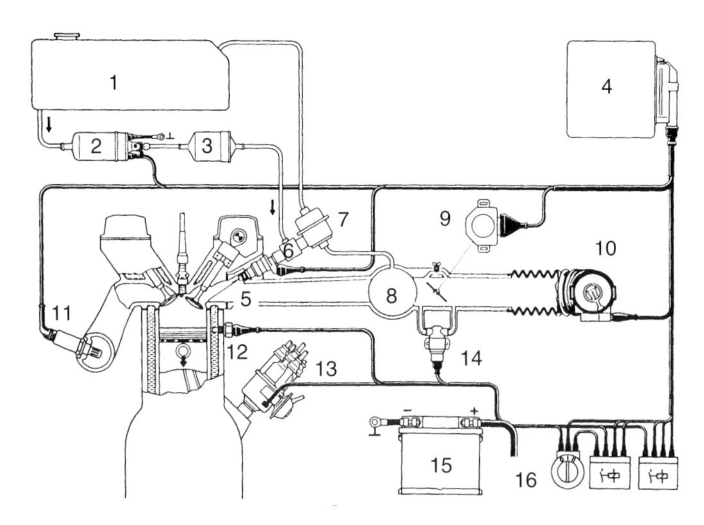

1. Fuel tank
2. Electric fuel pump
3. Fuel filter
4. Control unit
5. Fuel injector
6. Fuel distribution manifold
7. Fuel-pressure regulator
8. Intake manifold
9. Throttle-valve switch
10. Hot-wire mass flowmeter
11. Lambda sensor
12. Engine-temperature sensor
13. Ignition distributor
14. Idle-speed actuator
15. Battery
16. Ignition switch

Figure 3.10 *Electronic multipoint fuel-injection system. (Source: Bosch)*

A pressure regulator, which is located at one end of the distribution manifold, maintains a constant difference between the supply pressure (2 to 6 bar) and the air intake pressure.

The injectors are operated electromagnetically by a control unit. The control unit collects and processes operating information from a number of sensors. The control unit then sends electrical impulses that open and close the injectors. The frequency of injection is directly related to engine speed, but it can also depend on the type of electronic system. The most rudimentary systems provide only one simultaneous injection for all cylinders; in this case, an injection occurs with every turn of the crankshaft, which is twice per complete engine cycle.

Other more sophisticated systems actuate the injectors as a **group**: for a four-cylinder engine, two injectors would open and close at the same time, then the other two injectors would be actuated in turn after one complete engine rotation.

There is also **sequential** multipoint injection where the injectors operate individually once per cycle according to the crankshaft position. This technique is the best performing one of the three, but it is also the most expensive and therefore the least used.

The use of a **memory map** enables the control unit to determine the injection duration for each operating condition. Therefore, a driver pressing on the accelerator sets the operating condition, and as a consequence, the load and speed of the vehicle. For these operating conditions, there is a unique fuel flowrate for which the equivalence ratio will be 1.00. Therefore, because the gasoline supply pressure is low and constant with respect to the intake-air pressure, the injector flowrate is proportional to the opening duration. The injection time can thus be determined directly from the mass flowrate of the intake air and the engine speed.

The system described in Fig. 3.10 shows that the mass of the air ingested for each cycle is measured by a hot-wire mass flowmeter, which is located upstream of the intake manifold; the hot wire, which is maintained at a temperature higher than the intake air, is cooled by the air flow; the electrical current required to maintain the temperature of the wire provides a measure of the intake-air mass. This value is combined with the engine speed (provided by a signal from the electronic ignition) and both values are compared with an air-fuel ratio map that is stored in computer memory. The map information provides the optimal injection duration.

There are, of course, other methods of measuring air flow. For example, there are “flap type” volume flowmeters or indirect systems based on intake-manifold pressure or throttle position (see 3.2.5.2.b). These devices provide volume flowrates; the control unit must then determine the intake air temperature to determine the correct injection duration.

Furthermore, the control unit also receives operating information from other sensors. Signals can be received from the lambda sensor (see 3.2.5.3), the coolant temperature sensor, the battery voltage, ... The information received from these sensors enables the control unit to further refine its output to match each engine operating condition.

b. Single-point injection

Single point electronic fuel injection systems are used on low– to medium-priced automobiles that carry 3-way catalytic converters. Their low cost in comparison to multipoint systems derives from the use of a single injector. As shown in Fig. 3.11, there is little difference between the fuel supply circuits of the two systems: with the single-point device, an electric fuel pump draws fuel from the tank and supplies it through a filter to the injector.

The fuel is atomized above the throttle valve where maximum air flow occurs. A pressure regulator, which is linked to the injector, maintains a constant difference between the fuel pressure (0.7 to 1.1 bar) and the air pressure at the injection point, ensuring that the fuel flowrate depends solely on the opening duration of the injector.

As with multipoint injection systems, the injection timing and duration is controlled by a microprocessor. The opening-command processes are relatively easy to manage for the single injector: the control unit sends an electronic signal to open the injector once per engine cycle; the signal is based on the ignition pulses; when the signal stops, the injector closes.

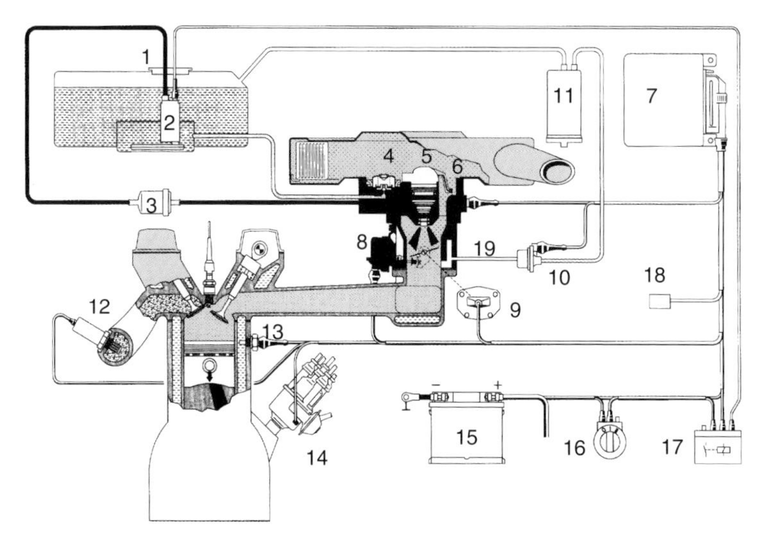

1. Fuel tank
2. Electric fuel pump
3. Fuel filter
4. Fuel-pressure regulator
5. Electromagnetic fuel injector
6. Air-temperature sensor
7. Electronic control unit
8. Throttle actuator
9. Throttle-valve potentiometer
10. Exhaust gas recirculation (EGR) valve
11. Carbon canister
12. Lambda sensor
13. Coolant-temperature sensor
14. Ignition distributor
15. Battery
16. Ignition switch
17. Relay
18. Diagnostic connector
19. Control injection unit

Figure 3.11 *Single-point electronic fuel injection system. (Source: Bosch)*

Controlling the duration of the injection is based on the same principles as the multipoint system: the microprocessor controller receives information on the volume of air ingested per cycle, the air temperature, and the engine speed; based on the mixture map stored in its memory, the controller determines the injection time required to provide the optimal combustion mixture.

The process used to determine the air intake volume is unique to the single-point system: a potentiometer is used to determine the position of the throttle valve, which represents the volume of air ingested per cycle. Also, intake-air temperature is essential to determine the mass flow of air and thereby derive the exact quantity of fuel to be injected for the cycle.

c. Injectors

Injectors ensure a precise mixture of the fuel as well as its proper atomization. Injectors are actuated by an electromagnet, which is controlled by a microprocessor.

The geometry of the needle and its support determine the shape of the fuel spray. A **pintle** is usually located at the end of the injector needle; when the pintle lifts from its seat, the spray hole is opened and the fuel is injected (see Fig 3.12). Multipoint systems are often equipped with injectors that provide **two conical-shaped fuel sprays**; the center hole is replaced by a nozzle tip with two holes in it (see Fig. 3.12). This type of injector works well with engines equipped with two intake valves per cylinder, since a spray can be directed to each intake valve. The atomization of the fuel spray can be improved with **air washing**: air is blown in and mixed with the fuel just before injection, which reduces the size of the droplets.

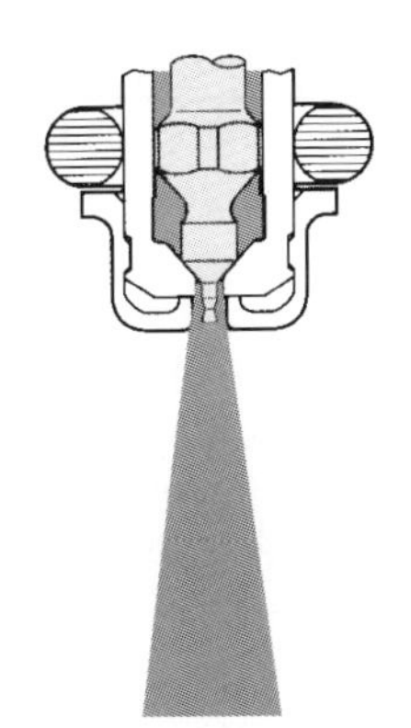

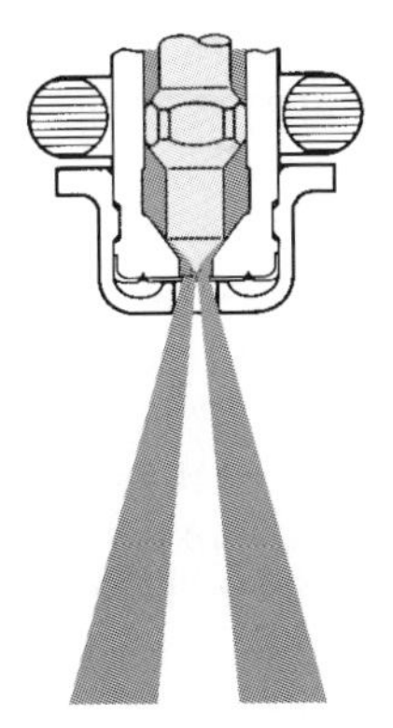

Figure 3.12 *Pintle and nozzle-hole injectors. (Source: Bosch)*

3.2.5.3 Mixture control by lambda sensor

Vehicles equipped with 3-way catalytic converters require effective mixture control. Maximum effectiveness is achieved when the fuel-air mixture is exactly

stoichiometric and the slightest deviation from this can result in a significant increase in polluting emissions (see Chapter 5).

Meeting these requirements necessitates the use of a **closed-loop** injection system, which consists of an oxygen sensor (or **lambda sensor**) that continuously monitors the composition of the exhaust gas and sends signals to the control unit. The control unit changes the injection duration to ensure that a stoichiometric mixture ($\varphi = 1.00$) is maintained. The schematic for the system follows:

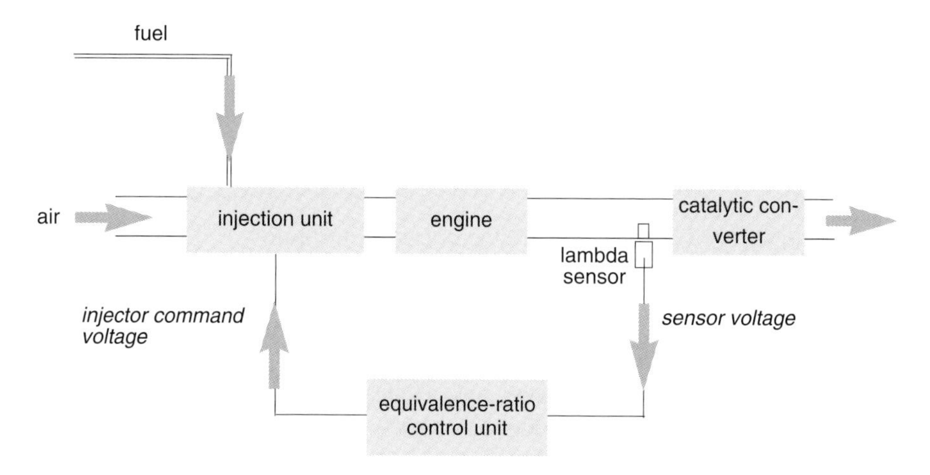

The lambda sensor is installed in the exhaust pipe, between the engine and the catalytic converter; that location results from the following **compromise**: the sensor must be located close enough to the engine to begin operating as soon as possible after starting, when the exhaust gases are relatively cold, but it also must be far enough from the engine to survive the high temperatures produced at full engine load. Vehicles with twin exhaust pipes are equipped with dual lambda sensors.

Fig 3.13 depicts this type of sensor: it consists of two electrodes, which are separated by a solid electrolyte; one electrode is in contact with the exhaust gases and the other is in contact with air. A voltage is generated as soon as there is a difference in oxygen content between the electrodes; in other words, when a mixture that is lean to begin with becomes rich—which means the exhaust gases are low in oxygen—it causes a “step change” in the voltage curve of the sensor.

The electrolyte usually consists of a zirconium dioxide based ceramic stabilized with yttrium oxide: the replacement of a few zirconium ions with yttrium ions makes the electrolyte a perfect conductor of oxygen ions (from about 300°C). Some manufacturers prefer to use titanium dioxide based electrolytes, which are also stabilized with yttrium oxide.

The electrodes are also covered with a layer of microporous platinum, whose catalytic effect amplifies the electrical signal, thus improving the detection of deviations from mixture stoichiometry.

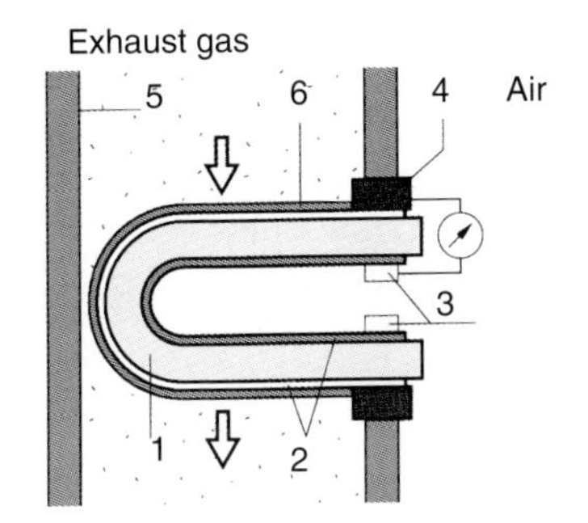

1. Ceramic
2. Electrodes
3. Contacts
4. Body contact
5. Exhaust pipe
6. Protective ceramic coating (porous)

Figure 3.13 *Diagram of a lambda sensor. (Source: Bosch)*

The most common lambda sensors are not heated. Unheated lambda sensors become effective when the exhaust gas temperature reaches 350°C, the temperature at which the electrolyte becomes conductive. **Heated sensors** become effective more quickly because they do not depend on the exhaust gas for heating the electrolyte; they use internal electric resistance heaters instead (see Fig. 3.14). Heated sensors become active when exhaust gas temperatures are low (200°C).

It is important to clearly differentiate between the method used to control injection duration using maps (see 3.3.5) and lambda-sensor control method. The latter system is used to adjust the mixture when it varies from stoichiometric and not to determine an optimal injection duration.

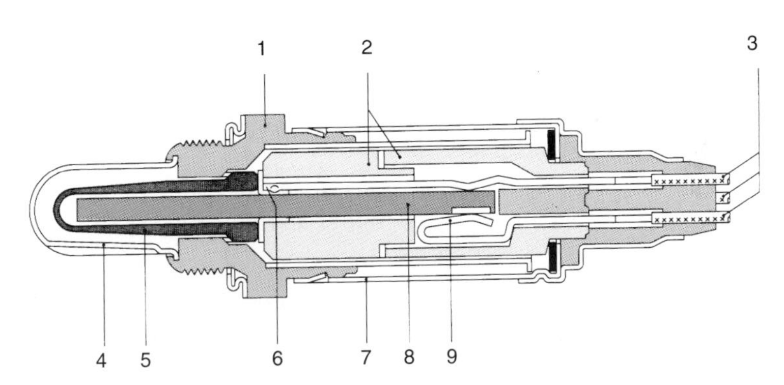

1. Sensor body
2. Ceramic support tube
3. Wiring connections
4. Louvered protection tube
5. Active ceramic portion of the sensor
6. Contact point
7. Protective cover
8. Heating element
9. Heating element connectors

Figure 3.14 *Heated lambda sensor. (Source: Bosch)*

There are also systems that have **two lambda sensors per exhaust pipe**: the first sensor, which is located upstream of the catalytic converter, detects the oxygen content of the exhaust gas while the second one, which is located downstream of the converter, monitors the effectiveness of the upstream sensor and the converter. This system was developed to meet new California regulation OBD II (On Board Diagnostic II), which is explained in Chapter 5.

3.2.5.4 OBD self-diagnostics

Since the beginning of the 1980s, American manufacturers have been using **on-board diagnostics** on vehicles equipped with closed-loop control systems and 3-way catalytic converters. The system is intended to alert drivers to operational defects and to aid repair technicians in problem diagnosis.

The California Air Resources Board (CARB) began to legislate, in 1985, a requirement for all vehicles sold in California to have an on-board diagnostic (OBD I) system beginning with the 1988 model year. Using special procedures, the system monitors each component related to the electronic engine-control system. The types of components that can be monitored are the controller itself, the injection system, the exhaust gas recirculation (EGR) system, the mixture sensor, ... If a default is detected, a dashboard telltale alerts the driver that repairs are required. The defect is also memorized by the control system to aid in diagnosing and repairing the problem.

CARB has since added to the 1985 measures by requiring OBD II in 1989, which applies to all vehicles sold in California beginning with the 1994 model year. This legislation requires the monitoring of all pollution related components (the catalytic converter for example). A major consequence of this legislation is the installation of a second lambda sensor downstream of the catalytic converter. Since this new legislative requirement is directly related to pollution, it is addressed in Chapter 5.

When this book went to press, there was no legislation in Europe requiring on-board diagnostics, but projects are underway and the results should be appearing soon.

3.3 Performance and control

Two major criteria are used to measure engine performance:

- **power**, which enables an engine to meet application requirements, for example, to provide tractive effort for vehicles
- **efficiency**, which is the measure of an engine's ability to convert the chemical energy contained in the fuel into mechanical work

This section covers some common definitions and formulas related to power and efficiency, as well the influence that some major operational parameters have on engine performance

3.3.1 Power, torque, mean pressure

3.3.1.1 Effective power

The **effective power (P_e)** developed at the output shaft (brake output) is the product of the torque (T) and the angular velocity (ω) according to the following equation:

$$P_e = T \cdot \omega$$

where P_e is in watts, T is in newton meters, and ω is in radians per second.

If N is the speed in revolutions per minute (rpm), then

$$\omega = \frac{2\pi}{60} N$$

and $$P_e = \frac{\pi}{30} TN \qquad \text{(where } P_e \text{ is in W)}$$

or $$P_e = \frac{\pi}{30\,000} TN \qquad \text{(where } P_e \text{ is in kW)}$$

Automobile engines usually develop effective power outputs of between 30 and 140 kW. The power of truck engines varies from 150 to 400 kW.

Fig. 3.15 shows an example of the power and the torque developed at full throttle, as a function of engine speed, for a modern engine.

The torque increases with the amount of energy released, which is related to **volumetric efficiency**. This parameter, which is referred to as V_e, represents the ratio of the mass of the air ingested for each engine cycle to the mass of

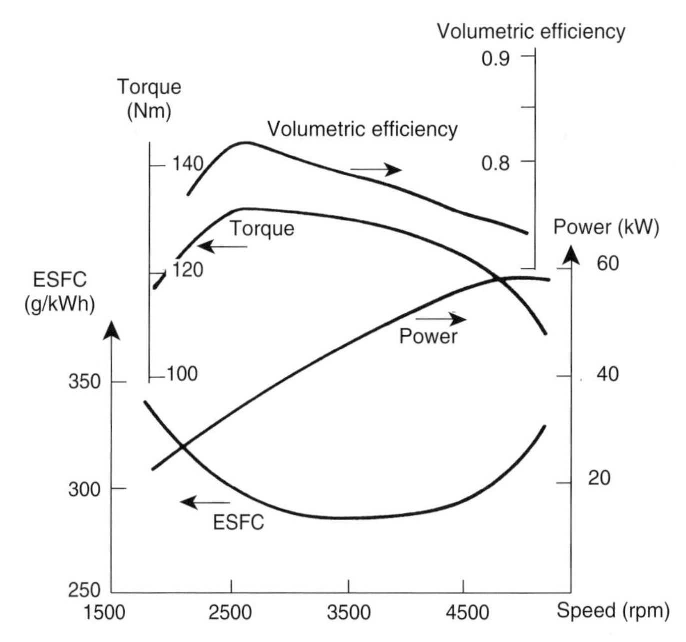

Figure 3.15 *Full-throttle characteristics of maximum power, maximum torque, volumetric efficiency and specific fuel consumption as a function of speed in a typical spark-ignition engine.*

air that could be contained in the swept volume at standard temperature and pressure (20°C and 1 bar).

An example of the change in volumetric efficiency as a function of speed is also shown in Fig. 3.15. By working on the intake flow dynamics, manufacturers seek to obtain a high volumetric efficiency at low speed to enhance driveability, and at high speed to produce maximum power.

With current engines, maximum torque occurs between 2000 and 3500 rpm. Torque decreases quickly at around 5500 to 6000 rpm; power also reaches its peak at this point and it also begins to decrease.

3.3.1.2 Specific power

Specific power is the maximum power developed per unit of cylinder volume. Specific power is usually expressed in kW/dm³ and it represents engine performance regardless of engine size. Modern engines in series production have specific power ratings between 30 and 50 kW/dm³. Racing engines, however, can attain much higher figures: for example, 300 kW/dm³ in Formula 1.

3.3.1.3 Administrative or taxable power

This characteristic is not directly related to engine performance; its sole purpose is to establish a basis for classifying vehicles for **taxation** and other tariffs. The rules governing its use are defined independently by each country. In France, the following formula was used before July 1998 :

$$P_f = m\left(0.0458\ \frac{V}{K}\right)^{1.48}$$

where

P_f = taxable (fiscal) horsepower in CV (*CheVaux*)
m = coefficient equal to 1, except for diesel engines when it is 0.7
V = cylinder volume in cm³
K = arithmetic average of the following coefficients:

$$K = \frac{k_1 + k_2 + k_3 + k_4\ (\text{or } k_5)}{4}$$

where k_1, k_2 ... represent the speed in km/h that correspond to 1000 rpm in each gear 1, 2 ... . Five speed gearboxes use k_5 instead of k_4, provided that k_5 is less than 1.25 k_4. The value of P_f is rounded to the nearest whole number, which is usually between 4 and 20 CV.

Since 1st July 1998, a new simpler formula is used:

$$P_f = \frac{[CO_2]}{45} + \left(\frac{P_{max}}{40}\right)^{1.6}$$

where

P_f = taxable horsepower in CV
$[CO_2]$ = carbone dioxide emissions in g/km (ECE + EUDC cycle)
P_{max} = maximum power in kW

3.3.1.4 Mean pressure

In following expression for specific power

$$P_s = \frac{T}{V}\omega$$

V is the engine displacement and the term T/V, which has units of pressure, is introduced.

This infers that the brake output power can be linked to a value referred to as the **mean effective pressure or MEP**. The MEP represents the virtual constant pressure that, if applied to the piston, would result in the same work as that delivered by the engine.

For one engine rotation, the average work would be

$$W_e = \frac{1}{2}\,\text{MEP} \cdot V$$

(1 power stroke for 2 revolutions)

The engine rotates at N rpm, or $N/60$ rev/s.

Therefore, the effective power is

$$P_e = \frac{1}{2}\,\text{MEP} \cdot \frac{VN}{60}$$

and

$$\text{MEP} = \frac{120\,P_e}{VN}$$

(P_e is in W; MEP is in Pa; V is in m^3; N is in rpm)

The following representation is another, more practical, formula that is often used:

$$\text{MEP} = \frac{1200\,P_e}{VN}$$

(P_e is in kW; MEP is in bar; V is in dm^3; N is in rpm)

The MEP, like the volumetric efficiency, is a variable that can be used to express engine load. For example, MEP is usually 0 to 1 bar at low loads. At wide open throttle, the MEP reveals an engine's compactness and its **intrinsic performance** in terms of power output. Typical MEP values are 8 to 10 bar for production engines, 12 to 14 bar for supercharged engines, and 30 to 40 bar for Formula 1 engines.

3.3.2 Efficiency

The **overall or effective efficiency** of an engine is equal to the ratio of the energy measured at the output shaft to the energy released by the complete combustion of the fuel, that is

$$\eta_e = \frac{\text{Effective work}}{\text{Chemical energy of the fuel}}$$

The energy ratio η_e is obtained directly from the effective specific fuel consumption (ESFC) and the net heating value (NHV) of the fuel, according to the following formula:

$$\eta_e = \frac{3.6 \cdot 10^6}{\text{ESFC} \cdot \text{NHV}}$$

(ESFC is in g/kWh; NHV is in J/g)

The η_e ratio translates the overall efficiency of the chemical-energy conversion process of the fuel into mechanical work—this is of primary interest to the user. However, η_e neither explains the source of any loses nor their relative value: thermodynamic limitations, friction, auxiliary drives, ...

One of the first limitations to efficiency is purely thermodynamic. This limitation appears when the **efficiency η_0 of the constant volume cycle** is derived as shown schematically in Fig. 3.16. As shown in the P-V diagram, there are two constant volume (isochoric) processes involving combustion heat (Q_1) and rejected heat (Q_2). The work done is represented by W.

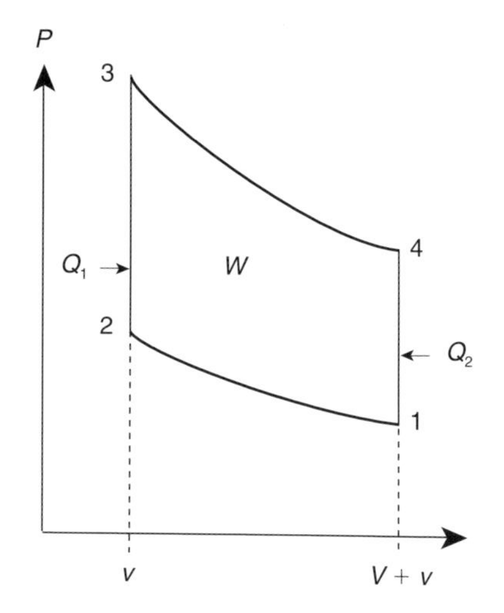

Figure 3.16
Thermodynamic representation of a 4-stroke cycle.

Applying the first principle of thermodynamics yields

$$Q_1 = W + Q_2$$

The efficiency is

$$\eta_0 = \frac{W}{Q_1} = \frac{Q_1 - Q_2}{Q_1} = 1 - \frac{Q_2}{Q_1}$$

On the other hand

$$Q_1 = C_v\,(T_3 - T_2)$$

and

$$Q_2 = C_v\,(T_4 - T_1)$$

where C_v is the specific heat at constant volume and T is the temperature; the subscripts 1,2,3,4 indicate the temperatures at various points in the cycle.

Therefore:

$$\eta_0 = 1 - \frac{T_4 - T_1}{T_3 - T_2} = 1 - \frac{T_1}{T_2}\,\frac{\left(\frac{T_4}{T_1} - 1\right)}{\left(\frac{T_3}{T_2} - 1\right)}$$

Assuming adiabatic processes between 1 and 2, and 3 and 4, yields

$$\frac{T_3}{T_4} = \frac{T_2}{T_1} = \varepsilon^{\gamma-1}$$

or

$$\varepsilon = \frac{V + v}{v}$$

In these formulas, ε is the compression ratio, V and v are the swept volume and the combustion chamber volume respectively, and $\gamma = C_p/C_v$ is the ratio of the specific heats of the working fluid, which is assumed to be constant.

This finally yields

$$\eta_0 = 1 - \frac{T_1}{T_2}$$

or

$$\eta_0 = 1 - \frac{1}{\varepsilon^{\gamma-1}}$$

This relationship, which is derived from an unrealistic hypothesis, has no other purpose than to involve the mechanical (ε) and thermodynamic (γ) parameters.

In effect, the term γ represents, in an overall sense, the thermo-aerodynamic phenomena of gas composition, combustion, and heat exchanges; γ changes according to the operating conditions, but its value is estimated to be about 1.22 by modeling and experimentation.

For example, by using this numerical value the following results are obtained,

$$\eta_0 = 0.383 \text{ for } \varepsilon = 9$$

$$\eta_0 = 0.397 \text{ for } \varepsilon = 10$$

which show a gain of 4% for a one point rise in the compression ratio. The influence of this parameter is really more complex because it modifies the rate of combustion and heat exchange; this is dealt with in subsequent sections.

Modern data acquisition methods (see 3.4.2.2) have enabled the plotting of engine pressure-volume diagrams (see Fig. 3.19) and the calculation of the efficiency of the thermodynamic processes from the work relationship $W = \int P\,\mathrm{d}V$. This calculation leads to the **indicated efficiency** η_i, which is greater than η_e (which includes mechanical losses). Fig. 3.17 shows an example of the change in η_i and η_e as a function of MEP for a series-production engine operating at 2000 rpm with an equivalence ratio of 1.00. Both values are smaller at low loads, but the change is more pronounced for η_e than for η_i, which indicates that mechanical losses are greater at part throttle. Effective efficiency can be broken down as

$$\eta_e = \eta_i \, \eta_m$$

where η_m is the **mechanical efficiency**. This representation is problematic because is supposes that η_m is effectively a constant characteristic. This is not so as shown in the example in Fig. 3.17; η_m varies from 63 to 92% and the MEP rises from 1.5 to 10 bar. On the other hand, at idle and at zero load: $\eta_m = 0$ and MEP = 0.

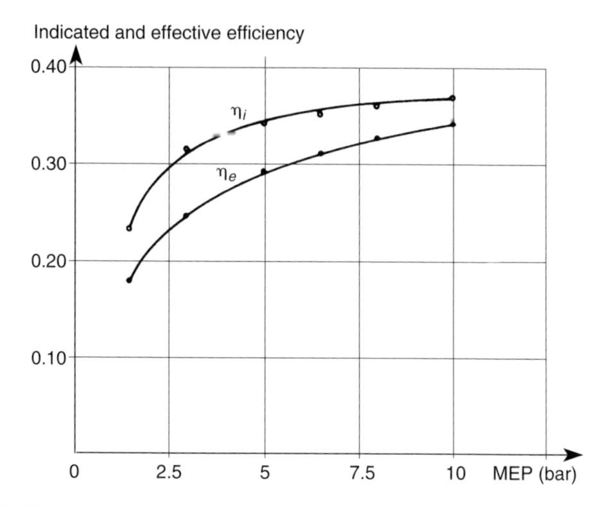

Figure 3.17 *Change in indicated and effective efficiency as a function of MEP. (speed = 2000 rpm; equivalence ratio = 1.00)*

Another approach to efficiency consists of establishing a balanced summation of the total energy input to the engine in terms of its distribution as effective work, mechanical losses, exhaust gas enthalpy, and heat transfers to the engine walls. Fig. 3.18 shows the results of such an approach based on model simulations that have been validated with precise experimental data. The loss distribution is shown as a function of equivalence ratio, but it could be shown as a function of other parameters such as load, compression ratio, speed, etc.

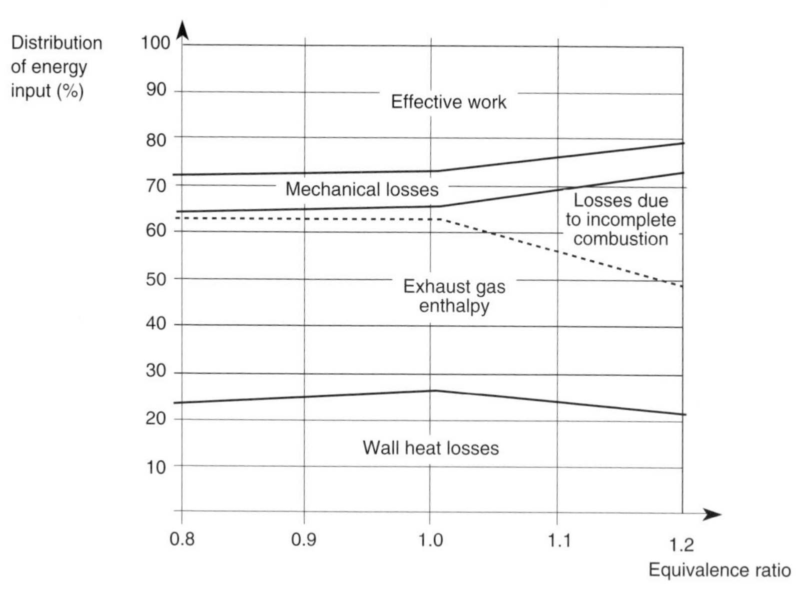

Figure 3.18 *Input energy distribution as a function of equivalence ratio. (speed = 2000 rpm; part load)*

3.3.3 Effective and indicated performance

It is interesting to extend the fundamentals of average pressure and specific fuel consumption to indicated performance. The following relationship defines **indicated specific fuel consumption (ISFC):**

$$\text{ISFC} = \frac{3.6 \cdot 10^6}{\eta_i \cdot \text{NHV}}$$

Mean indicated pressure (MIP) is defined as

$$\text{MIP} = \frac{\int P\text{d}V}{V}$$

The MIP itself includes two components of the P-V diagram (see Fig. 3.19): one is referred to as MIP_H, which corresponds to positive work represented by the high-pressure envelope and the other is referred to as MIP_L, which corresponds to the negative work required to evacuate exhaust gases and ingest the air-fuel mixture.

Therefore, according to Fig. 3.19,

$$\text{MIP} = \text{MIP}_H - \text{MIP}_L$$

MIP_L increases with manifold vacuum. Fig. 3.20 shows an example of the change in MIP_L with load for a series-production engine running at 2500 rpm. The MIP_L can reach minus 0.8 bar at low loads and for a MEP representing 75 km/h, pumping losses represent 30% of the effective output.

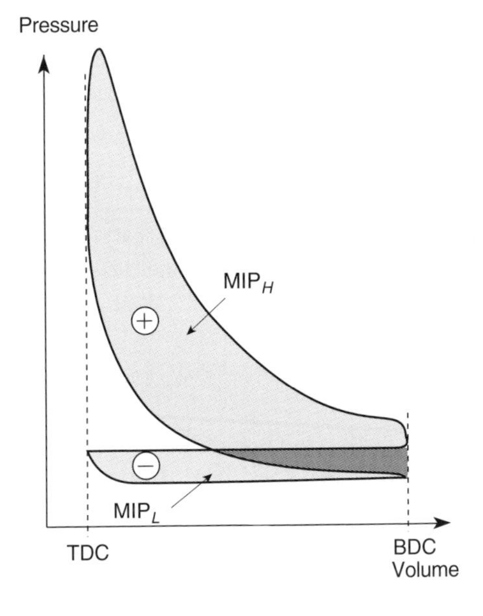

Figure 3.19 *Pressure-volume diagram.*

Therefore, excluding the other parameters, the transfer of mass contributes in large measure to the deterioration of efficiency a low loads.

Mechanical losses can be expressed as the **mean friction pressure** (**MFP**), which is defined by the following relationship:

$$\text{MFP} = \text{MIP} - \text{MEP}$$

The derivation of MFP from the mean indicated and effective pressures shows that this value is around 1 bar for a production engine and varies very little with load. This enables a fairly simple determination of the part played in the deterioration of effective efficiency by mechanical losses at low loads.

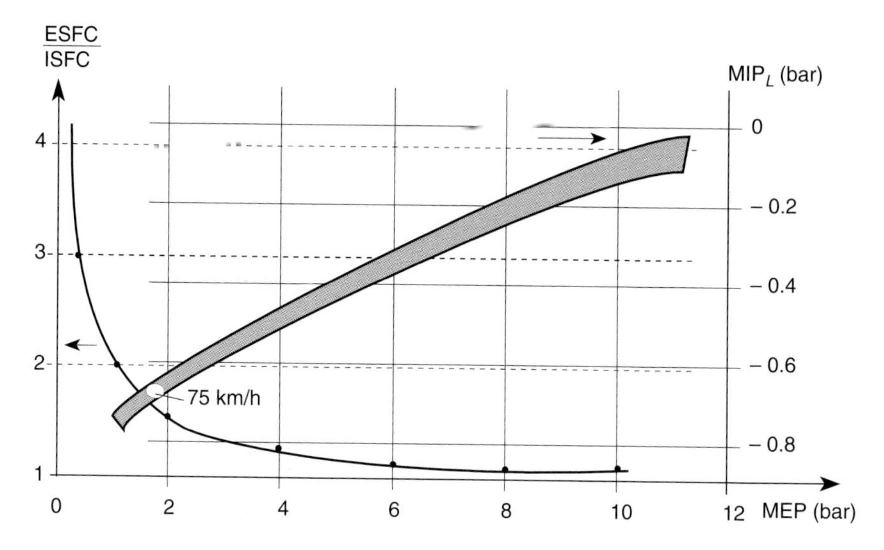

Figure 3.20 *Change in pumping losses and ESFC/ISFC ratio as a function of MEP. (assumes MFP = 1 bar)*

Because the fuel flowrate does not depend on the fact that indicated or effective characteristics are considered, the following relationship is justified,

$$\text{MIP} \cdot \text{ISFC} = \text{MEP} \cdot \text{ESFC}$$

which yields:

$$\frac{\text{ESFC}}{\text{ISFC}} = 1 + \frac{\text{MFP}}{\text{MEP}}$$

Fig. 3.20, which is based on an MFP of 1 bar, shows the hyperbolic relationship of the ESFC/ISFC ratio as a function of MEP.

Thus, for a MEP of 0.5 bar, the indicated efficiency is three times the value of the effective efficiency.

3.3.4 Influence of operating and design parameters

The previous analysis shows that it is possible to represent the overall efficiency by the following expression,

$$\eta_e = A(\varepsilon\ , \gamma, \varphi)\frac{\text{MEP}}{\text{MEP} + \text{MFP} + \text{MIP}_L} = A \cdot B \tag{1}$$

where the first term A is a function of several variables—compression ratio, γ coefficient, mixture air-fuel ratio—and A represents the contribution of the physical and chemical processes, whereas the second term B represents the losses due to friction and pumping. A simple and quick analysis shows that construction and control parameters exert their influence exclusively on either A or B, or in a more complex manner on all the mechanical and thermodynamic processes. This section describes some of the clear relationships.

3.3.4.1 Compression ratio

The compression ratio affects the thermal efficiency, but it does not affect the losses related to MFP and MIP_L.

The formula

$$\eta_0 = 1 - \frac{1}{\varepsilon^{\gamma-1}}$$

does not take into direct account the quantitative effect of the compression ratio, because γ varies with ε. In particular, the increase in heat exchanges due to a higher compression ratio results in a reduced γ value, which tempers the potential gains in efficiency.

From the previous relationships, the following relationship can be derived:

$$\frac{d\eta_0}{d\gamma} = \varepsilon^{1-\gamma} \ln \varepsilon$$

Applying this formula shows that if the heat exchanges are high—γ close to 1—efficiency becomes very sensitive to variations in these heat exchanges. This explains why, in practice, an increase in compression ratio beyond 9 or 10 results in only modest increases in efficiency of roughly 3%, or only 2% per unit of compression ratio.

3.3.4.2 Volumetric efficiency

Volumetric efficiency has a predominant effect on systemic losses, which is shown by applying formula (1).

The following numbers can be adopted as orders of magnitude,

Parameters	Load	
	Full load	Low load
MEP (bar)	10.0	0.2
MIP_L	0.2	0.7
MFP	1.0	1.0
therefore:		
B	0.893	0.541

which translate into a reduction in efficiency of 40% due to systemic losses alone.

Quality of combustion is another cause of reduced efficiency at low loads. This effect, which is described in subsequent sections, can be represented in the formula for η_0 by a lower γ value of 1.18 for low loads, instead of the 1.22 value used for higher volumetric efficiencies. Solving the η_0 formula results in a 15% decrease in efficiency. Using this example, the overall efficiency is almost one half when the MEP changes from 10 to 2 bar. This estimate agrees with the experimental results presented in Fig. 3.17.

3.3.4.3 Engine speed

The effect of engine speed results mainly from friction losses. The order of magnitude values chosen are an MFP increase from 0.7 to 1.4 bar due to a speed increase from 1000 to 5000 rpm. The corresponding efficiency losses associated with these changes depend on the expected operating range. The losses in the previous numerical examples are 6% at full load and 17% at low load.

Changes in engine speed also affect thermodynamic efficiency, but to a lesser degree, by modifying cylinder-wall heat transfer and the speed of combustion. This change can be evaluated using model studies, but it does not have a significant effect on engine performance.

3.3.4.4 Air-fuel ratio

Air-fuel ratio has no direct effect on systemic losses, but it can have a significant effect on power and indicated efficiency, with direct repercussions on effective performance (see Fig. 3.21). There are two possible functional zones on either side of stoichiometry.

A. *Lean mixture settings*

Indicated efficiency reaches maximum values at very lean mixtures. This results from the following two effects:

- A reduction in equivalence ratio at constant combustion speed improves thermodynamic efficiency by modifying the characteristics of the reaction environment, which translates into an overall increase in the γ coefficient.
- Conversely, combustion speed decreases with a lean mixture, which leads to a reduction in efficiency.

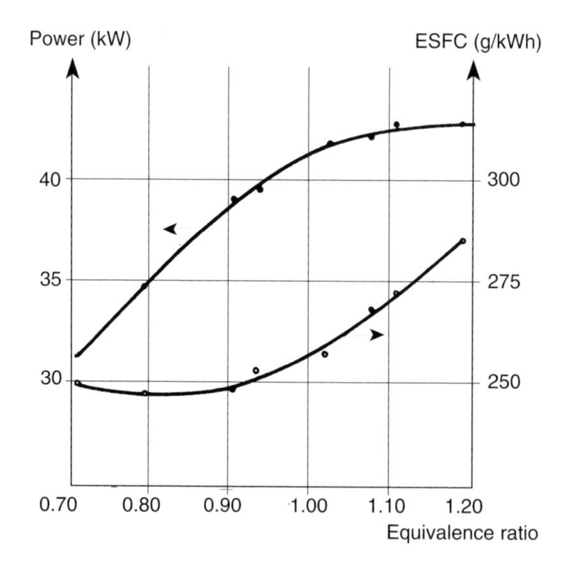

Figure 3.21 *Effect of equivalence ratio on power and specific fuel consumption. (full load at 4000 rpm)*

These effects explain the thrust of all the research done on lean combustion systems since the beginning of engine development. In 1996, lean combustion was applied to **Japanese engines** in particular (Honda, Mazda, Toyota, Mitsubishi, ...) and it yielded efficiency increases of 10 to 15% when compared to stoichiometric-mixture combustion (see Chapter 8).

However, to increase maximum power at full load, mixtures must approach stoichiometry because MEP increases gradually with the quantity of energy introduced into the combustion cycle.

B. Rich mixture settings

Efficiency decreases almost linearly as a function of the equivalence ratio once it exceeds 1.00. Excess fuel does not provide any particular efficiency advantage.

The MEP for a production multicylinder engine reaches a reasonably flat maximum value at a slightly rich mixture (1.05 to 1.10). This is probably related to the differences in the air-fuel ratio between cylinders. For example, at an average mixture of 1.05 the leanest cylinder is operating at a mixture of 1.00, thus providing maximum power. In contrast, an overall stoichiometric mixture ($\varphi = 1.00$) means one or several cylinders may be receiving a lean mixture resulting in a slight power loss.

3.3.4.5 Ignition advance

Ignition advance determines the timing of the combustion diagram during a cycle and it has a direct effect on indicated performance. For a given operating condition, the optimal ignition advance is the one that produces maximum torque or efficiency. Finding the **optimum advance** when setting up an engine is a critical step in obtaining both the best possible efficiency at part load and the maximum power at full load. This must be accomplished while including the limiting criteria required to ensure proper combustion (see 3.4.3.2) and to meet anti-pollution legislation (see 5.4.1.1).

Curves that relate changes in torque as a function of ignition advance are flat in the immediate vicinity of the optimum point and then they drop off sharply as the advance is decreased. This change is particularly acute at low loads. Hence, a reduction of 1°CA in this area can lead to a 1% loss in efficiency.

3.3.5 Engine maps

An engine map is a plot of all engine operating parameters (speed, load, mixture, specific fuel consumption, power, emissions) throughout the useful operating range.

A graphical representation that is now frequently used consists of drawing curves of **constant specific fuel consumption** against MEP and engine speed. This series of curves enables the immediate location of ranges of optimal efficiency (see Fig. 3.22) and it indicates the maximum MEP available at each engine speed.

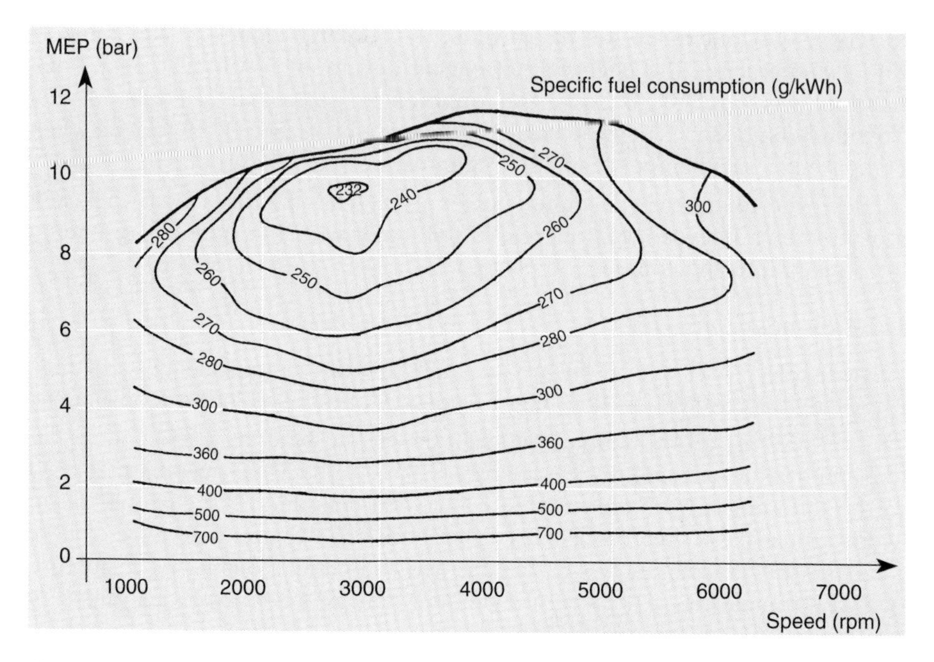

Figure 3.22 *Example of a constant specific fuel consumption engine map (Opel engine). (Source: Breitwieser, 1993)*

Furthermore, this map helps in the work of setting up new engines, by locating performance points with respect to reference values. This is done by comparing specific fuel consumption points with particular combinations of engine speed and MEP. Taking a medium sized European vehicle as an example, the following combinations are considered: MEP of 1 bar and a speed of 2500 rpm; MEP of 3 bar and a speed of 3500 rpm. These are conditions that are achieved in actual operation during a standard fuel consumption and emission measurement test (see Chapter 5).

3.4 Combustion processes

Describing engine combustion often results in a flowering of imprecise and ambiguous terms, which create confusion about processes that are often very distinct. This section helps to clarify this complex terminology problem by proposing precise definitions that are now accepted by specialists.

This is approached by explaining some theoretical considerations and then the various types of combustion—both normal and abnormal—are described along with explanations of how they affect engine performance.

3.4.1 Definitions and theoretical considerations

The theoretical study (Van Tiggelen et al. 1968; De Soete 1976) of various combustion mechanisms is the subject of numerous specialized studies, but these studies are not repeated herein. Instead, this section covers some basic fundamentals and, where possible, how they apply to engines.

3.4.1.1 General characteristics

A fuel-air mixture participates in a combustion reaction by either slow oxidation or rapid combustion. In the first case, a small increase in temperature is all that is required to see the progressive appearance of intermediate oxidation products (aldehydes, carbon dioxide, etc.). The speed of the transformation is always limited and after reaching a maximum it decreases progressively as the reactants are consumed.

Rapid combustion, which is used in all energy-consuming applications (propulsion, heating), is clearly distinguishable from slow oxidation by the presence of **non-equilibrium** concentrations of transitory atomic species—atoms and radicals—in the reaction zone. These atomic species emit light according to the **chemiluminescence** process, and, as a consequence, they are responsible for the appearance of flame. The reaction speed is very high and the transformation completes without the opportunity of analyzing the active species or stopping the reaction at an intermediate stage.

Depending on the characteristics of the reaction environment, rapid combustion can follow vastly different ways and means. Often, at the beginning stages, rapid combustion appears to be the same as slow oxidation, then an extremely rapid acceleration takes place in the process: this is **spontaneous ignition**. This process can affect the entire reaction mixture or it can be localized to an area where the conditions of temperature, pressure, and concentration are more favorable, or where they have been artificially enhanced. The latter case refers to an **ignition**. This is obtained in practice by bringing in outside energy such as an electric spark, a match, a glowing body, ...

If the initial composition of the mixture is correct—referred to as an **inflammable** mixture—combustion can continue point-by-point from the point of ignition. This process is called **propagation**. The flame front is very thin—about a few tenths of a mm—and it separates the mixture at each instant into two separate zones containing reactants (fresh charge) and products (burned gases). If the air-fuel mixture is well mixed before ignition, the flame is referred to as "**pre-mixed**."

In the opposite case, **diffusion flames** occur. The speed with which the reactants diffuse into each other can have a overriding effect.

The speed of flame propagation depends heavily on the nature of the mixture and the experimental conditions. A subsonic propagation speed is referred to as **deflagration** and it is distinguished from **detonation,** which occurs at supersonic speed. Deflagration phenomena are the only things that occur in engines and the term "detonation", which is frequently used to refer to abnormal combustion (**knock**) is incorrect.

From a kinetic standpoint, rapid combustion brings into play a series of chain reactions involving molecules, atoms, and especially free radicals. The latter items are combinations of atoms with electrons that are not linked in covalent chemical bonds. The following species are free radicals: $CH_3^{\bullet}$, $OH^{\bullet}$, $HO_2^{\bullet}$.

If the **reactants** (fuel and oxygen) are designated as A and B, the **reaction products** as C and D, and the **free radicals** as X and Y, the overall reaction can be described by the following equation

$$A + B \rightarrow C + D$$

and the **propagation of the chain** reaction includes at least the following elementary stages:

$$A + X \rightarrow C + Y$$

$$B + Y \rightarrow D + X$$

According to this schema, the propagator radical X is regenerated at each link in the chain, and the reaction can theoretically continue indefinitely, transforming one molecule of reactant into one molecule of product at each step. The $OH^{\bullet}$ radical often plays the role of chain propagator.

In reality, there are other elementary reactions that cause either the disappearance of free radicals (**break away**) or their multiplication (**branching**). Break-away reactions can occur in a homogeneous phase or on the chamber walls. These reactions are always compensated for in large measure by the appearance of new active atomic species during branching reactions. These new species can occur as the result of the collision between the initial reactants and a radical or they can involve an intermediate combustion product (E), which is formed in the propagation chain. The corresponding reaction schemas are

$$A + Y \rightarrow 3 \text{ radicals}$$

(**direct** branching)
or

$$E + Y \rightarrow 3 \text{ radicals}$$

$$E + A \rightarrow 2 \text{ radicals}$$

(**indirect** branching)

An example using the combustion of methane, which is the simplest hydrocarbon, illustrates these theoretical concepts. In this case, one of the propagation chains leads to the formation of an intermediate oxidation product, formaldehyde CH_2O:

$$CH_4 + OH^{\bullet} \rightarrow CH_3^{\bullet} + H_2O$$

$$CH_3^{\bullet} + O_2 \rightarrow CH_2O + OH^{\bullet}$$

The CH_2O is then transformed according to the following type of schema:

$$CH_2O + OH^{\bullet} \rightarrow COH^{\bullet} + H_2O$$

$$COH^{\bullet} + O_2 \rightarrow CO_2 + OH^{\bullet}$$

However, CH_2O can also enter into an indirect branching reaction:

$$CH_2O + O_2 \longrightarrow COH^{\bullet} + HO_2^{\bullet}$$

Consequently, after each branching reaction, new radicals appear and reenter other propagation chains. The transformation speed of the reactants becomes faster and faster. This acceleration leads to a **runaway** reaction that is characteristic of rapid combustion.

3.4.1.2 Auto-ignition

A. Limiting conditions

Spontaneous combustion can only occur within certain **pressure** and **temperature** limits. For a conventional hydrocarbon, Fig. 3.23 shows the shape of the curve that separates slow oxidation and auto-ignition. In general, the minimum spontaneous-ignition temperature decreases with increasing pressure. However, the curve shows that situations exist where several critical temperatures can occur at the same pressure. Some hydrocarbons can also spawn **cold flames**. These differ from conventional spontaneous ignitions by their fugitive character and the fact that they lead to incomplete oxidation compounds (formaldehyde) instead of complete combustion products (CO_2). Cold flames may or may not be followed by auto-ignition; cold flames have been observed in engines, in the instants just before the appearance of knock.

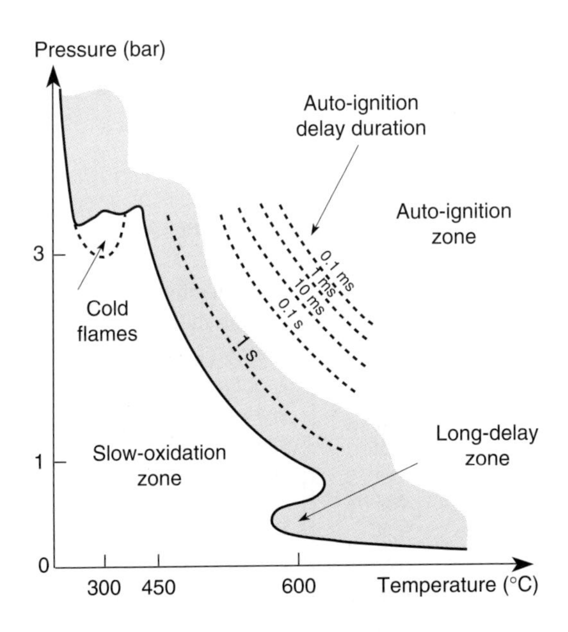

Figure 3.23 *Schematic representation of the slow oxidation and spontaneous ignition zones for a hydrocarbon-air mixture. (Data for* n*-heptane)*

Table 3.2 lists the minimum auto-ignition temperature (T_i) in air at atmospheric pressure for some organic compounds. These temperatures are about 500 to 600°C for hydrogen, carbon monoxide, alcohols, and light hydrocarbons, and they drop to 200 to 250°C for long chain paraffinic or olefinic hydrocarbons (C_8-C_{10}). These values are not precise, but they can serve as order-of-magnitude values, given the difficulty of obtaining accurate measurements. In fact, it is very difficult to maintain the reaction environment at a constant temperature under limit conditions.

Also, T_i varies with other parameters like pressure (see Fig. 3.23), mixture composition (air-fuel ratio and dilution), and the shape of the container enclosing the air-fuel mixture. For example, when plotted as a function of the equivalence ratio, T_i reaches a minimum value that varies with the type of hydrocarbon; dilution by inert gases (CO_2, N_2) always tends to increase T_i. As a final note, T_i increases as the diameter of a cylindrical enclosure decreases, due to the increasing level of break-away reactions on the walls.

Knowing the minimum spontaneous-ignition temperatures is especially valuable in safety planning and in the prevention of **fires** and **explosions**. The values provided in table 3.2, despite their lack of precision, constitute valuable

Table 3.2 *Minimum auto-ignition temperatures for some compounds. Estimated of derived in air at atmospheric pressure. (Source: Van Tiggelen 1968)*

Compound	T_i (°C)	Compound	T_i (°C)
Hydrogen	570	Naphthalene	575
Carbon monoxide	630	Methanol	510
Methane	580	Ethanol	490
Ethane	490	*n*-Propanol	480
Propane	480	Isopropanol	540
n-Butane	420	*n*-Butanol	430
Isobutane	480	Acetaldehyde	230
n-Hexane	260	Acetone	560
n-Heptane	285	Methylamine	430
n-Octane	220	Dimethylether	350
n-Decane	240	Diethylether	190
Isooctane	670	Ammonia	650
Ethylene	520	Hydrazine	270
Propylene	460	Carbon disulfide	130
But-1-ene	385	Hydrogen sulfide	290
But-2-ene	435		
Butadyene	420	Eurosuper*	450
Acetylene	320	Diesel*	260
Benzene	620	Kerosene*	240
Toluene	585		
o-Xylene	520		

* Orders of magnitude.

information because they indicate the minimum temperatures below which spontaneous ignition will not occur, even under the most favorable laboratory conditions. However, temperature T_i is not a determining parameter in engines; in fact, during the compression phase the air-fuel mixture rapidly exceeds this minimum temperature level. The critical factor is therefore the **time** required to initiate spontaneous ignition.

B. Auto-ignition delay

Delay is the time that passes between the instant that the mixture is placed in temperature and pressure conditions that favor auto-ignition, and the time that it actually occurs.

Several experimental techniques have been developed to measure or estimate the orders of magnitude of spontaneous ignition delays. One of these, referred to as the **rapid compression machine**, is of some interest to an engine designer because it enables the study of spontaneous ignition under conditions that are very close to those encountered in an engine. The mixture to be evaluated is compressed in an almost **adiabatic** manner using a piston and the mixture remains enclosed up to the moment that spontaneous ignition occurs. The difficulty with this method consists of obtaining a compression time that is negligible with respect to the delay being measured. Also, the pressure and the temperature must be held constant during the entire compression period after the piston stops. Ignition is followed by an abrupt pressure rise, which is followed by oscillations.

Figure 3.24 shows how the interpretation of the pressure-time diagram enables the delay (θ) to be derived.

Following up on the results for a given hydrocarbon shows that θ can be expressed as a function of temperature and pressure in the following way,

$$\theta = AP^{-n} \exp\left(\frac{B}{T}\right) \quad (1)$$

where P and T represent pressure and temperature, and A, n, B are positive constants, which vary with the nature and concentration of the fuel.

Also worth mentioning is the coefficient B, which is linked to the **activation energy** (E) of the spontaneous ignition kinetic process by the following relationship,

$$B = \frac{E}{R}$$

where R is the universal gas constant ($R = 8.314$ J mol^{-1} K^{-1}).

The validity of relationship (1) is proven by a number of numerical results (De Soete 1976) that show a linear change of log θ as a function of $1/T$ at constant pressure for some hydrocarbons. The values of E derived from these calculations vary between 2000 and 6000 kJ/mol, depending on the hydrocarbon.

The universal formula (1) actually represents the simultaneous and successive involvement of several different and highly complex kinetic mechanisms, depending on the pressure and temperature zones being investigated.

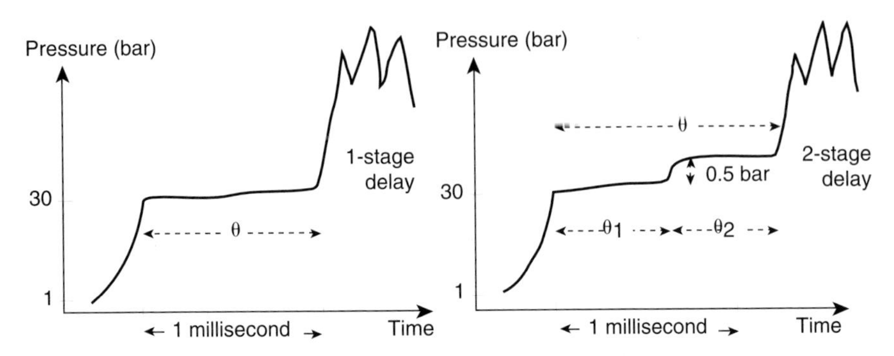

Figure 3.24 *Measurement of auto-ignition delay using a rapid compression machine. An example of the change in pressure as a function of time.*

It is therefore not surprising that the values for A, n, and B quoted by various authors do not concur.

The involvement of several reaction mechanisms results in a **two-stage** delay for some hydrocarbons, like *n*-heptane for example (see Fig. 3.24).

In this case, there is a slight increase in pressure before auto-ignition, which corresponds to the exothermic production of intermediate atomic species that subsequently involve themselves in new chain reaction processes.

One last note concerns the auto-ignition delay that varies as a function of the **equivalence ratio**; this delay reaches a minimum value around stoichiometry (φ between 0.90 and 1.00) and it increases rapidly on either side of this zone.

3.4.1.3 Flame propagation

A. Upper and lower inflammability limits

For a mixture of fuel, air, and eventually a diluant, experience demonstrates that the propagation of a flame can only occur within certain limits of composition. The **upper (L_u) and lower (L_l) inflammability limits** are by definition the extreme percentages of fuel (generally expressed by volume) at which inflammability can occur. The occurrence of these limits can be easily demonstrated based on the general schema of chain reactions that involve thermal and kinetic criteria.

In fact, the respective concentrations of the reactants must ensure that the speeds of the propagation and branching steps remain faster than the breakaway reactions.

Furthermore, the overall thermal balance of these various elementary reactions must remain positive.

To illustrate, Table 3.3 lists the inflammability limits obtained at rest in air at 25°C for a number of current products. The results are expressed in terms of both composition (volume%) and equivalence ratio. Inflammability varies considerably between fuels; some compounds such as hydrogen and acetylene have a very broad range.

Table 3.3 *Approximate inflammability limits for various organic compounds, measured in air at atmospheric pressure and at 25°C. (Source: Van Tiggelen 1968)*

Fuel	L_l*		L_u**	
	Volume %	Equivalence ratio	Volume %	Equivalence ratio
Methane	5.0	0.50	15.0	1.7
Ethane	3.0	0.51	12.5	2.4
Propane	2.2	0.53	9.4	2.5
n-Butane	1.9	0.59	8.5	2.8
n-Pentane	1.5	0.58	7.8	3.2
n-Hexane	1.2	0.54	7.0	3.4
n-Heptane	1.1	0.58	6.7	3.7
Isooctane	1.0	0.60	6.0	3.8
Ethylene	3.1	0.46	32.0	6.7
Propylene	2.2	0.48	10.5	2.5
Benzene	1.4	0.50	7.1	2.7
Toluene	1.4	0.60	6.7	3.0
o-Xylene	1.0	0.50	6.0	3.1
Methanol	6.7	0.51	36.0	4.0
Ethanol	3.3	0.48	19.0	3.3
Isopropanol	2.0	0.46	12.0	3.0
Butanol	1.45	0.42	11.2	3.6
Diethylether	1.8	0.52	36.5	16.4
Hydrogen	4.0	0.1	74.5	6.9
Acetylene	2.5	0.30	81.0	9.6
Carbon disulfide	1.25	0.18	50.0	14.2
Carbon monoxide	12.5	0.34	74.0	6.8
Ammonia	15.0	0.63	27.0	1.3
Hydrogen sulfide	4.3	0.32	45.5	6.0
Dimethylether	3.4	0.50	27.0	5.3
Acetone	2.6	0.51	12.8	2.8
Acetaldehyde	4.0	0.50	56.0	15.2

* L_l: lower limit, ** L_u: upper limit.

The relative data on inflammability limits have always been subject to a degree of imprecision due to the ignition source. In fact, the initiation of the flame always requires a minimum quantity of energy, which becomes greater toward the limits. The experimenter is thus left to ascertain the impossibility of igniting the mixture with the experimental device instead of the impossibility of a flame being able to propagate.

Inflammability limits vary equally as a function of the experimental conditions: temperature, pressure, and oxygen dilution. An increase in temperature always results in an increase in L_u and a reduction in L_l. The effect of pressure is more complex because the zone of inflammability can increase or decrease at high pressure, depending on the fuel. Dilution—for example, the change from pure oxygen to air—always restricts the inflammability.

The lower inflammability for a mixture of several hydrocarbons can be estimated using the semi-empirical **Le Chatelier** relationship

$$L_{lM} = \frac{100}{\frac{\varphi_1}{L_{l1}} + \frac{\varphi_2}{L_{l2}} + \frac{\varphi_3}{L_{l3}} + \cdots} = \frac{100}{\sum_1^j \left(\frac{\varphi_j}{L_{lj}}\right)}$$

where

L_{lj} = lower inflammability limit for the constituent j in air

L_{lM} = lower inflammability limit of the mixture

φ_j (%) = content by volume of the constituent j in the mixture

Theoretical considerations enable the lower inflammability limit to be connected with the combustion reaction's variation in molar enthalpy (ΔH_C). Therefore, the following equation can be chosen from among several semi-empirical equations

$$\Delta H_C \cdot L_l = 4680$$

where

L_l = is the lower inflammability limit (% vol)

ΔH_C = molar enthalpy of combustion (kJ/mol)

The relationship has been reasonably well verified for a certain number of hydrocarbons and organic compounds (diethylether, acetone, alcohols, ...)

B. *Ignition energy*

To ensure that an ignition attempt is successful, a sufficiently intense burst of energy must be released in a minimum volume and in a relatively short time. These qualitative criteria can be translated to mathematical terms that can be used to calculate the initiation conditions required for a flame. From an experimental standpoint, the minimum ignition energy varies depending on the nature of the fuel, the air-fuel ratio, the pressure, and the aerodynamic conditions. It also depends on the ignition device, especially the distance between the electrodes for an electric spark.

Fig. 3.25 shows that the ignition energy is at a minimum at stoichiometry or at a slightly rich mixture. The ignition energy required in an engine surpasses—by a factor of 10 to 100—the minimum amount derived in the laboratory for a mixture at rest. The energy required is about 30 to 100 mJ.

This overcapacity is required in a turbulent milieu where intense gas movements tend to dissipate the ignition-system energy outside the critical zone of the ignition source.

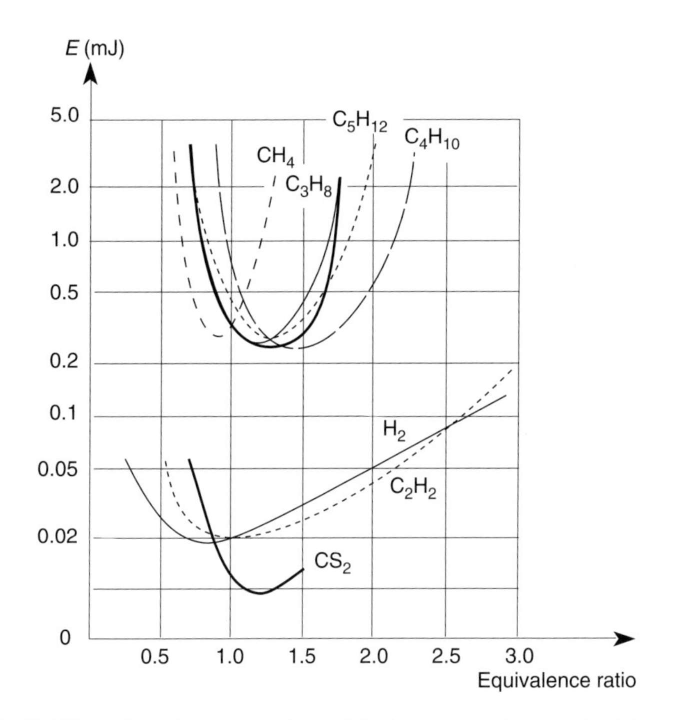

Figure 3.25 *Effect of equivalence ratio and fuel type on minimum ignition energy. (Source: Van Tiggelen 1968)*

C. Propagation speed

The following two propagation situations are considered: the first situation is an inflammable mixture in a laminar flow environment and the second situation is the same mixture in a turbulent environment.

a. Laminar flow

Consider the propagation of a flame front in an air-fuel mixture at rest or in laminar flow. The flame's spatial speed of displacement (V_s) with respect to a fixed axis can be expressed as

$$V_s = V_g + V_n$$

where

V_g = fresh gas speed or displacement (which is sometimes nil)

V_n = relative speed of the reaction zone with respect to the fresh gases

V_n is referred to as **laminar normal speed of propagation** or of **deflagration**; it is an intrinsic characteristic of the inflammable mixture.

Using the classic example of a stabilized Bunsen burner flame, V_n is easily determined from the gas flow (F) and the flame front surface (S), which is measured photographically:

$$V_n = \frac{F}{S}$$

V_n is usually expressed in cm/s, F in cm^3/s, and S in cm^2.

Table 3.4 shows the V_n values for various hydrocarbons when burning in air at 25°C under a pressure of 1 bar. The change in V_n as function of air-fuel ratio is also provided. For most organic compounds, V_n values are between 30 and 50 cm/s; higher values (up to 2 to 3 m/s) occur with hydrogen, acetylene, ethylene, ethylene oxide, propylene oxide, carbon disulfide.

Maximum laminar speed is usually reached near stoichiometry and it decreases markedly with lean mixtures ($\varphi = 0.80$). There does not appear to be a systematic variation of V_n as a function of the chemical structure of hydrocarbons. However, V_n is usually low with paraffins, from methane to isooctane, and higher for some aromatics like benzene, and likewise for methanol.

The laminar propagation speed increases along with the initial temperature of the air-fuel mixture (see Fig. 3.26) and it decreases when dilution with inert gases is increased (see Fig. 3.27). Increasing pressure also tends to decrease flame propagation speed (De Soete et al. 1976).

Table 3.4 *Normal laminar speed V_n of flame propagation in air for various mixtures of some fuels*.*

Fuel	V_n (cm/s)					Maximum V_n (cm/s)	φ for V_n maximum
	$\varphi = 0.8$	$\varphi = 0.9$	$\varphi = 1.0$	$\varphi = 1.1$	$\varphi = 1.2$		
Methane	30.0	38.3	43.4	44.7	39.8	44.8	1.07
Propane	....	42.3	45.6	46.2	42.3	46.8	1.06
n-Butane	38.0	42.6	44.8	44.2	41.2	44.8	1.02
n-Heptane	37.0	39.8	42.2	42.0	35.5	42.5	1.05
Isooctane	37.5	40.2	41.0	37.2	31.0	41.0	0.98
Cyclohexane	....	41.3	43.5	49.9	38.0	49.9	1.10
Ethylene	50.0	60.0	68.0	73.0	72.0	73.5	1.13
Propylene	....	48.4	51.2	49.9	46.4	51.2	1.00
Benzene	39.4	45.6	47.6	44.8	40.2	47.6	1.00
Methanol	34.5	42.0	48.0	50.2	47.5	50.4	1.08
Diethyl oxide	37.0	43.4	48.0	47.6	40.4	48.2	1.05
Hydrogen	120.0	145.0	170.0	204.0	245.0	325.0	1.80
Acetylene	107.0	130.0	144.0	151.0	154.0	155.0	1.25
Carbon disulfide	58.0	59.4	58.8	57.0	55.0	59.4	0.91
Carbon monoxide	....	....	....	28.5	32.5	52.0	2.05
Isopropanol	34.4	39.2	41.3	40.6	38.2	41.3	1.02
Acetaldehyde	26.6	35.0	41.4	41.4	36.0	41.4	1.10
Acetone	40.4	44.2	42.6	38.2	....	44.2	0.90
Ethylene oxide	70.7	83.0	88.8	89.5	87.2	89.5	1.10
Propylene oxide	53.3	62.6	66.5	66.4	62.5	66.5	1.00

* Initial conditions: 25°C and 1 bar.

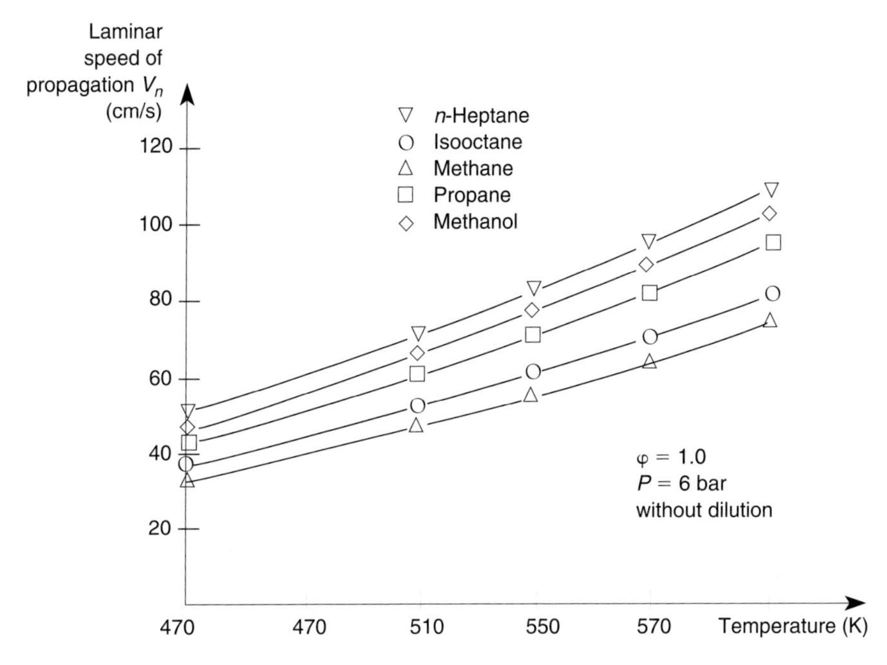

Figure 3.26 *Effect of mixture temperature on the laminar normal speed of flame propagation.*
(Source: Ryan 1980)

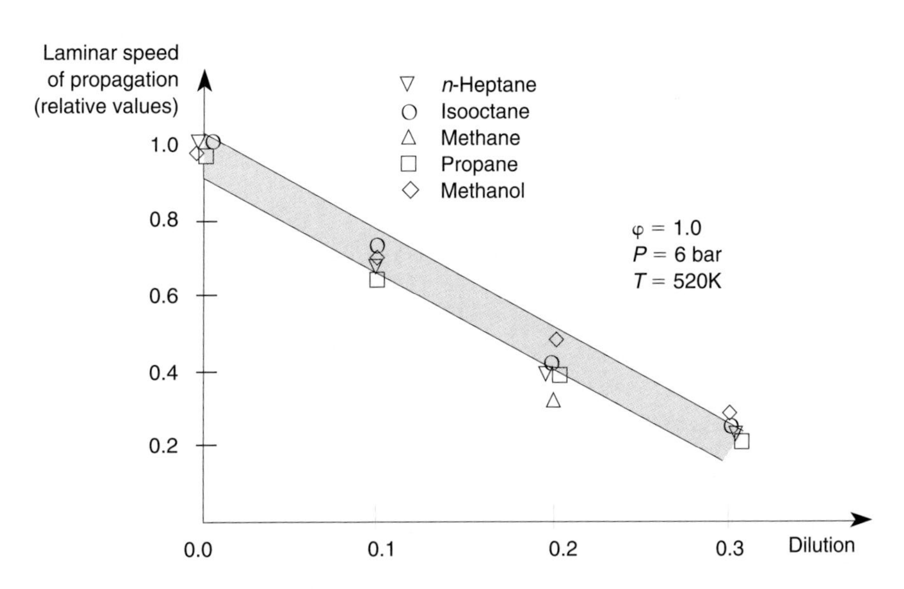

Figure 3.27 *Relative effect of dilution on the laminar normal speed of flame propagation. (The dilution is the ratio of the mass of the diluant to the total mass of the charge. The diluant is a 15% by volume mix of CO_2 in N_2.)*
(Source: Ryan 1980)

b. Turbulent flow

In most energy applications that use combustion and in all cases involving engines, flame propagation takes place in a turbulent rather than a laminar environment. This means that various volumes of air-fuel mixture are subjected to **fluctuating speeds** that are superimposed on the average speed of the flow.

Speed fluctuations occur in a disorderly way over time and cannot be described mathematically except by statistical methods. In a very simplistic way, the instantaneous speed (v_i) at a given point in the fluid is equal to the algebraic sum of the average value $(\bar{v})$ is constant in time and imposed by the gas flow, and the fluctuation (u), which can take on all values in all possible directions:

$$v_i = \bar{v} + u$$

The intensity of the turbulence u' is the quadratic average speed, which is the square root of the squares of u:

$$u' = \sqrt{\bar{u}^2}$$

The factor u' is expressed in the same units as the speed.

Flame propagation speed increases considerably with increasing turbulence as illustrated by the results in Fig. 3.28. This effect is especially pronounced in engines, since the flame propagation speed can reach **several tens of meters per second** (up to 50 m/s). The values however, remain well below the values that characterize detonation (500 to 1000 m/s).

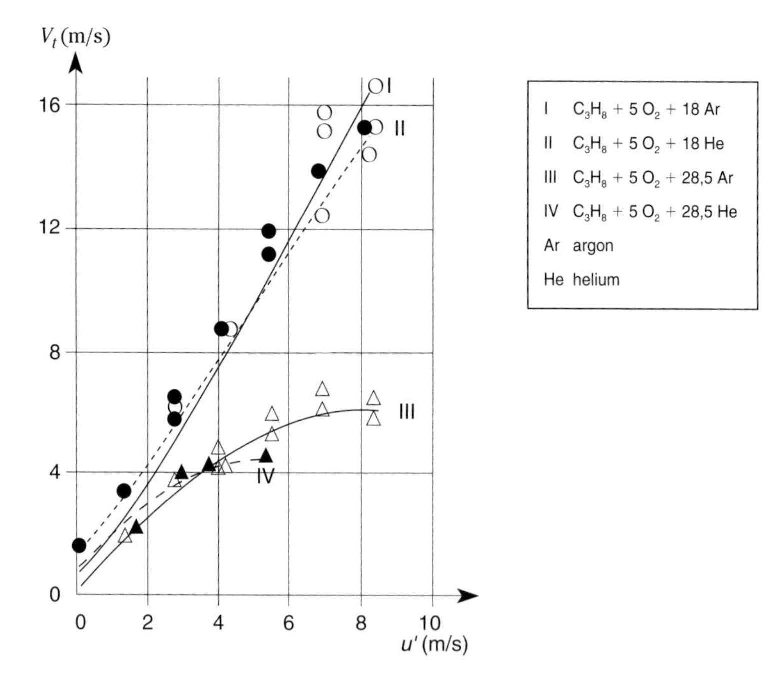

Figure 3.28 *Change in flame propagation speed (V_t) as a function of turbulence intensity (u').*

To explain the rapid deflagration in a turbulent environment, a simplified model of the **rippled flame front** is used (Fig. 3.29).

According to this schema, the reaction zone undergoes a large number of disordered fluctuations about the median position, due to instantaneous variations in gas flow. The increase in the speed of the flame front results solely from the increase in the rippled flame surface S_r with respect to the apparent surface S_a.

Therefore,

$$\frac{V_t}{V_l} = \frac{S_r}{S_a}$$

where V_t and V_l are the propagation speeds for turbulent and laminar environments respectively.

This hypothesis concurs with a number of experimental results, especially with the fact that the relative differences in flame propagation speeds between different fuels, which occur in laminar flow, are preserved in turbulent flow, regardless of the intensity.

In any case, the flame front in engines can always be considered as a continuous zone. Swirling combustion spreads in all directions, especially toward the unburned gases where it causes localized spontaneous combustion. The mathematical representation of such a process is evidently very difficult.

Turbulence not only accelerates flame front propagation but it also restricts the limits of inflammability and makes ignition more difficult. This explains the fact that the combustion of a homogeneous mixture in an engine becomes erratic and incomplete at equivalence ratios of 0.70 to 0.80, while corresponding mixtures easily ignite in a laminar environment.

c. Flame front temperature and burned gases

Accounting for all the possible reactions of the combustion products enables the determination of their composition under conditions of thermodynamic equilibrium. This was described in a previous section (see 1.3.3.4). This procedure also presents the possibility of calculating the theoretical temperature of

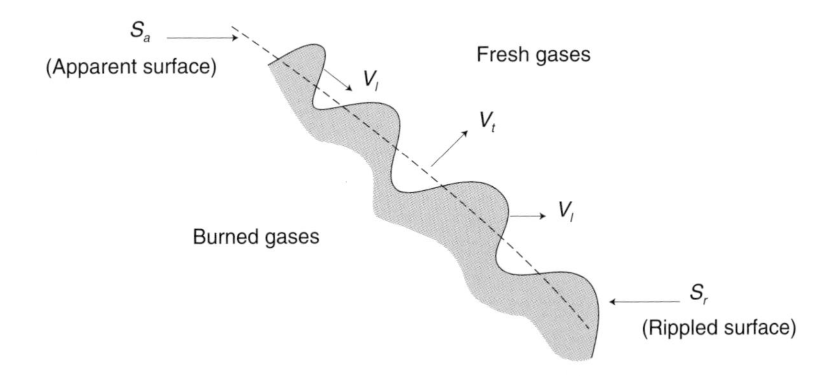

Figure 3.29 *Rippled flame front model.*

the flame front, using the enthalpy of combustion to bring the products to equilibrium temperature. This is often referred to as the **adiabatic flame temperature**. The word adiabatic is a reminder that thermal losses are not accounted for, nor are the out-of-equilibrium conditions that exist within the flame front.

Table 3.5 lists adiabatic flame temperatures for various fuels during stoichiometric combustion at 25°C under a pressure of 1 bar. Note the small differences in temperature between the standard hydrocarbons. However, some products have high flame temperatures (acetylene, hydrogen, nitromethane, ...) and some have only moderate temperatures (methanol, ethanol, ...).

Even though the differences in flame front temperatures are small between the various fuels, they are important because they can change the level of nitrogen oxide emissions (see 5.4.1.1).

As a function of air-fuel ratio, flame temperatures reach maximum values at stoichiometry or slightly above and decrease rapidly with leaner mixtures (Fig. 3.30).

Table 3.5 *Adiabatic flame temperatures* (T_f) *of some fuels**.

Fuel**	Chemical formula	T_f (K)
Cyanogen (g)	C_2N_2	2596
Hydrogen (g)	H_2	2383
Ammonia (g)	NH_3	2076
Methane (g)	CH_4	2227
Propane (g)	C_3H_8	2268
n-Octane (l)	C_8H_{18}	2266
Pentadecane (l)	$C_{15}H_{32}$	2269
Eicosane (s)	$C_{20}H_{40}$	2291
Acetylene (g)	C_2H_2	2540
Naphthalene (s)	$C_{10}H_8$	2328
Methanol (l)	CH_3OH	2151
Ethanol (l)	C_2H_5OH	2197
Nitromethane (l)	CH_3NO_2	2545

* Initial conditions: temperature 25°C, pressure 1 bar, equivalence ratio 1.00.
** The organic compounds are either solid (s), liquid (l) or gas (g).

d. Flame extinction—wall effect

In a great number of applications, especially in engines, as the flame propagates it quickly comes into contact with metallic structures like the cylinder walls and the top of the piston. In proximity to the walls, the reaction environment is highly perturbed by the loss of heat and the retention of free radicals. It is believed that endothermic branching is weaker and that break-away processes are intensified. Lastly, the propagation speed slows to a stop before reaching the walls.

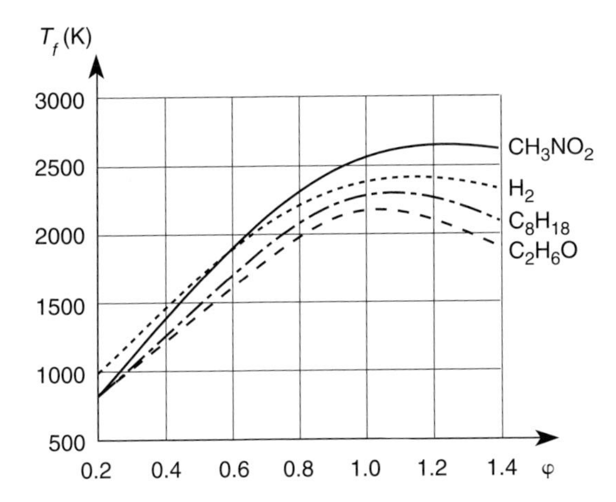

Figure 3.30 *Adiabatic flame temperatures (T_f) as a function of equivalence ratio (φ) for some fuels. Initial conditions: temperature = 25°C; pressure = 1 bar.*

The flame-wall distance that corresponds to the point of extinction is referred to as the **quenching distance**; it depends on several parameters: temperature, pressure, equivalence ratio, turbulence intensity, and combustion chamber geometry. In engines, the distance is often a few tenths of a millimeter and it seems to increase with lean mixtures.

The phenomenon that was just described qualitatively has often been cited to explain the emissions of unburned hydrocarbons from engines. It is now understood that it is definitely involved, but that it is not a major player (see 5.4.1.1).

3.4.2 Normal combustion

Normal combustion in a spark-ignition engine results from **deflagration in a turbulent environment**, which is initiated by the spark from a spark plug, and which is not interfered with by parasitic or uncontrolled phenomena.

3.4.2.1 General description

Fuel, which can still be in the form of droplets dispersed in air at the induction point, continues to vaporize during compression. The spark occurs in a **supposedly homogeneous environment**, but this environment is also a very turbulent one when engine speeds are between 600 and 6000 rpm. The spark occurs at a point 10 to 40°CA before TDC and a flame forms from an initial node and then propagates throughout a volume that is changing instantaneously as the piston moves.

The flame, whose shape is controlled by the walls of the chamber, consists of a surface that separates the zone containing the products of combustion (burned gases) from the zone containing the initial charge (fuel and air).

Fig. 3.31 shows, in a schematic sense, the relationship between the flame propagation and the pressure-volume or pressure-time diagrams. From TDC, the downward movement of the piston tends to reduce the pressure in the chamber while the energy released tends to have the opposite effect. The pressure reaches its maximum value after TDC (at 10 to 15°CA). The combustion takes place over 60 to 90°CA, which corresponds to a duration of 3 to 5 ms for an engine speed of 3000 rpm. This explanation clearly demonstrates that the term "explosion", in its current sense, is not a suitable description for this type of energy release.

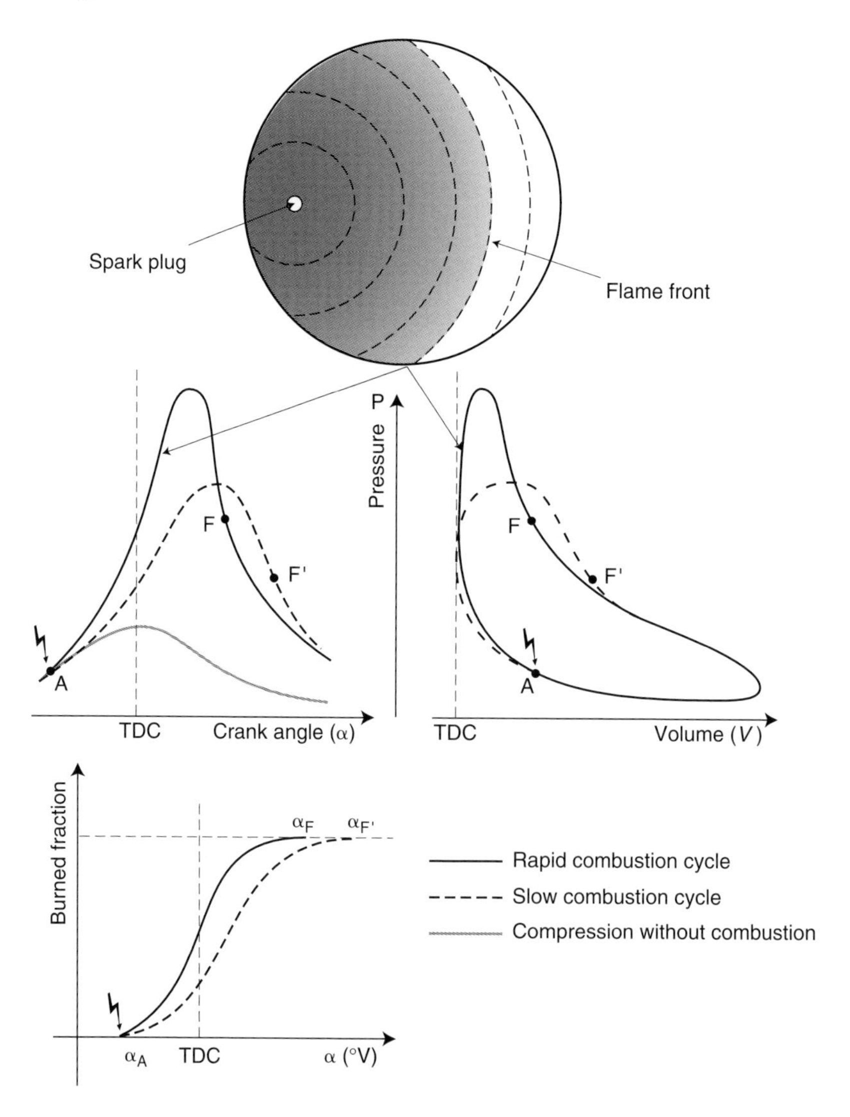

Figure 3.31 *Normal combustion.*
A = ignition point
F or F' = combustion endpoints

Another characteristic of the spark-ignition engine is the existence of **cyclic dispersion**. Provided that imperfections in the ignition or fuel systems are not the cause, significant variations in pressure-time curves have been observed during the combustion phase. The examples shown in Fig. 3.32 show that the variation in amplitude is considerable. This phenomenon occurs with flame propagation under all **turbulent conditions**. The random movement of gases results in statistical fluctuations in the surface of the flame front, which have inevitable repercussions on the speed of energy release and the pressure rise in the chamber. Statistical methods—choice of population size, determination of mean values and standard deviations—are therefore necessary to classify the combustion processes of the spark-ignition engine.

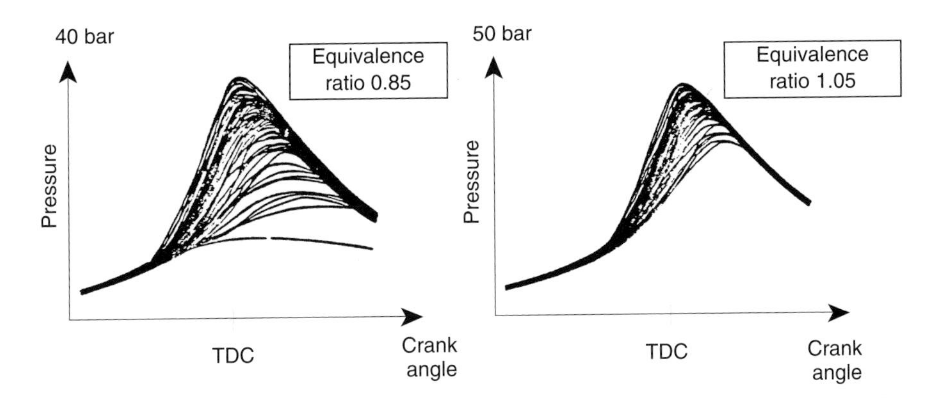

Figure 3.32 *Examples of cyclic dispersion.*

3.4.2.2 Experimental investigative methods

Methods for directly observing combustion have recently been developed. Many of these methods are based on the use of **engines with optical access** to the combustion chamber: quartz viewports are installed in pistons and heads, or both, to make combustion-chamber activities visible.

Some measurement techniques like **laser induced fluorescence** have been adapted for the study of air-fuel mixtures: carefully chosen substances are mixed with the fuel, which then fluoresces in a laser beam. This enables the equivalence ratio of the fuel mixture to be measured at any point in the combustion chamber. Fluorescent substances can also be mixed with air.

Other techniques, like **Mie scattering** or **doppler laser anemometry**, are used specifically to study gas flows and flame-front propagation. With both techniques, particles in air scatter when passed through a laser beam; the scatter signal reveals the speed of the particles.

Still other techniques enable the qualitative observation of the flame, either by **shadowgram** (the flame is illuminated and the shadow is viewed) or **schlieren interferometry** (refracted light is analyzed), or by **direct observation**. There are studies currently underway on the development of "transparent" engines with larger viewports and geometry that is closer to standard engines; also, transparent cylinders of synthetic sapphire have been developed.

Other methods for studying combustion use **fiber optics** placed in the top of the piston and in the head. The results are not as precise, but these methods provide, on the one hand, a good estimate of the location of the flame as a function of time, and on the other hand, an easy-to-use technique that is relatively inexpensive.

All these optical techniques (Eckbreth 1988) are an important support for the basic research being done in this area; however, these techniques are not covered in detail here because they are highly specialized. On the other hand, **conventional data acquisition systems** are described in detail.

Conventional systems provide data that enables the very-precise study of any given condition of engine operation. Different types of information are collected by the judicious placement of various sensors within the engine. The results are then fed to computers where they are stored and processed if required. Using these techniques, the combustion process during each engine cycle can be analyzed.

For example, methods for measuring combustion pressure and crank angle are described in this section.

Pressure is normally measured using a piezo-electric quartz crystal imbedded in the engine head. This sensor provides an electrical charge in proportion to the pressure it receives. An adjustable-gain amplifier transfers the electrical charge into a voltage that is proportional to the pressure. Static or dynamic calibration is used to ensure that the response is linear and stable over time.

The **timing registration**, or crankshaft angle measurement, is done using a disc with 360 slots at 1°CA spacing. The disc is mounted on the crankshaft. An infrared light-emitting diode placed in front of the slots directs its beam through the slots onto a phototransistor on the opposite side. The 0°CA reference point is located by a second light-emitting diode that shines through a single slot located on another track on the disc.

This provides a **synchronization signal**. To ensure an exact representation of TDC, the precise positioning of the disc requires careful attention and meticulous verification to record accurate results, especially for indicated work.

Data acquisition systems are used on engines operating in transitory or steady-state conditions. These systems are not limited to the measurement and recording of the combustion analysis parameters; they can be used to record any type of information that sensors can provide. The systems can also process all recorded data, either mathematically or statistically. In a general sense, data acquisition systems can be divided into two categories. The first type, which are referred to as **rapid systems**, operate at high frequency (from 0.1 to 1°CA); for example, they measure intake, exhaust, and combustion chamber pressures as well as fuel-injector-needle lift off. The second type are referred to as **slow systems** because they measure parameters about once per second. These measurement channels are used to measure data with low variability per cycle and from cycle-to-cycle; these measurements are speed, power, intake and exhaust temperatures, emission concentrations, … .

There are several types of data acquisition systems. Their major differences are in the frequency of parameter measurement, the number of fast versus slow channels, the price, … . Also, the degree of mathematical and statistical data processing varies from system to system. Lastly, ergonomic characteristics and ease-of-use, in both data gathering and results analysis, varies greatly between systems. The best known and best performing system is made by AVL and sold under the name Indimaster 670 (it can be used to acquire data under transitory conditions). Other systems worth mentioning are the Klepcat (developed by IFP), Fevis, and Superflow systems, which are less well-known but offer an interesting compromise of cost and performance.

3.4.2.3 Thermodynamic evaluation

Thermodynamic evaluation essentially consists of determining the speed of the energy release and the temperature change over time of the burned and unburned gases. This information is derived from the pressure diagram and data that characterize the operating condition being investigated.

Two thermodynamic systems exist during combustion. One system contains the **unburned gases** and the other contains the **burned gases**, which are represented by u and b respectively (u meaning unburned); the gases are separated by a flame front of negligible thickness. Experience has shown that the pressure is homogeneous throughout the combustion chamber and that (given the high temperatures involved) each system obeys the universal gas law. The various equations, differentiated with respect to time,[1] are

- The conservation of mass

$$\dot{M}_u + \dot{M}_b = 0 \tag{1}$$

- The conservation of volume

$$\dot{V}_u + \dot{V}_b = \dot{V} \tag{2}$$

The change in total volume as a function of crank angle is given by the following relationship,

$$V = v + \frac{\pi D^2}{4}\frac{L}{\lambda}\left[(1 - \cos\alpha) + \lambda\left(1 - \left[\frac{1 - \sin^2\alpha}{\lambda^2}\right]^{0,5}\right)\right]$$

where

v = the volume of the combustion chamber (at TDC)
D = cylinder diameter (bore)
L = connecting rod length
λ = rod-to-crank ratio ($\lambda = 2L/S$, where S is the engine stroke)

1. The notation $\dot{X}$ represents the derivative of the variable X with respect to time. $\dot{X} = \frac{dX}{dt}$.

- The universal gas constant for system u and system b,

$$\frac{\dot{P}}{P} + \frac{\dot{V}_u}{V} = \frac{\dot{M}_u}{M_u} + \frac{\dot{r}_u}{r_u} + \frac{\dot{T}_u}{T_u} \tag{3}$$

$$\frac{\dot{P}}{P} + \frac{\dot{V}_b}{V} = \frac{\dot{M}_b}{M_b} + \frac{\dot{r}_b}{r_b} + \frac{\dot{T}_b}{T_b} \tag{4}$$

where r is the mass constant for universal gases, which depends on the composition.

- The conservation of energy

$$\overline{M_u U_u}^{\cdot} = -P\dot{V}_u + \Sigma_{ui}\dot{Q}_{ui} + H_u \dot{M}_u \tag{5}$$

$$\overline{M_b U_b}^{\cdot} = -P\dot{V}_b + \Sigma_{bi}\dot{Q}_{bi} + H_u \dot{M}_b \tag{6}$$

In the last two equations, the internal energy of each system is represented by the product of its mass (M) and its internal specific energy (U), which are both functions of the temperature and the composition. Also, the term H refers to the specific enthalpy. The Q expression represents the heat exchanged between the gas and the walls. This heat is represented by the following equation,

$$Q_{ji} = \frac{dQ_{ji}}{dt} = A_{ji}\, k_{ji}\, (T_{ji} - T_j)$$

where

A_{ji} = surface area of the walls i in contact with system j

k_{ji} = transfer coefficient at the wall/gas interface

T_{ji} = temperature of the wall i in contact with the system

T_j = temperature of the gas in system j

The combustion chamber is divided into various zones: valves, piston, head, cylinder, etc. The temperature of these zones can be estimated with reasonable precision. The transfer coefficients are determined from semi-empirical equations, of which the best known ones are those of Woschni and Trapy (Woschini 1967; Trapy 1985).

There are now six equations, (1) to (6), with six unknowns; M_u, V_u, T_u, M_b, V_b, and T_b. The system can be solved by conventional numerical methods and it provides all the needed information including the **heat release rate**. This is expressed in terms of the burned fraction as a function of time.

$$x(t) = \frac{1}{M} M_b(t) \qquad 0 < x(t) < 1$$

Fig. 3.33 shows, for an average-pressure diagram, the standard $x(t)$ curve. This curve has the characteristic S shape, which represents the heat release rate during combustion and the maximum energy release during the median phase.

Using x (t), the combustion is described by two parameters that are displayed in Fig. 3.33 and they are referred to respectively as

CA50 = angle of rotation when 50% of the total energy has been released

HRR = heat release rate, expressed in°CA^{-1}. This is the maximum value of the curve that represents the derivative of the burned fraction, which is the speed of the energy release.

CA50 describes the position of the combustion in the cycle, whereas HRR supplies a method of expressing the speed of the combustion. Usually the CA50 point occurs 5 to 10°CA after TDC; the maximum value of HRR is about 0.04, which means that at the most rapid combustion phase, the flame consumes 4% of the total charge per degree of crankshaft rotation.

Fig. 3.33 also shows the variations in pressure diagrams resulting from cyclic dispersion.

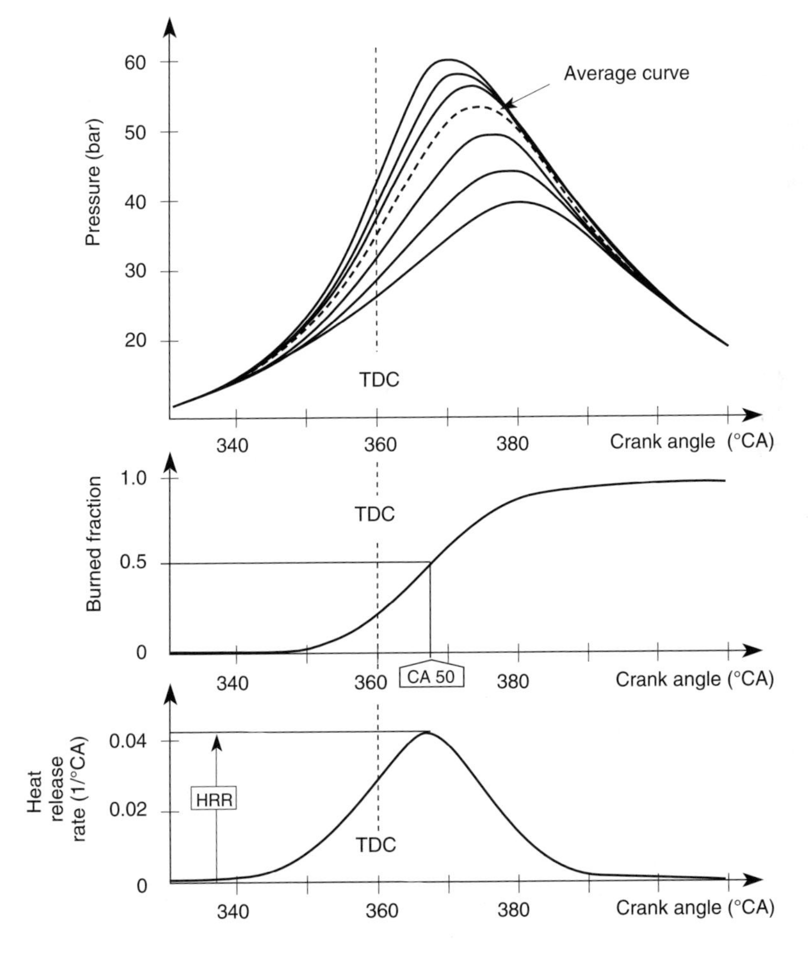

Figure 3.33 *An example of the rate of combustion that is based on a pressure versus crank angle diagram.*

Another interesting point is the speed of the initial energy release, which occurs at the CA5 point; that point corresponds to the crank angle when 5% of the energy has been released.

3.4.2.4 The effect of operating conditions

Engine design and control parameters, as well as fuel variables (composition, equivalence ratio), can change the rate of combustion, and they can affect engine performance as a result. Modeling techniques can be used to determine the separate effects of various parameters, and to rank their possible consequences in terms of performance or MEP for example.

Prior to initiating such an analysis, an estimated ranking of the possible variations can be done. Fig. 3.34 shows the effect of the variations in CA50 and HRR on the indicated efficiency (η_i), which is based on model simulations.

The optimal value of CA50 is found a few °CA after TDC. The engine designer can optimize this parameter by adjusting the spark advance. This does not work for HRR however, because it depends on a number of parameters that are rarely controllable by external intervention.

An increase in HRR normally indicates an increase in performance because it is directly related to the speed of the energy release. For example, this change occurs when HRR increases from 0.02 to 0.04°CA. Above this value however, any increase in HRR does not result in increased performance. Therefore, very rapid combustion does not result in a net performance gain.

It is more effective to attempt to increase the speed of combustion in the type of operation where it is very slow, such as low-load conditions.

Cyclic dispersion is a limiting phenomenon in terms of performance, especially at low loads (Douaud 1981).

The following sections describe how some operating and control parameters act on the normal combustion process.

A. *Engine speed*

Increased turbulence at higher engine speeds helps to increase the speed of flame propagation. Turbulence increases almost **in proportion to engine speed** in such a way that the parameters of combustion (P, x, HRR), which are expressed in degrees of crankshaft rotation, are not greatly affected by a change in engine speed. At high speeds however, the initial and final combustion phases spread out, which requires a greater spark-advance.

B. *Volumetric efficiency*

A drop in volumetric efficiency tends to slow combustion, due to a reduction in the temperature and the pressure of the charge to be ignited, but especially due to dilution by residual exhaust gases trapped in the dead volume (v). Fig. 3.35 shows an example of the change in HRR for volumetric efficiencies between 0.4 to 0.8; within this range, HRR still retains an acceptable value (0.03°CA^{-1}). The effect of volumetric efficiency on combustion is not a major factor, except at very-low loads, or when artificial dilution occurs due to exhaust-gas recycling.

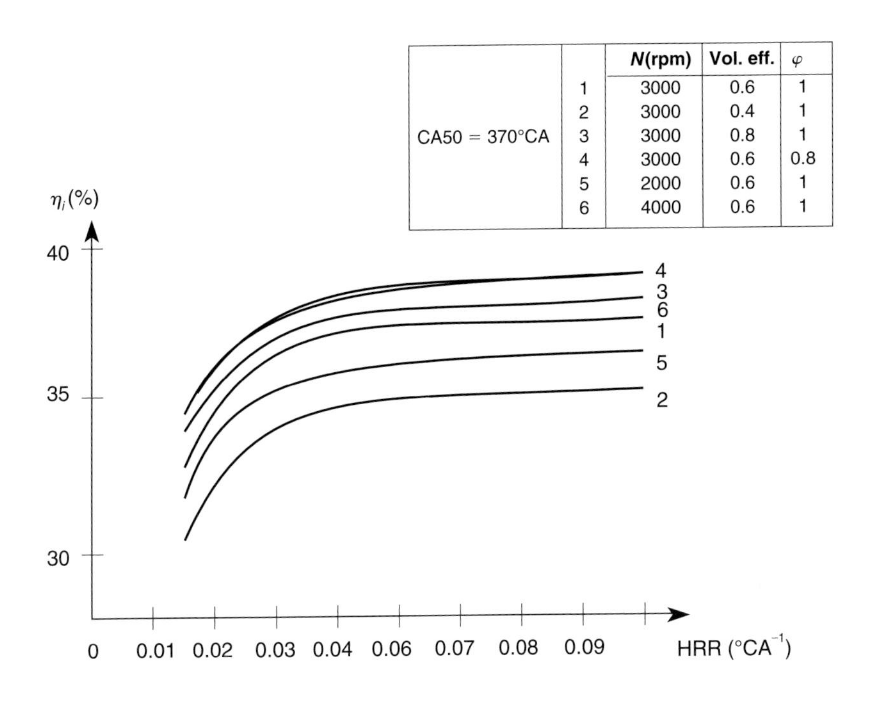

		N(rpm)	Vol. eff.	φ
CA50 = 370°CA	1	3000	0.6	1
	2	3000	0.4	1
	3	3000	0.8	1
	4	3000	0.6	0.8
	5	2000	0.6	1
	6	4000	0.6	1

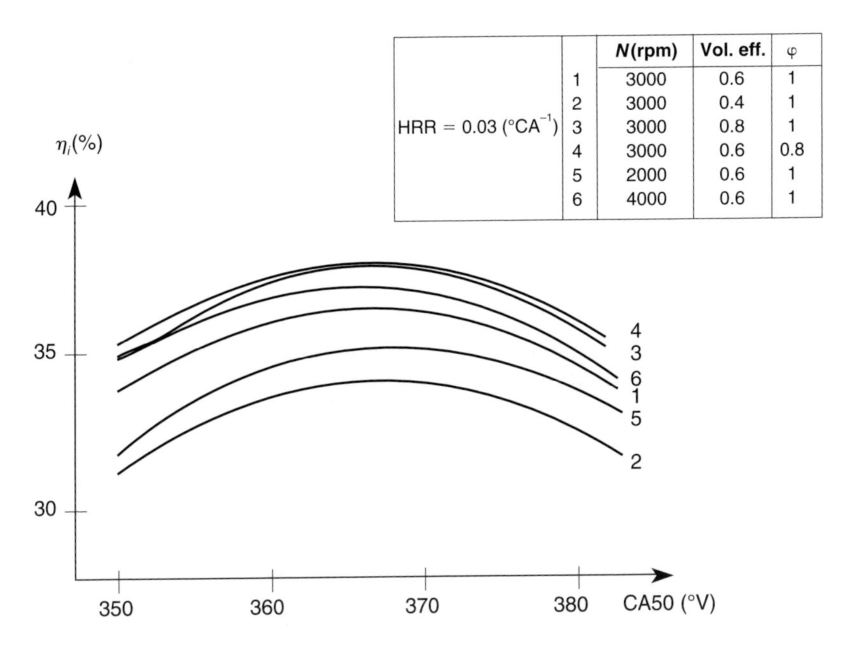

		N(rpm)	Vol. eff.	φ
HRR = 0.03 (°CA^{-1})	1	3000	0.6	1
	2	3000	0.4	1
	3	3000	0.8	1
	4	3000	0.6	0.8
	5	2000	0.6	1
	6	4000	0.6	1

Figure 3.34 *Effect of the HRR and CA50 parameters on indicated performance. Simulated results for a single cylinder engine.*

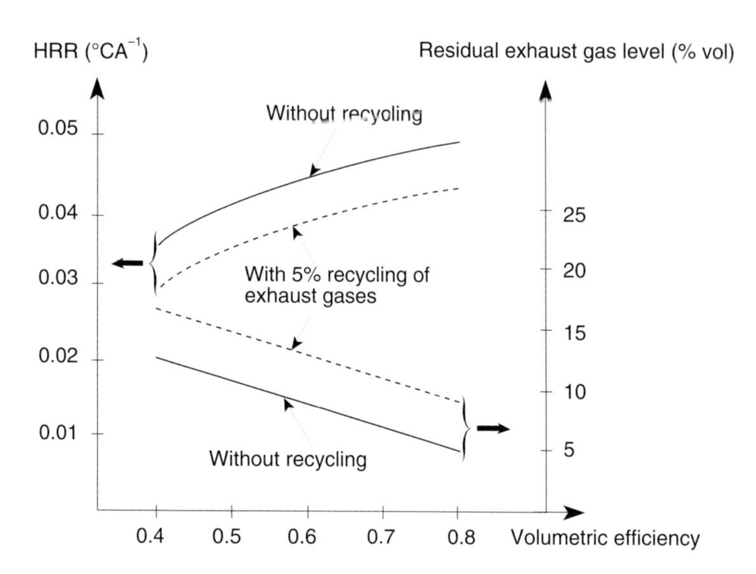

Figure 3.35 *Effect of volumetric efficiency on the speed of combustion.*

C. Engine geometry

Geometric parameters affect the intensity of turbulence and the surface area of the flame front. This results from the flame's contact with the walls and the resulting heat transfer. Therefore, it is logical that the HRR and CA50 parameters will change with the shape of the combustion chamber. Also, changes to the aerodynamics of the intake system can contribute to the creation of a directed gas flow, which changes the **surface area** and the **path** of the flame. These geometric factors are considered at the design stage to optimize energy release and spark advance. The shape of the combustion chamber also affects other criteria—volumetric efficiency, knock tendency, pollutant emissions—and the chosen solution is always an overall compromise.

D. Equivalence ratio

The air-fuel ratio of the mixture is of **paramount** importance. As shown in Fig. 3.36, combustion time is short with a rich mixture ($\varphi = 1.10$ to 1.20) and it increases markedly at less than stoichiometry ($\varphi = 0.80$ to 0.90) as cyclic dispersion increases. If the mixture's air-fuel ratio is further reduced, unstable operation ensues with misfiring and complete extinction of combustion. This confirms the existence of inflammability limits within the engine, which are like those observed in the laboratory, and which are affected by perturbing influences: turbulence, the presence of exhaust gases, and limited ignition energy. As a result, the operational limit of the air-fuel ratio depends on the type of operation being studied.

With a conventional engine, the minimal value of the equivalence ratio ranges from about 0.70 to 0.80 at high load and low speed, and it approaches stoichiometry under more difficult combustion conditions, especially at very-low volumetric efficiency.

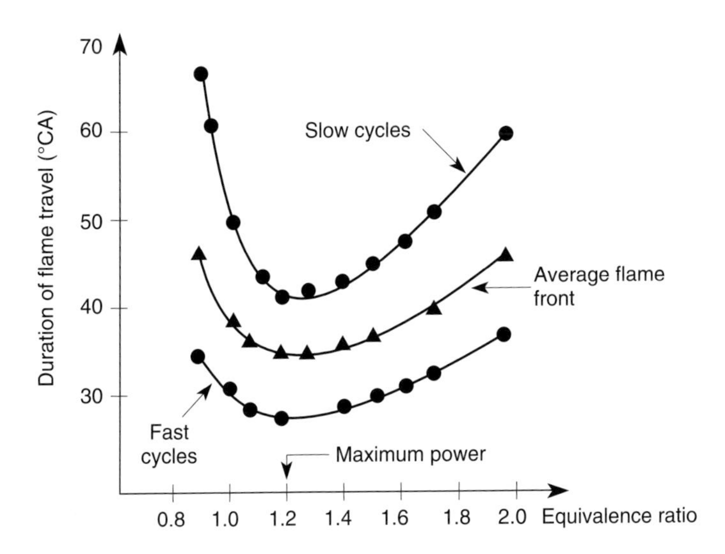

Figure 3.36 *Effect of air-fuel ratio on the duration of flame propagation. (Source: Young 1981)*

A number of studies are currently underway to develop engines that run on lean mixtures, because these engines provide higher efficiency than conventional engines. To obtain satisfactory combustion with overall lean mixtures, automobile manufacturers are developing engines that operate with a **stratified charge** (see 8.2.2).

E. Fuel composition

A number of engine tests with pure hydrocarbons, all at the same equivalence ratio, have shown that the speed of combustion varies slightly depending on the chemical structure of the fuel and it retains the **imprint** of the laminar propagation speed, as shown, for example, in Table 3.4. The variations are not very great (Table 3.6); the difference in flame speed between benzene, which burns quickly in an engine, and isooctane, which burns slowly, is no greater than

Table 3.6 *Relative combustion speeds of various hydrocarbons. Experimental single-cylinder engine operated at 4000 rpm, under full load, and at 1.10 equivalence ratio. The flame duration times were obtained using an ionization detector.*

Hydrocarbon	Relative combustion speed	Hydrocarbon	Relative combustion speed
Isooctane	1.02	Cyclohexane	1.21
Xylene	1.00	Cumene	1.13
Diisobutylene	1.12	Ethylbenzene	1.17
Toluene	1.07	Benzene	1.27

25%. The difference between the laminar flame speeds of these two hydrocarbons is 20% at an equivalence ratio of 1.10 (Table 3.4). For conventional fuels, which consist of a large mixture of compounds, variations of this magnitude would not be attained.

In practice, the values of HRR and CA50 are not heavily dependent on the details of fuel composition, such as the aromatic and olefin content. Very-specialized products must be used to see any change in this area; only in extreme cases would a change in fuels imply a small readjustment in the spark advance.

There is no known additive that can increase the laminar propagation speed of hydrocarbons, and as a consequence the speed of energy release in engines. Lead alkyls, in particular, have no effect on the propagation process of normal combustion.

3.4.3 Abnormal combustion

The term abnormal combustion applies to a large number of possible parasitical phenomena in combustion. Some are completely harmless and others can lead to the destruction of vital engine components. This section provides some essential definitions and describes the most common types of abnormal combustion: knock, run-on, and preignition.

3.4.3.1 Definitions

Occurrences of abnormal combustion are phenomena that can affect the whole fuel-air charge, or just a fraction of it, and they do not result from the propagation of the flame front initiated by the spark plug. From a reaction-mechanism standpoint, these phenomena are the

- **auto-ignition** of all or a part of the fuel-air mixture (knock, run-on)
- **propagation** of one or more flame fronts that originate from abnormal sources such as hot points on the spark plug or the exhaust valve (preignition)

Abnormal combustion can be temporary, episodic, or on the contrary, it can occur continuously with potentially severe results.

3.4.3.2 Knock

Knock has been known almost since the beginning of engines and it has spawned (without exaggeration) several thousand publications. In France, the first work in this area was done by P. Dumanois[1] and his associates in 1925. Researchers and experimenters rapidly reached an accord on the description of the predacious nature of knock, but they spent a long time discussing the exact nature of the process that puts it into play. This explains why a multitude of terms have been used to describe it—detonation, auto-ignition, ping-

1. Paul Dumanois (1885-1964): Ingénieur Général de l'Air, father of the *École Nationale Supérieure des Moteurs* in 1924.

ing, … . These terms are not precise and some are completely misapplied. The term knock is used here because it is a well-accepted term by specialists and the general public.

A. The nature of the phenomenon

Under certain operating conditions, part of the unburned charge located in advance of the flame front auto-ignites and burns instantaneously, or at very-high speed.

The result is a local increase in pressure followed by **vibrations** of the **gaseous mass,** which continue until the pressure in all areas of the combustion chamber is equalized. The pressure wave creates a **characteristic noise** that sounds like metallic ringing, which is distinct from the muted sound of combustion. The sound made by knock has a fundamental frequency of about 5000 to 8000 Hz.

Knock is auto-ignition that occurs in a homogeneous environment and it is different from the process in diesel engines, where auto-ignition starts in one or several very-localized areas (see 4.2). On the other hand, the term "detonation", which is often used to describe knock, is not correct because detonation spawns a supersonic flame propagation that is accompanied by shock waves.

Knock is easily detected by examining a pressure diagram, as shown schematically in Fig. 3.37. A normal combustion curve is disrupted at a point in the cycle by intense pressure oscillations that continue for part of the downstroke. This type of change occurs with only the **small fraction** of the total

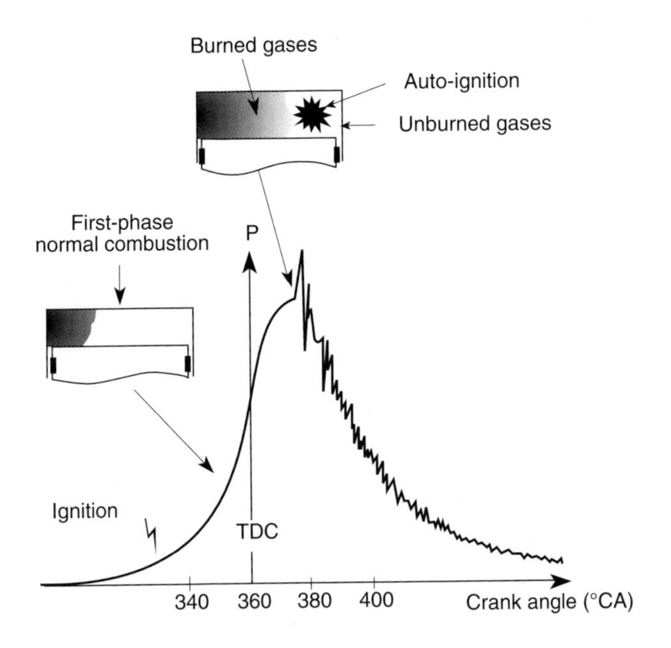

Figure 3.37 *Pressure diagram with intense knock.*

cycles that have the highest pressure levels, due to cyclic dispersion. Therefore, when trace knock does occur, it appears in less than 1% of the engine cycles.

It appears that knock is not very dangerous if it occurs on an episodic basis, for a very short time (a few seconds), and if it involves only a small fraction of the charge (for example, less than 10%).

However, with intense and sustained knock, various types of events can occur. These can range from minor symptoms (distinct noise, increase in coolant temperature) to a complete and immediate engine failure (blown head gasket; seizing or partial melting of the piston; valve and head deterioration). The destructive mechanisms behind this phenomenon are still not completely understood. For the engine manufacturer, the best course of action is to avoid the occurrence of knock—even trace knock.

B. Detection methods

For many years, the only instrument used to detect knock, both on the test bench and in vehicles, was the human ear, which was sometimes aided by a stethoscope.

One precise and easy-to-use technique consists of placing an **accelerometer** on the head and selecting the one frequency among many that corresponds to knock.

When a **pressure sensor** is implanted in the engine head, the detection and quantification of knock is no longer a problem. One can then study the cylinders in which knock preferentially occurs. Rapid data acquisition techniques also enable more precise knock intensity levels to be chosen as criteria. For example, pressure oscillations can be classified by the quantitative value $\Sigma(\Delta P)$:

$$\Sigma(\Delta P) = \sum_{i=1}^{n} (P_{i+1} - P_i)$$

The term i represents the angle of rotation for the phase of the cycle that involves knock, and it continues for n°CA. The expression $\Sigma(\Delta P)$ represents the intensity of the knock.

C. Prediction and modeling

Knock will occur if the unburned gases are susceptible to auto-ignition before being consumed by the flame front during its normal propagation. **Time** plays an important role. The duration of the normal combustion phase is obtained from the pressure diagram. With unburned gases, the change in pressure is known; the temperature can be estimated by computation or measured by experimental means. Calculated values are used here. To illustrate, Fig. 3.38 shows that the temperature and pressure can exceed 900 K and 30 bar respectively for unburned gases.

In this situation, auto-ignition occurs when the molar fraction of the free radicals (y) reaches a critical value y^*, which is probably independent of temperature and pressure—according to kinetic theory.

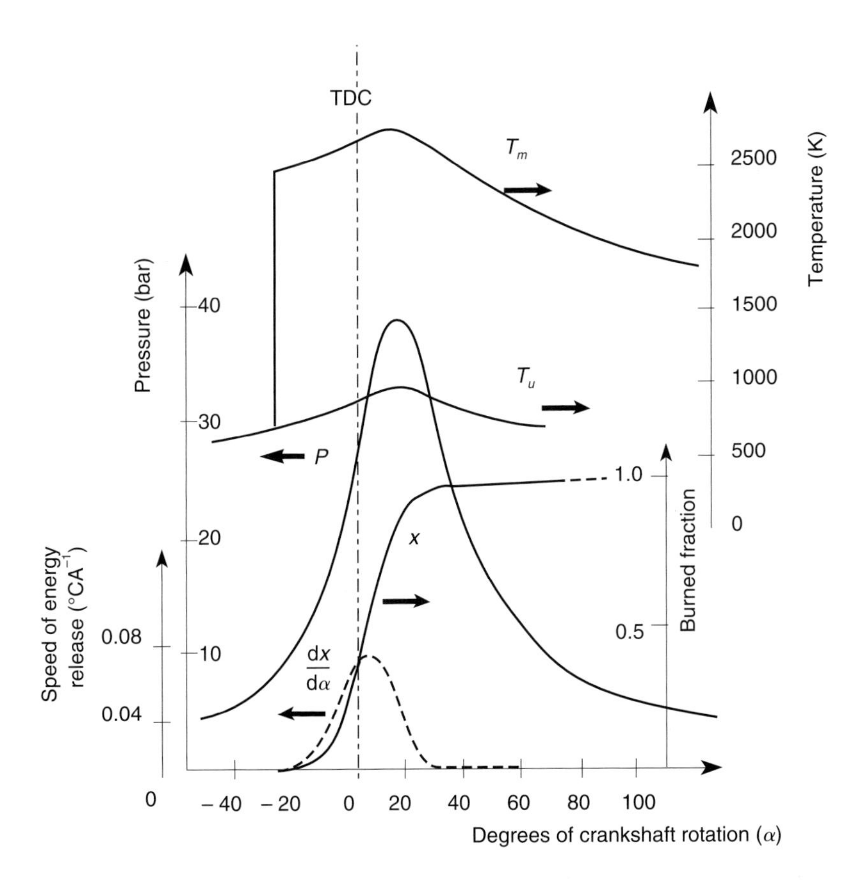

Figure 3.38 *An example of the change in temperature and pressure of unburned gases. (T_u = temperature of the unburned gases; T_m = average temperature)*

The speed at which reaction species appear is

$$\frac{dy}{dt} = Ky^a$$

where a is a constant of value 0, 1, i, ... , which depends mainly on the nature of the initial stages; K is a function that varies with temperature and pressure according to the conventional Arrhenius model,

$$K = kP^n \exp\left(-\frac{B}{T}\right)$$

where k, n, B are positive constants.

Auto-ignition will occur at instant t^* according to the following equation:

$$\int_{y_0}^{y^*} \frac{dy}{y^a} = \int_{t_0}^{t^*} K\,dt$$

The initial times (t_0, y_0) represent conditions when the temperature and pressure are relatively far removed from those occurring at the point of auto-ignition—times when the chemical activity is negligible.

The preceding expression can also be written as

$$\int_{t_0}^{t^*} \frac{K}{C}\,dt = 1 \qquad (7)$$

where C is a constant.

If the gases are at constant temperature and pressure, the temperature difference t^*-t_0 is by definition the auto-ignition delay θ. Integrating the general equation (7) for this particular case leads to the following relationship:

$$C = K\theta$$

Thus, in a real situation where fresh gases are constantly evolving over time, the auto-ignition condition becomes

$$\int_{t_0}^{t^*} \frac{dt}{\theta} = 1$$

This results in the following well-known equation for auto-ignition:

$$\theta = \frac{C}{K} P^{-n} \exp\left(\frac{B}{T}\right) = AP^{-n} \exp\left(\frac{B}{T}\right)$$

For some hydrocarbons or special fuels, parameters A, n, and B have been measured or estimated in the laboratory. Based on the evolution of the pressure and temperature of unburned gases, the **theoretical instant of ignition** (t^*) can be calculated in these cases.

Otherwise, the flame front ends its travel normally at the known time t_f.

Knock occurs if $t^* < t_f$ and the knock intensity is proportional to the amount of chemical energy contained in the unburned gas at time t^*.

Conversely, knock cannot occur if

$$t^* > t_f$$

or

$$\int_{t_0}^{t^*} \frac{dt}{\theta(P,T)} < 1$$

The three possible scenarios (absence of knock, trace knock, and severe knock) are shown schematically in Fig. 3.39. The validity of this model has been verified in circumstances where the values of the A, n, and B parameters were known. In these cases, the instant that knock appeared on the diagram was exactly the time predicted by the calculations.

This method of predicting the occurrence of knock is only of theoretical interest, since the numeric values for A, n, and B are not available for commercial fuels. In practice, engine **modeling** would be used to establish a general formula for the delay. Therefore, the operational method is the following one: for a given number of combustion cycles in which knock occurs, the change in temperature and pressure of the unburned gases is determined experimentally, as well as the critical time t_c. Then, coefficients A, n, B in the following equation

$$\int_{t_0}^{t_c} \frac{dt}{AP^{-n} \exp\left(\frac{B}{T}\right)} = 1$$

are determined to verify it, at least for the range of the gathered data.

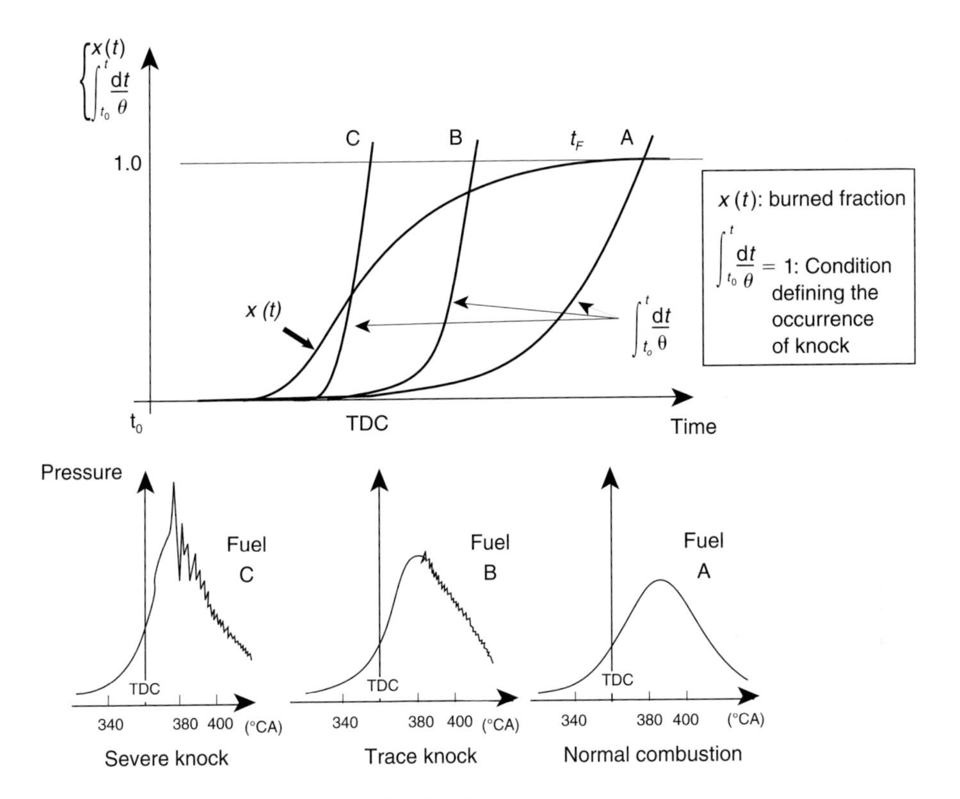

Figure 3.39 *Conditions under which knock occurs.*
(Fuel A = no knock; Fuel B = trace knock; Fuel C = severe knock)

The test conditions must be chosen to ensure maximum decoupling of the pressure and temperature variables. The validity of this approach can be verified, because the delay formula obtained accounts for tests done on several types of engines under very-different operating conditions at speeds from 600 to 5000 rpm.

After completing the verification, the formula can be introduced in simulation programs that enable the prediction of the risk of knock under all possible operating conditions.

Thus, for isooctane, the following relationship

$$\theta = 1.931 \cdot 10^{-2} P^{-1.7} \exp\left(\frac{3\,800}{T}\right)$$

(θ in s, P in bar, T in K)

describes the knock behavior of this hydrocarbon in various engines.

The preceding formula appears to be numerically different of the formulas derived from experiments on the rapid compression machine. However, this does not mean that the models presented herein are irrelevant. Experiments done on rapid compression machines and those done on engines do not always agree. It would be naive and risky to assume, given the complexity of the kinetic mechanism of auto-ignition, that general and semi-empirical formulas can be used outside of their experimental context.

D. Factors affecting knock

The builder who designs the engine and the refiner who produces the fuel must contribute to the creation of an environment free of knock. Engine-fuel matching requires a means to express the tendency to knock. The conventional concept is to use octane ratings and octane requirements; this would take too long to describe in this introductory section, which is devoted exclusively to engine operation. These fundamentals are covered in subsequent chapters and this section only provides some qualitative characteristics that are logically related to the overall problem of auto-ignition.

Everything that tends to increase the duration of flame propagation and to reduce the auto-ignition delay of the unburned gases promotes knock. In reality, the involvement of all parameters—mechanical or chemical—changes the delay, which is more sensitive to pressure and temperature changes than the propagation speed.

All increases in the temperature, pressure, or life-span of the unburned gases favors the occurrence of knock. Therefore, knock tendency tends to increase with compression ratio, volumetric efficiency, ignition advance, and intake-air pressure and temperature.

In cases where the pressure-crankangle diagram does not depend on engine speed, knock usually shows up at **low engine speeds**. In fact, if α (crank angle) is expressed in°CA, t in s, and N in rpm, the following relationship is derived

$$\alpha = 6Nt$$

and the auto-ignition condition, which is expressed as a function of crankangle becomes,

$$\int_{\alpha_0}^{\alpha^*} \frac{d\alpha}{\theta\,(P,T)} = 6N$$

where α_0 and α^* represent the initial calculated angle and the angle when knock occurs.

Observe that α^* increases with N and the tendency to knock decreases as the speed increases. This relationship holds true, except when certain parameters—volumetric efficiency, heat transfers, etc.—cause an increase in the temperature and pressure of the auto-ignitable mixture at high engine speed. This causes another type of knock—known as **high-speed knock**—that is described in subsequent sections.

The knock behavior of fuels, particularly hydrocarbons, is narrowly dependent on their **chemical structure**.

Hence, within the *n*-paraffin series, the resistance to auto-ignition is high for some of the lighter constituents (methane, ethane) but it decreases very rapidly as the molecular chain lengthens. For the same number of carbon atoms, isoparaffins are more knock resistant than their isomers with straight-chain structures and this difference increases with the number and the complexity of the lateral chains. Olefins have the same general structure in terms of carbon chain length and the number of branches. For the same type of structure, olefins can be more conducive or more resistant to auto-ignition than paraffins, depending on the pressure-temperature environment in which they evolve. The position of double bonds within the chain also has an effect. The aromatic group also has very-good knock resistance.

The knocking tendency of all fuels in terms of octane rating is discussed in subsequent sections (see 3.6).

3.4.3.3 Other types of abnormal combustion

Other types of abnormal combustion do not have knock's technical and economic consequences, because the methods used to control them do not limit engine performance nor do they impose severe constraints on product formulation. These types of abnormal combustion manifest themselves in a number of different ways, which has led to a very complex repertoire of terms and classifications. This section describes two such phenomena that can happen in practice, but rarely occur: **run-on** and **preignition**. Fundamentaly, the first involves auto-ignition and second involves the propagation of one or several flame fronts.

A. Run-on

Run-on, which is otherwise known as "after running" or "dieseling", occurs when an engine continues to operate at low speeds (50 to 500 rpm) for a limited time after the ignition system is turned off. Run-on in a conventional (carburetor equipped) vehicle usually occurs after a period of operation at low load, or when the coolant temperature rises, and it continues for a few seconds or for a few minutes in exceptional cases. This type of combustion does not affect the engine mechanically, but it can be very irritating to the operator.

Run-on is caused by **auto-ignition** after the ignition system is turned off. Due to the slow rotation of the engine at idle, time is the determining characteristic that causes auto-ignition, despite the low levels of temperature and pressure.

As a function of the chemical structure of the fuel, knock and run-on tendencies are obviously similar. With some vehicles, it has been noted that run-on precedes or accompanies knock, which can be a useful warning sign.

Various engine control and design parameters have a considerable effect on run-on. As a result, its intensity decreases or it disappears entirely with decreases in compression ratio, inlet air temperature, coolant temperature, and volumetric efficiency at idle. This last parameter explains why high idle speeds contribute to run-on.

This type of malfunction has practically disappeared today; on modern engines with fuel injection, the fuel is cut off along with the ignition system.

B. Preignition

Preignition occurs when ignition and a more-or-less complete combustion of the charge occurs **before spark ignition**. When this phenomenon happens, the unwanted ignition occurs regularly with each cycle and it becomes more and more premature (Guibet et al. 1972). The evolution toward extremely-advanced ignition can be slow (tens of minutes) or very fast (a few seconds). Fig. 3.40 shows several examples of pressure diagrams taken during a progressive evolution to extremely advanced preignition. The maximum pressure, which always occurs after TDC with normal combustion, moves to TDC or slightly before it, once preignition begins. Pressures reach very-high levels—up to 73 bar in the example. Evidence of cyclic dispersion is also visible, not just around TDC, but very early in the cycle.

Preignition causes a considerable increase in the heat transfer between the combustion gases and the walls, which often leads to serious and spectacular incidents: major reductions in power, partial melting of valves, seizing or holing of pistons, and the return flow of burned gases into the intake manifold and the fuel system.

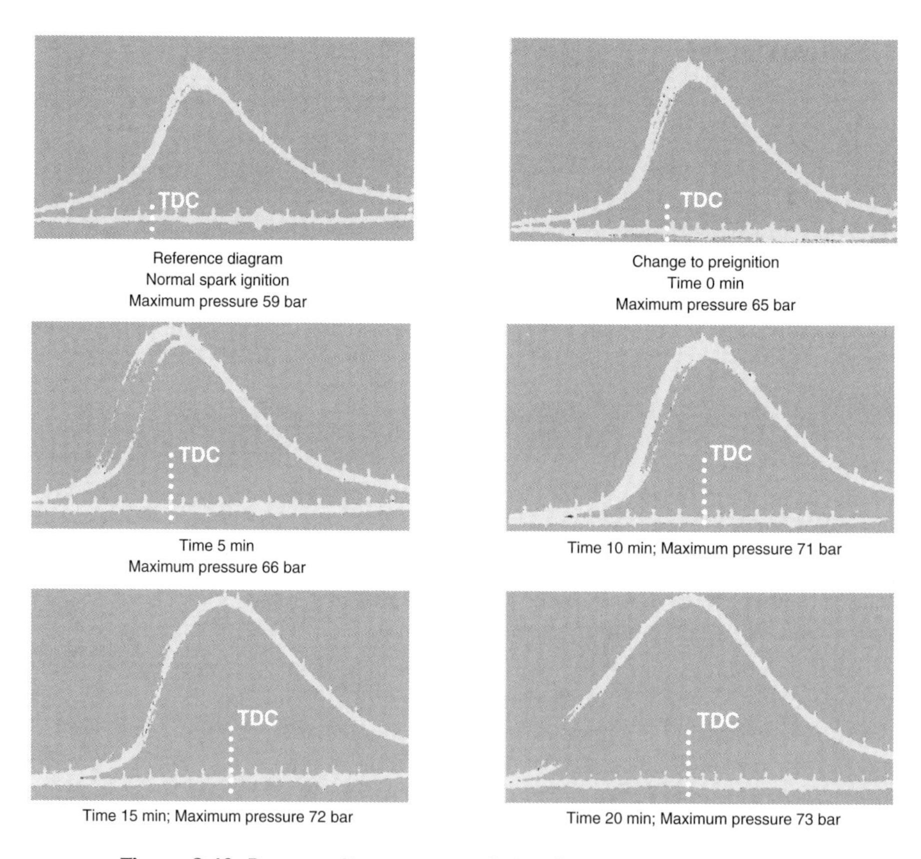

Figure 3.40 *Pressure diagrams recorded under preignition conditions.*

Several studies have shown that preignition can occur when a point or area in the combustion chamber reaches about 1000°C. In practice, it is usually an **exhaust valve** covered in deposits that acts as the initiating source of ignition. Following that, the **spark plug** may be next, if it does not have thermal characteristics that enable it to rapidly dissipate the flow of heat caused by the preignition.

In preignition conditions, the exhaust valve reaches a critical temperature due to the progressive accumulation of deposits; the deposits are almost exclusively mineral and they consist of contributions from the fuel (especially if the fuel contains lead) and the engine lubricant. With some engines, a mass of about 1g per valve is enough to initiate preignition.

Preignition has almost completely disappeared in service, thanks to a number of simultaneous occurrences: reduced lubricant consumption, control of ash content, and spark plugs with appropriate heat ranges.

Gasoline characteristics and specifications

An automobile's driveability and usability under all circumstances, its satisfactory performance, and its reliability and longevity depend on several technological factors including the fuel quality criteria (physical and chemical properties, octane rating) that are described in this section. The section also describes other gasoline characteristics that are required to satisfy regulations, and to meet the constraints of storage and distribution.

3.5 Physical properties

The physical properties of a fuel have a fundamental effect on the fuel-supply system and the combustion process, especially under difficult conditions: starting, operation, and cold and hot weather performance. The most important characteristics in this regard are the **density** and the **volatility**, which are represented respectively by the distillation curve and the vapor pressure. Other measures like viscosity and heat of vaporization are also involved, but to a lesser extent.

3.5.1 Gasoline density

Density is usually measured at 15°C using a **hydrometer** (ISO method 3675); it is expressed in kg/dm^3 with a precision of 0.0002 to 0.0005, depending on the type of hydrometer used. However, in practice only three decimal places are typically used.

Density varies with temperature according to the following relationship

$$\rho_t = \rho_{15} - k\,(t - 15)$$

where

t = temperature in°C

ρ_t and ρ_{15} = density at t°C and 15°C respectively

k = numerical coefficient, which for gasolines is about 0.00085

Therefore, when the temperature rises from 15 to 25°C for example, ρ decreases by 0.008, that means 1%. These changes, although very small in absolute value, must be taken into account in various commercial transactions related to the storage and distribution of gasoline.

The specifications for gasoline density are fairly uniform throughout the world. In Europe, conventional unleaded gasoline (Eurosuper) must have a density between 0.725 and 0.780 kg/dm^3. In the United States, the minimum and

Catalytic cracking unit. R2R process.

Idemitsu Corporation. Aïchi Japan.

Capacity: 40 000 bbl/d (2 Mt/yr).

(Photothèque IFP)

Hyvahl deep conversion unit, Ssang-Yong, Korea.
Capacity: 32 000 bbl/d (1.6 Mt/yr).
(Photothèque IFP – C. Dumont/Scopimag photo)

CFR engine lab, IFP – Solaize. (Photothèque IFP)

Hole-type injector
Gasoline pressure = 3 bar
Time = 4 ms

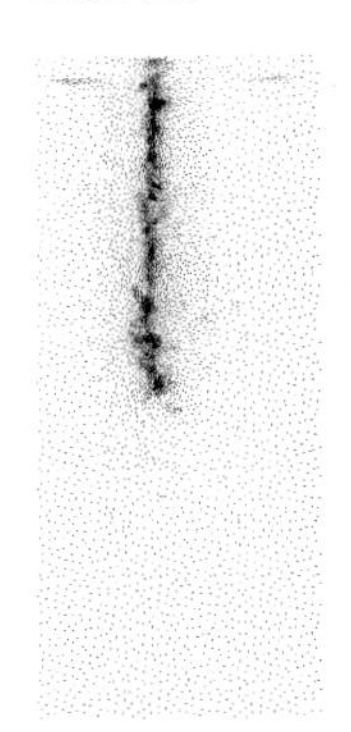

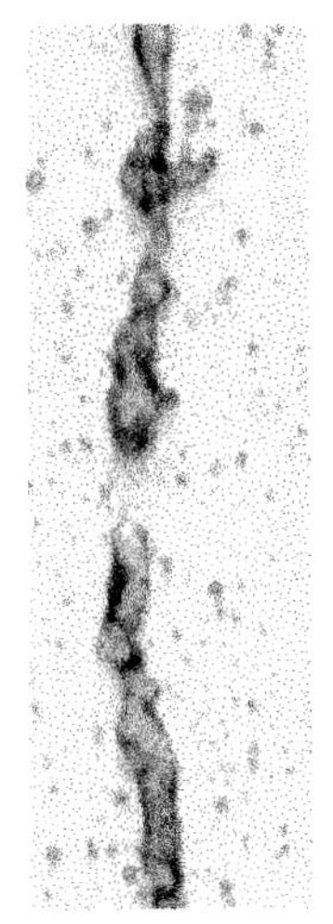

Close up of the spray
(liquid column 0.3 mm
in diameter)

Pintle-type injector
Gasoline pressure = 3 bar
Time = 4 ms

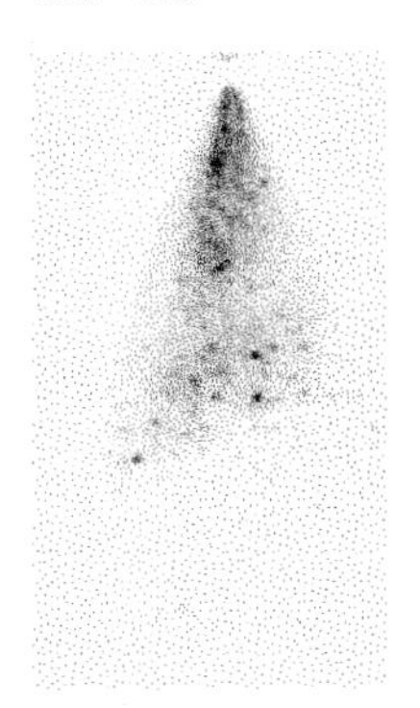

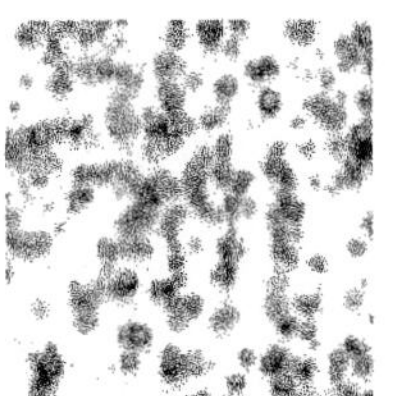

Close up of the spray
(droplets 0.1 to 0.2 mm
average diameter)

Spray patterns frome hole- and pintle-type injectors in a gasoline engine.

(IFP document)

Evaluation of intake valves after use with a gasoline without additives.

Total mass of the deposits = 5322 mg. Score = 5.40/10.

(Photothèque IFP)

Evaluation of intake valves after use with a gasoline containing a detergent additive.

Total mass of the deposits = 36 mg. Score = 9.83/10.

(Photothèque IFP)

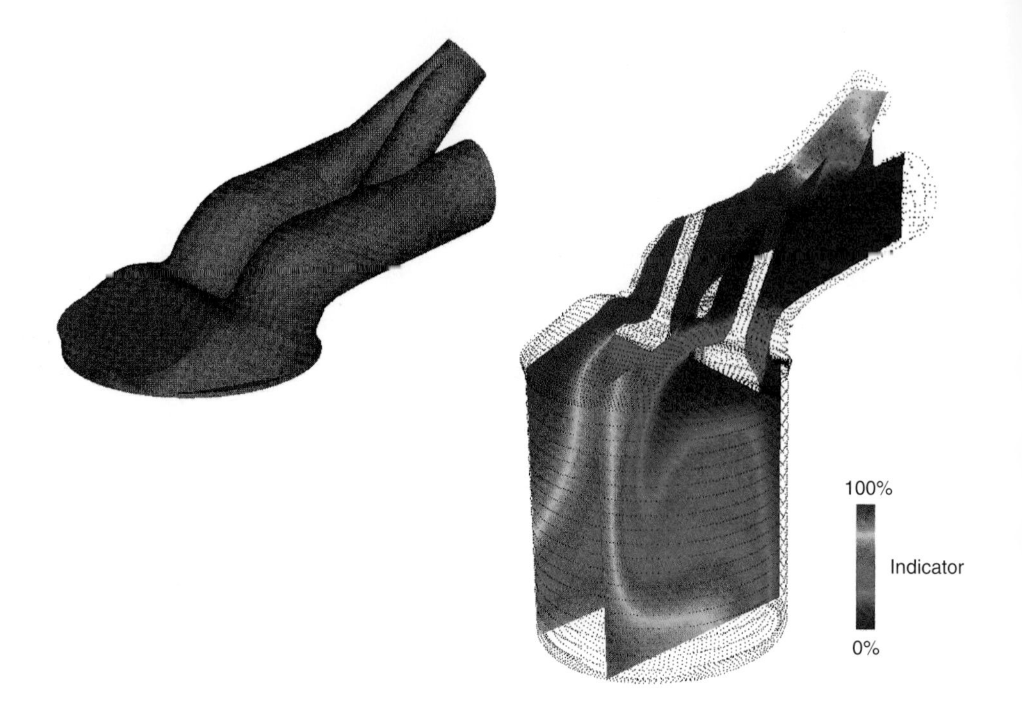

Model of the intake and compression phases in a 4-valve gasoline engine.
Spatial distribution of the indicator.

Defining the geometry of the intake system enables the optimization of the mixture stratification, turbulence, and speed at the point of ignition.

(IFP document)

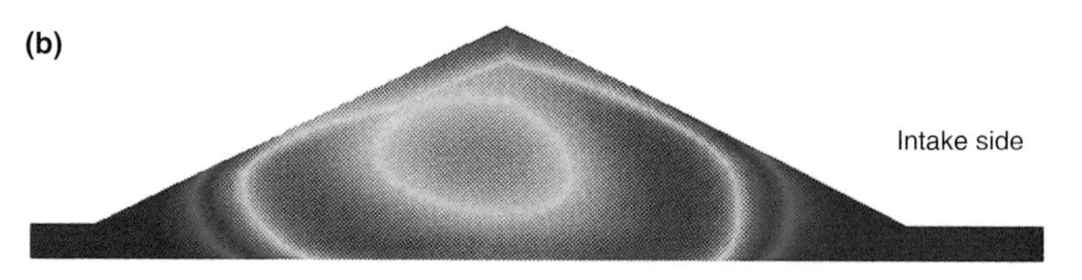

Model of the combustio
a 4-valve gasoline engin
Flame surface dens
(cm^2/cm^3) at a burn
fraction of 9%.

***(a)** Two intake valves*
cylinder.

***(b)** One intake valve*
cylinder.

(IFP document)

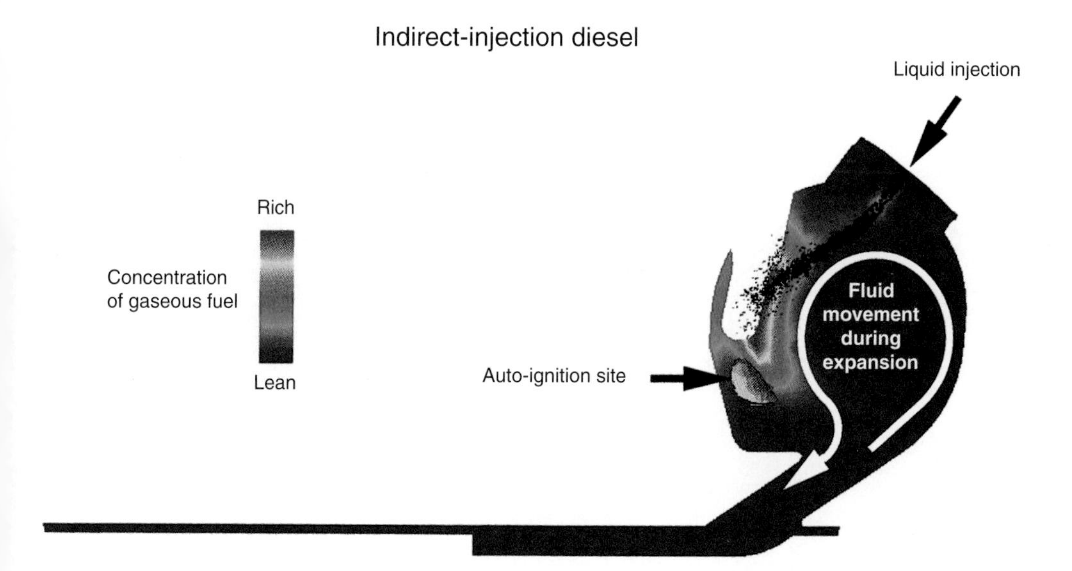

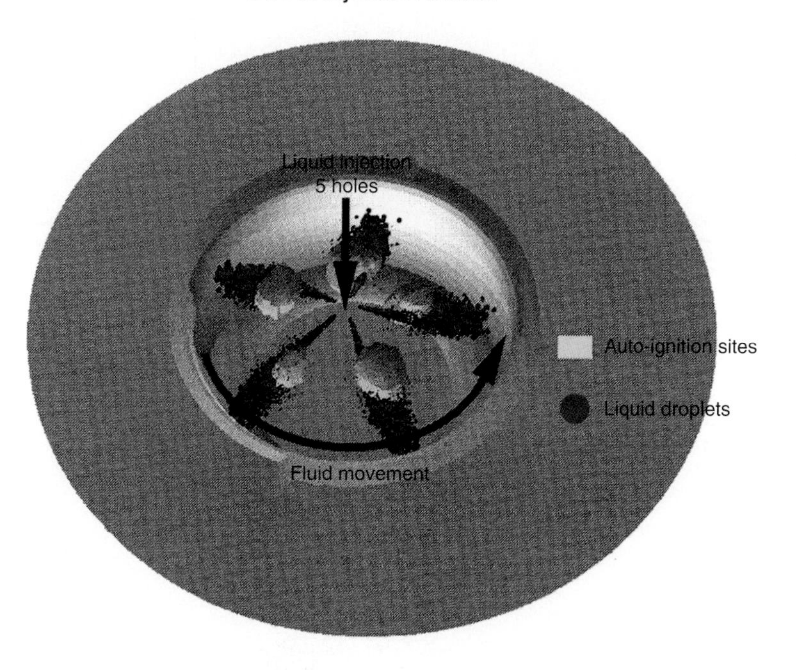

Simulation of auto-ignition phenomena in a diesel engine.
Indirect and direct injection.
(IFP document)

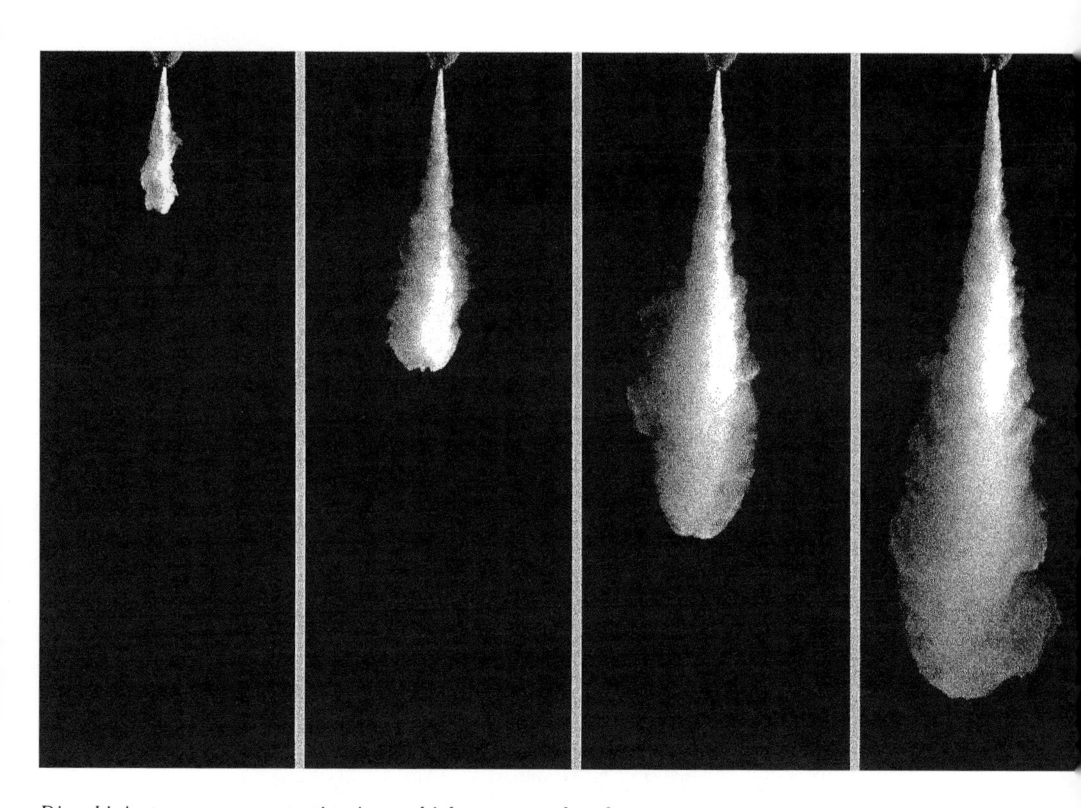

Diesel-injector spray penetration into a high-pressure chamber.

Injector supply pressure = 800 bar.
Fuel = dodecane.

Density of the receiving gas (nitrogen) = 25 kg/m^3.
Gas temperature = 400 K.
Instantaneous images taken from left to right at 0.20, 0.40, 0.70 and 1.02 ms after the start of injection.
(The distance between the injector tip and the base of the image is 68 mm).

(IFP document)

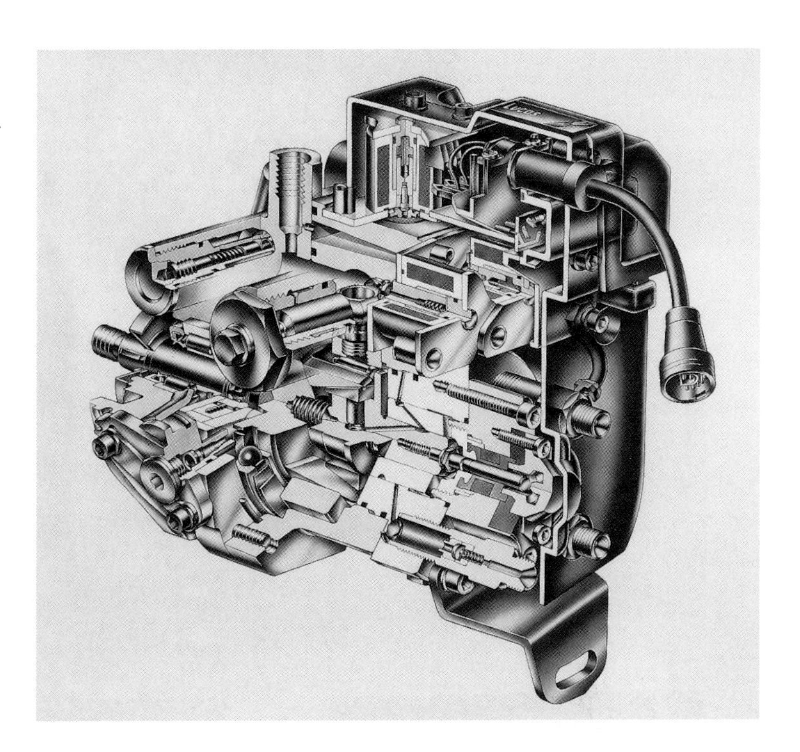

Section view of a Lucas-Epic injection pump. (Lucas Diesel document)

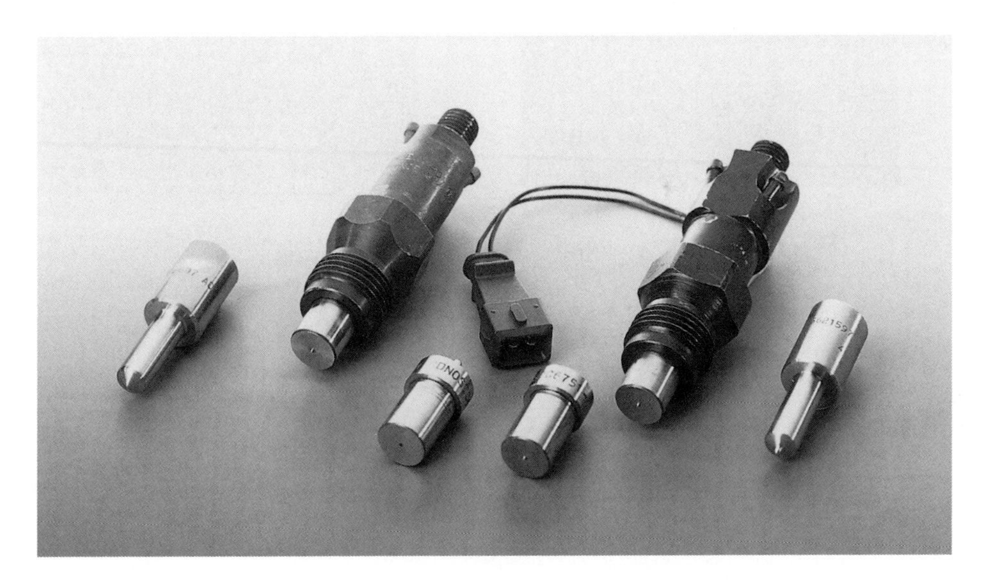

Complete assemblies: injectors and injector bodies. (Lucas Diesel document)

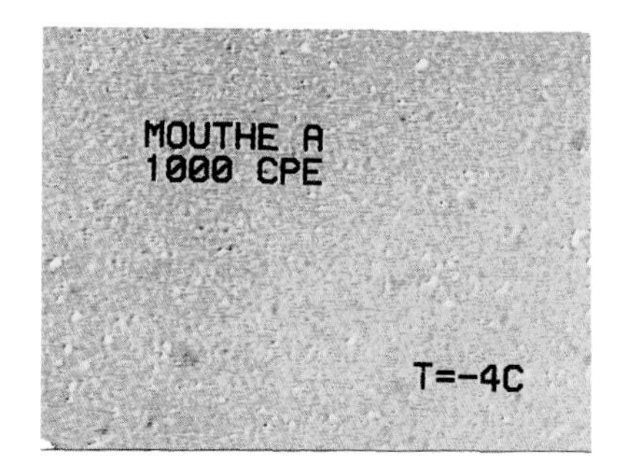

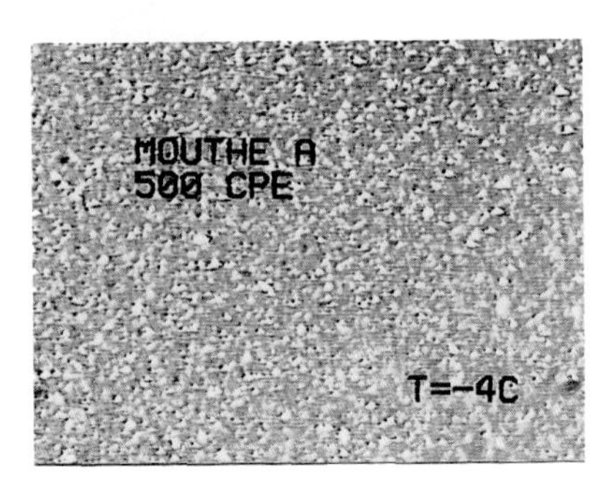

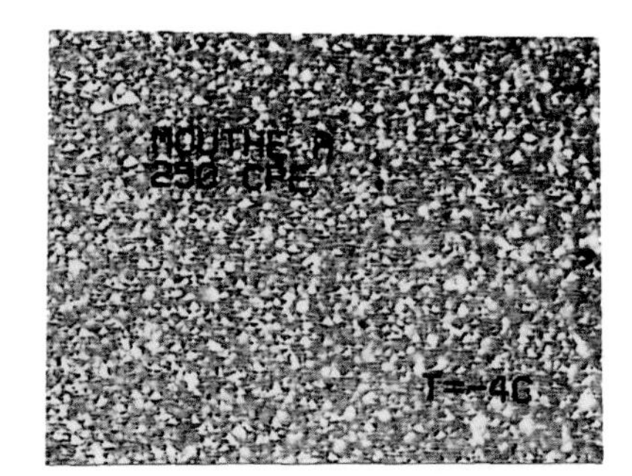

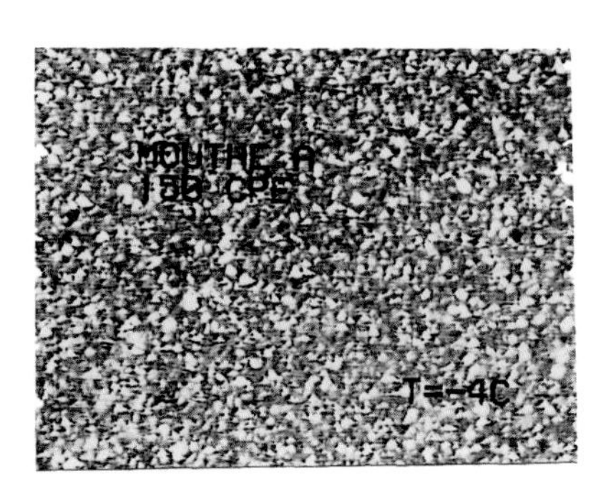

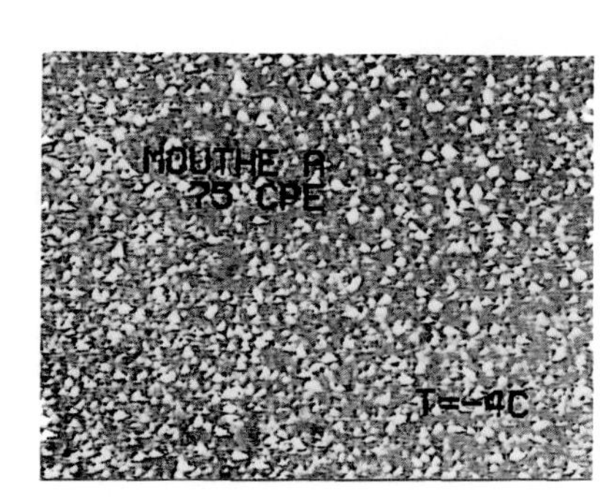

Paraffin crystals in virgin diesel fuel and in fuel containing various levels of flow improvers.

Temperature = –4 °C.

Each plate matches a specific additive dose expressed in mg/kg.

(Elf-Antar France document)

maximum values are 0.745 and 0.765 kg/dm^3 respectively. In Japan, there is only an upper limit of 0.783 kg/dm^3 for both qualities of unleaded gasolines.

Worldwide, more than 80% of the gasoline consumed falls within a density range of 0.735 to 0.765 kg/dm^3. In general, density increases with aromatic content in a such a way that high-octane gasolines usually have a higher density. However, adding oxygenated products like MTBE (ρ_{15} = 0.745 kg/dm^3) results in increased octane ratings without changing density.

A certain gasoline density range is necessary for satisfactory vehicle operation. In fact, automobile manufacturers take this into account when designing fuel inlet systems and when choosing the appropriate flowrates for the various components (nozzles, injectors).

In practice, the lambda sensor equivalence-ratio control system (see Chapter 5) is not affected by variations in fuel density. However, for older vehicles equipped with carburetors and for newer vehicles operating temporarily under conditions where the air-fuel ratio is not controlled (for example, cold starting), variations in density can have an effect on the **air-fuel ratio**. Therefore, on a carburetor-equipped vehicle, higher fuel density means a reduction in equivalence ratio, which can cause unstable operation in extreme cases.

In practice, the user wants the highest possible density that is compatible with the specifications because this provides the highest volumetric NHV and the lowest specific fuel consumption in liters per 100 km. It is estimated that an increase in density of 1% results in a consumption reduction of 1%.

The 50 kg/m^3 latitude in gasoline density is an acceptable compromise for the refiner, whereas any tightening of the requirement would be too constraining.

3.5.2 Gasoline volatility

3.5.2.1 Classification methods

Fuel volatility is expressed by one or several characteristics: distillation curve, vapor pressure, and less often, the *V/L* ratio. This section provides a brief description of the measurement methods and the significance of these values.

A. Distillation curve

The distillation curve represents the change in volume of the distilled fraction at atmospheric pressure as a function of temperature, when measured in a standard apparatus (ASTM D 86 and ISO 3405 standards). The method is often called "ASTM distillation." The change in temperature is plotted as a function of the quantity distilled, taking note of

- the **initial point** (IP), which is the temperature at which the first drop of distillates appears
- the **temperatures at various distillation percentages** (5%, 10%, 20%, … 90%, 95%, …)

- the **final point (FP) of distillation**, which is the temperature at which the last drop of distillate occurs
- the level of losses (in%) and eventually the level of residue

Fig 3.41 shows an example of a distillation curve for a gasoline along with the boiling points of some typical constituents. The initial point is usually about 35°C, the final point is about 180°C, and about 50% of the sample is distilled at 100°C.

Measurement accuracy depends on the area of the curve and the calculated value of the change in temperature per unit volume: $\Delta t/\Delta V$. For example, for a distillation range of 10 to 90% and for a $\Delta t/\Delta V$ value of 1°C/%, the temperature-based repeatability and reproducibility are 1.5 and 3.5°C respectively. The allowable dispersions are slightly more important for the beginning and particularly for the end of distillation. Therefore, differences of 2 or 3°C for the final boiling point of a gasoline often occur in the same laboratory. The determination of distillation curves is now totally **automated**.

To classify the volatility of a gasoline, the variable chosen is not the percentage distilled, but the percentage **evaporated**. The latter value includes, as a consequence, the losses related to the most volatile constituents.

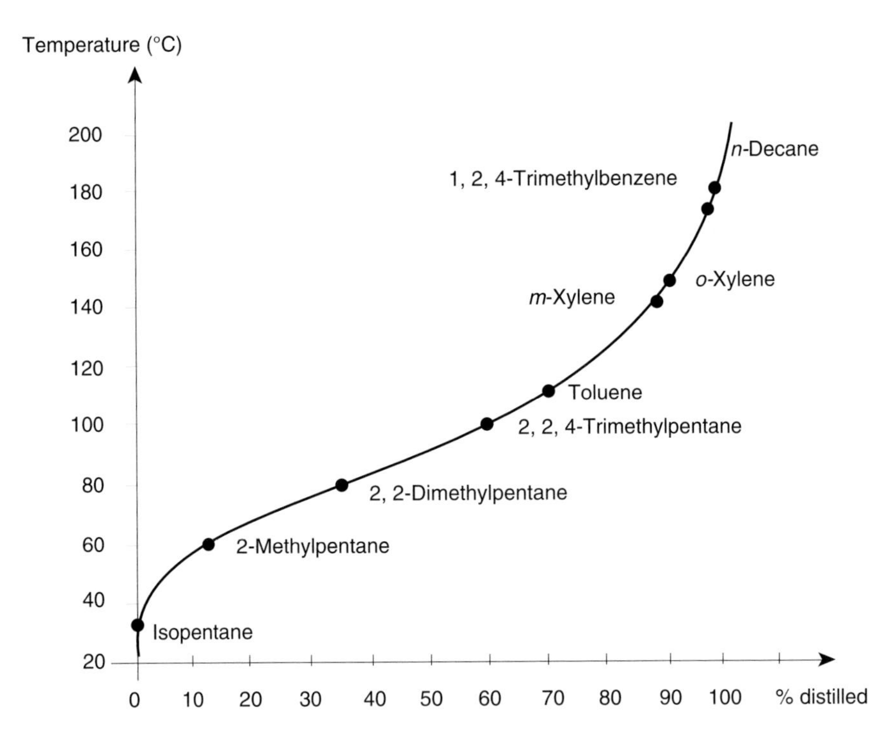

Figure 3.41 *Distillation curve for a gasoline showing the boiling points of some typical constituents.*

Evaporation criteria can be expressed as

- the temperature corresponding to a certain evaporation percentage; for example, T10 represents the temperature for 10% evaporation
- the evaporated percentage at a given temperature; in this case, E70 would be the percentage evaporated at 70°C

Specifications are generally established based on the second criteria (E70, E100, ...) because it enables the establishment of simpler mixing rules for the formulation of gasolines. To classify the end of the distillation curve, points T90 and T95 (temperatures that correspond to 90 and 95% evaporation) are preferred over the final point (FP), whose experimental determination is rather imprecise.

B. Vapor pressure

The vapor pressure of a complex mixture, at a given temperature, is the pressure at which **liquid-vapor equilibrium** is established. Volatility is directly related to vapor pressure.

There are three major methods of determining the vapor pressure of gasolines:

a. Reid method

Both European (EN 12) and American (ASTM D 323) procedures consist of determining the pressure developed by the vapors emitted from a sample of fuel contained in a standard metallic flask at a temperature of 37.8°C (100°F). The vapors are collected in another metallic reservoir, which is four times the volume of the first one. This volume is referred to as the air chamber.

Before the test, the air chamber is immersed in water and drained, but not dried. It is then connected to the chamber containing the fuel, which has been chilled to 0°C and saturated with air. When the test begins, the assembly is placed in a bath at 37.8°C and agitated until equilibrium is reached; the manometer reading at this point represents the pressure in the air chamber.

The new European standard applies to the group of low-viscosity products (crude oils included) with vapor pressures of less than 180 kPa (1.8 bar). It is also applied to fuels containing oxygenated products with precise content requirements (3% maximum for methanol, 5% for ethanol, 15% for MTBE, ...).

The American standard is more restrictive with respect to oxygenated products, because it cannot be applied to fuels containing MTBE.

These limitations result from the fact that some oxygenated molecules (especially methanol and ethanol) are water soluble. If these compounds are present in a fuel, the vapor pressure reading on the manometer is slightly lower than the real value. This is why the old "dry bomb" method was used, where the air chamber was dried before the test.

The result from this test is referred to as the **Reid Vapor Pressure (RVP).** This is a **relative** pressure, which means a pressure difference based on atmospheric pressure. The RVP of gasolines is generally within 350 to 1100 mbar. Typical dispersions for the E12 method are 21 mbar for repeatability and 49 mbar for reproducibility.

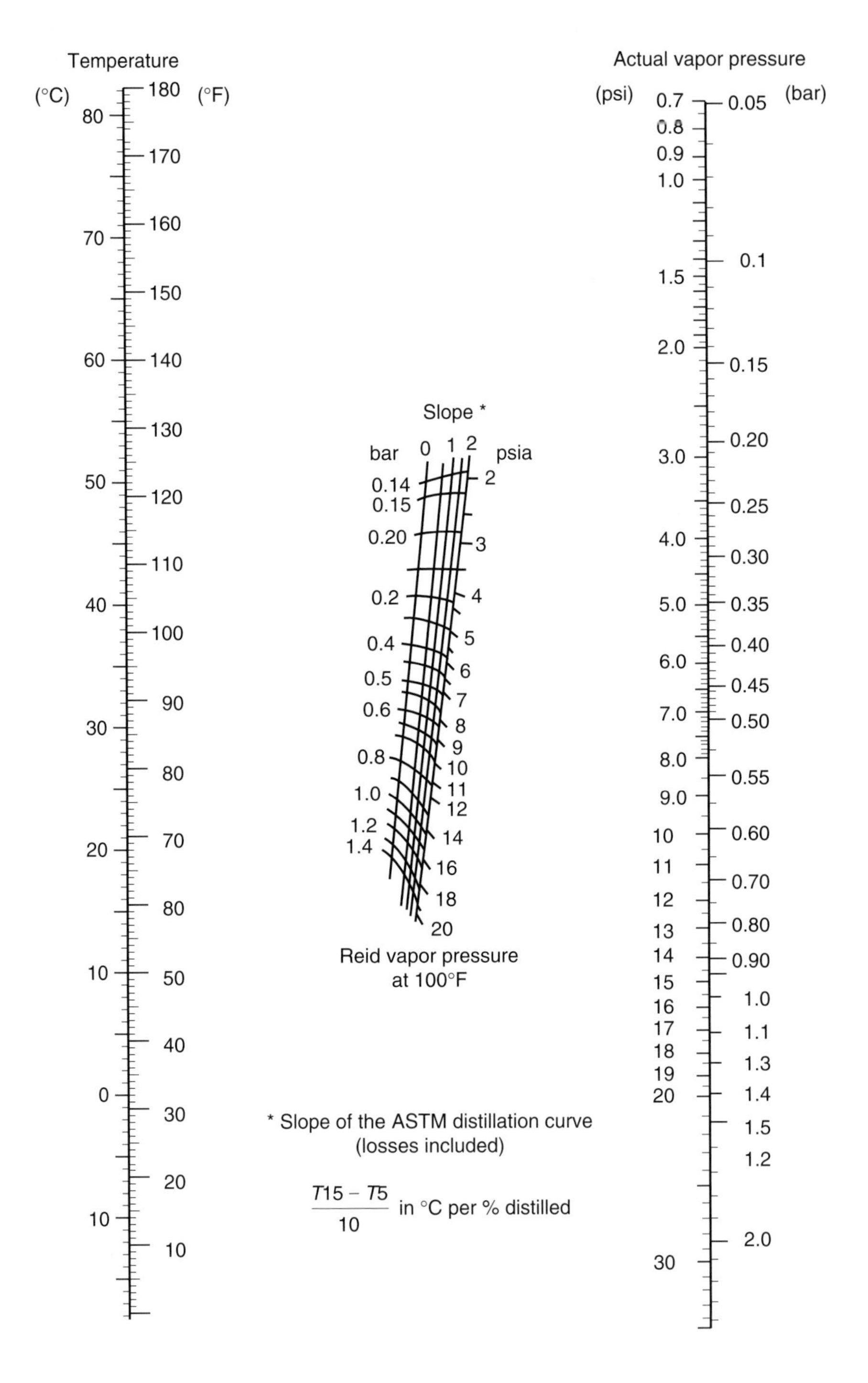

Figure 3.42 *Correlation of Reid vapor pressure (RVP) to actual vapor pressure (AVP).*

The RVP does not provide a true measure of a fuel's vapor pressure, because it includes the contribution of the gases dissolved in the fuel and released as the temperature increases from 0°C to 37.8°C. Fig. 3.42 is a nomograph that provides the actual vapor pressure (AVP) of a fuel at various temperatures based on the RVP and information about the first part of the distillation curve.

b. Grabner method

This recently developed and fully-automated technique (NF M 07-079 and ASTM D 5191) has several advantages: it only uses a very-small fuel sample (1 cm^3 instead of the 150 cm^3 used by the RVP); it provides a **direct reading** of vapor pressure at different temperatures; the test equipment is dry, allowing the test to be used for oxygenated fuels.

The Grabner method collects fuel vapors in a vacuum. The air dissolved in the sample remains an artifact, as in the RVP technique. The precision of this technique is acceptable with dispersions of 11 mbar for repeatability and 19 mbar for reproducibility.

Regardless of the technique chosen to measure vapor pressure (RVP or Grabner), samples must be collected according to precise criteria (ISO standard 3170) to ensure that lighter constituents are not lost to evaporation.

c. Vapor-liquid ratio

The vapor-liquid (*V/L*) ratio is a volatility measure that is not widely used in Europe, but it is popular in Japan and in the United States, where it has been standardized (ASTM D 2533). At a fixed temperature and pressure, the *V/L* ratio represents the **volume of vapor** formed for a given volume of liquid initially drawn at 0°C.

Therefore, the volatility of fuels is expressed as the temperature at given *V/L* ratios, for example *V/L* = 12, 20, or 36. There are correlations between the temperatures that correspond to the *V/L* ratios and standard volatility measurements (RVP, distillation curve). For example, the following relationships exist:

$$T_{(V/L)12} = 88.5 - 0.19 \text{ E70} - 42.5 \text{ RVP}$$
$$T_{(V/L)20} = 90.6 - 0.25 \text{ E70} - 39.2 \text{ RVP}$$
$$T_{(V/L)36} = 94.7 - 0.36 \text{ E70} - 32.3 \text{ RVP}$$

where

$T_{(V/L)x}$ = temperature (°C) at which $V/L = x$
E70 = percent evaporated at 70°C
RVP = Reid vapor pressure in bar

3.5.2.2 Vehicle requirements

If gasoline volatility were either too high or too low, it would not cause complete engine breakdown or a serious incident, except in extreme cases. However, inappropriate volatility can **deteriorate** vehicle performance in relatively disconcerting ways. Several operating conditions can be affected: starting and warm-up time at both hot and cold temperatures, idle stability, acceleration performance, and operation at cruising speeds.

This section describes vehicle behavior in relation to gasoline volatility at both warm and cold temperatures. The section also examines the effect that the heavy fractions in fuels have on engine life.

A. Cold temperature behavior

Vehicle operating problems at low temperatures (for example, below 0°C) are usually linked to incomplete fuel vaporization and an uncontrolled reduction in equivalence ratio.

Considerable progress has ensued in this area with the replacement of carburetors by single and multipoint fuel-injection systems. However, for any given technology, a vehicle's performance can vary widely depending on the volatility of the fuel used.

Several cold-temperature evaluation procedures have been developed by the Coordinating European Council (CEC) in Europe (Falk 1983) and by the Coordinating Research Council (CRC) in the United States (Benson 1971).

The European procedure consists of a series of 12 identical cycles with a duration of 104 s each, which are described in Fig. 3.43. Disruptions and operating problems are recorded during each cycle:

- starting time
- stalling on departure, at idle, or under high load (very unusual)
- the refusal or inability of the vehicle to attain or maintain a given speed
- abrupt hesitations during acceleration or at cruising speed

The procedure used in the United States involves a complex series of idles, accelerations, and decelerations (CRC 1978).

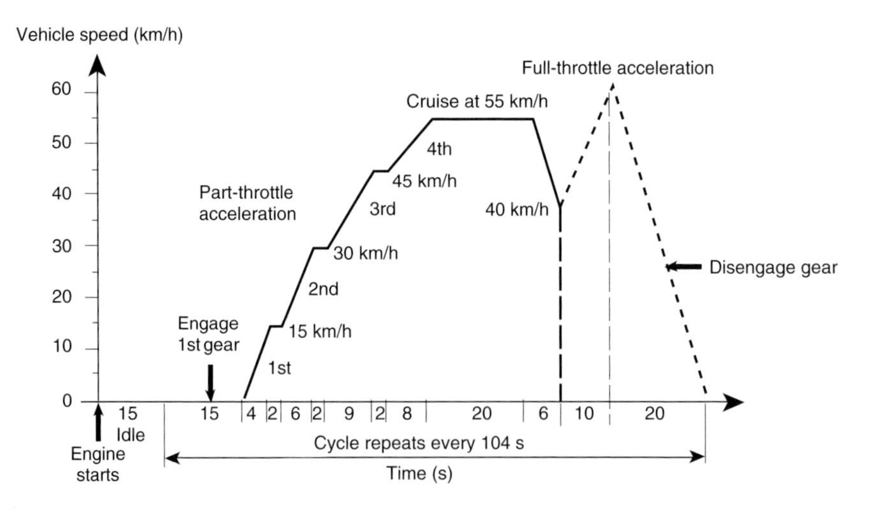

Figure 3.43 *The effect of fuel volatility on cold-temperature operation. (Cycle used in the GFC-CEC procedure)*

Regardless of the operating method chosen, defaults are evaluated and rated according to a pre-established scale. The combination of ratings leads to an overall **demerit** rating.

The fuel volatility parameters that have a significant effect on cold-temperature performance have been isolated by testing on fleets of vehicles.

In the United States (Barker 1988), the following Driveability Index (DI) has been established:

$$DI = 1.5\,T10 + 3\,T50 + T90$$

where T10, T50, and T90 are the temperatures corresponding to 10, 50, and 90% evaporation.

The following relationship has been proposed in Europe (Le Breton 1984),

$$\text{demerit} = a - RVP - 22\,E100$$

where RVP and E100 are the Reid vapor pressure and the percentage evaporated at 100°C respectively; a is a positive constant.

As expected, all these studies have confirmed that satisfactory cold-temperature performance depends on a **minimum fuel volatility**, of which the RVP and the E100 or T50 are the most significant parameters.

The test procedures described above correlate well with users impressions in actual service (Pearson 1985), which ensures the validity of these procedures.

In the past, **icing** was a phenomenon that often disrupted vehicle operations; icing occurs when **ice** forms inside the carburetor body and the fuel system (Owen et al. 1989) during low temperatures (0 to +4°C) and high humidity (hygrometer readings over 80%). Icing was caused by fuels that were too volatile and their rapid evaporation chilled the carburetor. Special additives such as isopropyl alcohol usually had a positive effect on icing.

B. Hot weather behavior

At high ambient temperatures, the disruptive phenomena most likely to affect vehicle fuel systems are mainly vapor lock and percolation.

Vapor lock is the occurrence of pockets of vaporized fuel between the fuel tank and the fueling device (carburetor or injection system). Under these conditions, the fuel pump cannot provide sufficient flow and the mixture equivalence ratio is reduced; in some situations, the engine stalls.

These risks can be avoided by making appropriate modifications: locating the fuel pump in the fuel tank, increasing fuel pressure, running fuel hoses through low-temperature areas, ... However, fuel volatility is still the predominant characteristic in any case.

Percolation is also the result of unwanted and uncontrolled vaporization of fuel, but it happens when the vehicle is idling or has been stopped, after operating for a long time at high temperatures (parking, toll booths). Once the cooling effect of the under-the-hood airflow provided by the vehicle's motion ceases, the underhood temperature rises quickly and the fuel vaporizes before it can be injected. The engine then becomes difficult or impossible to start, or instability occurs when idling, acceleration, or cruising.

The increased use of **fuel injection** is helping to solve vapor lock and percolation problems.

On the other hand, modern vehicles are becoming more intrinsically sensitive than their predecessors for the following reasons:

- **transverse mounted engines**, which have been chosen to increase passenger space, leave little room in the engine compartment and contribute to increased under-hood temperatures
- **intake-system heating,** which often results from the close proximity of the catalytic converter
- **high-speed operation**, which causes additional kinetic heating of the area around the engine

Several procedures to rate the hot-temperature performance of vehicles have been developed in both the Unites States (CRC 1976) and Europe (GFC 1995). To illustrate, Fig. 3.44 shows the driving cycle used in Europe; it consists of a series of high-load operating phases, interspersed with periods of idling or stopping.

The driving defaults and anomalies are scored in the same way as the cold-driving cycle described previously.

Tests on numerous vehicles, done in the United States as well as in Europe, show a clear relationship between demerits and the following two fuel-volatility parameters: RVP and E70.

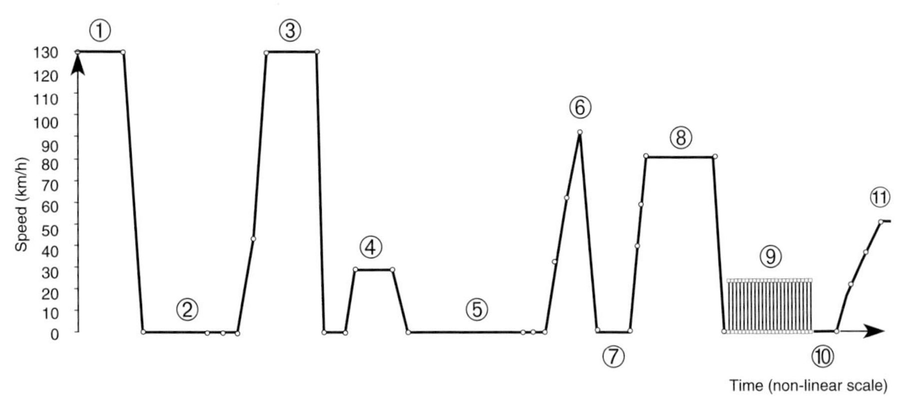

1 and 3: Temperature stabilization
2: Engine is stopped for 15 min, followed by starting and idling
4: Ascent of a hill with 10% slope at a constant speed of 40 km/h for a period of 7 min
5: Engine is stopped for 30 min, followed by starting and idling
6: Full-throttle acceleration to 4500 rpm through the first three transmission speeds (5% slope)
7: Engine is stopped; fresh fuel is added
8: Cruise at 80 km/h for 15 min in the highest gear (canister purge)
9: Repeat 20 “stop and go” cycles consisting of:

- 40 s at idle
- 20 s in first gear at 20 km/h

10: 5 min of idle
11: Cruise at 50 km/h for 5 min in third gear

Figure 3.44 *European driving cycle used to rate hot-temperature performance.*

Thus, an expression of the following form

$$\text{Demerit} = \text{RVP} + n\ \text{E70}$$

(where n is a whole number constant) accounts for the behavior of an automobile fleet.

The Fuel Volatility Index (FVI), which is used in Europe, is defined by the following formula,

$$\text{FVI} = \text{RVP} + 7\ \text{E70}$$

where

RVP is in mbar

E70 is the % evaporated at 70°C

The value of FVI is from 800 to 1200.

Vehicles now require more from fuels at hot temperatures than they do at cold temperatures, to such an extent that automobile manufacturers are hoping for a **reduction** in RVP instead of an increase.

In addition to these technological constraints, there is an increasing demand to limit **evaporative loses**, which are an important source of atmospheric pollution from gasoline engines (McArragher et al. 1990). This question is discussed in detail in Chapter 5.

C. Engine durability and longevity

Engine durability and longevity are usually not highly dependent on fuel volatility, but they can be affected if the fuel contains, for example, **heavy fractions** that distill at 200 to 220°C.

Improperly vaporized fuel constituents are deposited on the internal walls of the engine where they undergo incomplete transformation during combustion. This results in several **problems**: losses in combustion efficiency, higher emissions of unburned hydrocarbons at the exhaust, lubricant dilution caused by the fuel passing between the cylinder walls and the pistons, component wear caused by the presence of the acidic by-products of incomplete combustion, ... Engine deterioration of this type only occurs after long operating periods (several thousand kilometers) and it is accentuated by severe-service conditions such as door-to-door service, deliveries, short trips in winter, ...

The preceding conditions explain the need to limit the **final distillation point** of gasolines. Chapter 5 indicates that a similar control is desirable to reduce the rate and toxicity of exhaust emissions.

3.5.2.3 Gasoline volatility specifications

Table 3.7 lists the volatility characteristics required for "Eurosuper" unleaded fuel sold in various European countries. There are eight possible classes, which are defined by the following criteria: RVP, FVI, E70, E100, E180, and FBP. Each country chooses a class of fuel for each season, depending on their particular climatic conditions. Thus, fuel volatility characteristics can change **four times per year** in the same country, at the beginning of each season. Note that the characteristics specified for spring and fall are the same.

Table 3.7 *Volatility specifications for European unleaded gasoline. (Standard EN 228)*

Characteristics	Limiting values by class							
	1*	2	3*	4	5	6*	7	8
Minimum vapor pressure (bar)	0.35	0.35	0.45	0.45	0.55	0.55	0.60	0.65
Maximum Reid vapor pressure (bar)	0.70	0.70	0.80	0.80	0.90	0.90	0.95	1.00
E70 (%) min	15	15	15	15	15	15	15	20
max	45	45	45	45	47	47	47	50
Maximum FVI**	900	950	1 000	1 050	1 100	1 150	1 200	1 250
E100 (%) min	40	40	40	40	43	43	43	43
max	65	65	65	65	70	70	70	70
E180 (%) min	85	85	85	85	85	85	85	85
Maximum FBP (°C)	215	215	215	215	215	215	215	215
Maximum residue (vol %)		2	2	2	2	2	2	2

* Classes used in France.
** FVI = RVP (mbar) + 7 E70.

For example, France has chosen to use classes 1, 3, and 6 according to the following calendar:

- Class 1 from June 20 to September 9
- Class 3 from April 10 to June 19
- Class 3 from September 10 to October 31
- Class 6 from November 1 to April 9

In all European countries, the heavy-fraction specifications are identical: E180 must be greater than 85% and the final boiling point must not exceed 215°C.

In the United States as well, gasoline volatility must meet specific requirements defined by region and time of year. Thus, ASTM standard D 4814 lists a series of specifications of varying severity for each federal state, depending on the season. The requirements essentially consist of a Reid vapor pressure and a test temperature delivering a *V/L* ratio of 20. Table 3.8 lists the two series of volatility classes defined by these characteristics. Choosing classes provides a definition of the requirements.

The Environmental Protection Agency (EPA) has also promulgated a ruling. It is more restrictive than ASTM D 4814 and it can be applied in any state that requests it. The ruling only requires a maximum Reid vapor pressure limit, whose value varies depending on the geographic location of the state being considered (see Fig. 3.45).

One last observation concerns reformulated gasoline, which is controlled by very-specific regulations that are described in Chapter 5 (see 5.12.1).

Table 3.8 *Volatility specifications for American gasolines based in Reid vapor pressure and V/L ratio.*
(Based on ASTM D 4814)

Characteristics	Limiting values by class					
	AA	A	B	C	D	E
Maximum RVP* (bar)	0.54	0.62	0.69	0.79	0.93	1.03
T10 (°C) max	70	70	65	60	55	50
T50 (°C) min	77	77	77	77	77	77
max	121	121	118	116	113	110
T90 (°C) max	190	190	190	185	185	185
FBP (°C)	225	225	225	225	225	225
Maximum residue (vol %)	2	2	2	2	2	2

* Use ASTM D 4953, ASTM D 5190 or ASTM D 5191 standards if the fuel contains oxygenated compounds.

Characteristics	Limiting values by class				
	1	2	3	4	5
Test temperature at which a maximum value of 20 is attained for the *V/L* ratio** (°C)	60	56	51	47	41

** Changes can be applied to ASTM D 2533 if the fuel contains oxygenated compounds.

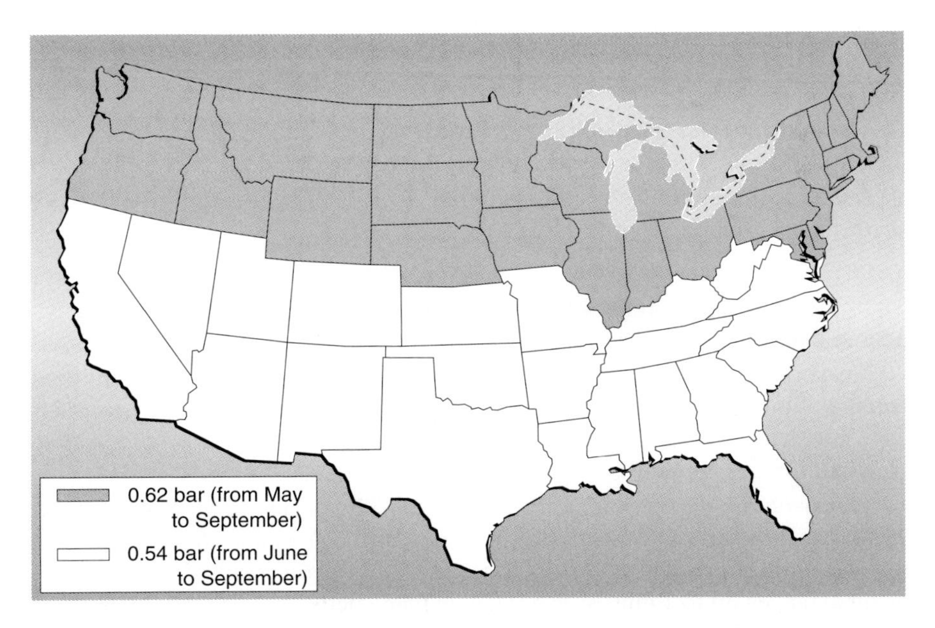

Figure 3.45 *Maximum Reid vapor specifications for American gasolines, stipulated by the EPA since May 1992.*
(A waiver of 70 mbar is allowed for gasolines containing ethanol.)

Almost everywhere in the world, the current trend is toward **reduced vapor pressure and volatility indexes (FVI)** for gasolines. This has been confirmed by the auto manufacturers in France who have indicated in their *Cahier des Charges Qualité* (quality requirements) that the maximum summer values for E70 and FVI should be 40 and 850 respectively, instead of the values of 45 and 900 listed in the official specifications.

The requirement to control gasoline volatility at the formulation stage is a major constraint because it implies a limit on the addition of light fractions. Each **1%** by mass of a C_4 fraction (butane, butenes) results in an average increase of about **50 mbar** in RVP. The addition of some alcohols (methanol, ethanol) also cause a noticeable increase in RVP (about 150 mbar with the addition of about 1% methanol; about 50 mbar for same quantity of ethanol).

Therefore, to meet volatility specifications, alcohol fuels must have a restricted light-hydrocarbon content.

3.5.3 Other physical characteristics of gasoline

Other gasoline characteristics have less effect on vehicle operation than the ones described previously. This section discusses viscosity and heat of vaporization.

3.5.3.1 Viscosity

Viscosity is the resistance that molecules exhibit when a force tries to move some molecules with respect to the others. If, within a fluid, an element with surface S slides at speed $\mathrm{d}V$ in opposition to force F over the same surface area, the **absolute dynamic viscosity** is expressed as

$$\mu = \frac{F}{S}\frac{\mathrm{d}x}{\mathrm{d}V}$$

The dynamic viscosity unit is the **Pascal-second**. The millipascal-second is the sub-multiple, which is also known as the centipoise.

Absolute kinematic viscosity is often used, which accounts for density according to the following formula,

$$\nu = \frac{\mu}{\rho}$$

where ν is in m^2/s, or preferably mm^2/s for normal liquids and petroleum products in particular.[1]

Measurements are taken using a viscometer (ISO standard 3104) and they consist of measuring the time it takes for a liquid to flow between two reference marks on a capillary tube. A previous calibration is used to provide a time constant for the apparatus.

The viscosity of petroleum products is usually measured at 20 or 50°C; at 20°C, viscosity varies from less than 1 mm^2/s for gasolines to 500 mm^2/s or more for heavy products. Results are given with a precision of two or three decimal places up to 1 mm^2/s, one decimal place between 1 and 4.5 mm^2/s, and whole numbers above 50 mm^2/s.

1. The old term "centistoke" is still used sometimes (1 centistoke = 1 mm^2/s).

Viscosity has an overriding effect on the use of medium – and heavy-weight petroleum products, and it has a noticeable effect on the operation of diesel engines and turbine engines at very-low temperatures. These questions are explored in more detail elsewhere (see 4.4.3 and 7.3.1). Fuel viscosity does not play an important role in the operation of the spark-ignition engine. As the values listed in Table 3.9 show, viscosity remains low, at less than 1 mm^2/s until – 20°C, and it does not much depend on the chemical composition of the gasoline. Aromatic hydrocarbons have generally higher viscosity than paraffins and olefins, but the differences do not have significant effects on engine intake and fuel systems, or on combustion.

However, it is possible that fuel viscosity might become important at **very-low temperatures** by changing the flow and the atomization in various parts of the intake system. This intrinsic effect is probably masked by other parameters and it is not generally accounted for.

Table 3.9 *Absolute kinematic viscosity of light hydrocarbons and fuels. (values are provided for illustrative purposes)*

Product	Temperature (°C)	Viscosity (mm^2/s)
Hexane	20	0.49
Heptane	20	0.60
Isooctane	20	0.73
Benzene	20	0.74
Toluene	20	0.68
Gasoline 1	20	0.55
Gasoline 2	20	0.58
Gasoline 3	20	0.61
Gasoline 3	0	0.76
Gasoline 3	– 20	0.97

3.5.3.2 Heat of vaporization

Heat of vaporization is the heat required to convert one mole of liquid to a vapor. It is usually referred to by the symbol L and it is expressed in J/mol. The value of L depends on the boiling temperature and the slope ($d\pi/dT$) of the change-of-state curve. The value is calculated from the **Clapeyron** formula,

$$L = T(V_2 - V_1)\frac{d\pi}{dT} \qquad (1)$$

where

T = absolute temperature

V_2 and V_1 = molar volumes of the vapor and liquid phases respectively (m^3/mol)

π = vapor pressure (Pa)

Note that L decreases as the temperature rises and reaches zero at the critical point where $V_2 = V_1$, whereas $d\pi/dT$ maintains a finite value.

Within the normal temperature range for changes of state, V_1 is negligible compared to V_2. If the vapor phase is related to a perfect gas, equation (1') leads to the following expression,

$$\ln \frac{\pi_2}{\pi_1} = \frac{L}{R}\left(\frac{1}{T_1} - \frac{1}{T_2}\right) \quad (2)$$

which supposes that the heat of vaporization stays reasonably constant within the narrow temperature range of $T_2 - T_1$ that corresponds to the vapor pressures π_2 and π_1.

Equation (2)—where R is the universal gas constant—is often used to estimate the change in vapor pressure with temperature, once the order of magnitude of L is known.

The following calculation provides a numerical example:

The vapor pressure of *n*-pentane is $\pi_1 = 533$ mbar at 18.5°C. At that temperature the heat of vaporization is 25 725 J/mol. The known value of R is 8.31 J/mol · K. Therefore, the calculation of the vapor pressure π_2 is given by the following calculation:

$$\ln \frac{\pi_2}{\pi_1} = \frac{25\,725}{8.31}\left(\frac{1}{291.5} - \frac{1}{301.5}\right) = 0.3522$$

$$\frac{\pi_2}{\pi_1} = 1.422$$

$$\pi_2 = 758 \text{ mbar at } 28.5°\text{C}$$

The data in Table 3.10 show that most of the conventional hydrocarbons that constitute gasoline have mass heats of vaporization of about 300 kJ/kg. Among the other types of fuels, oxygenated organic compounds and particularly light alcohols (methanol, ethanol) are differentiated by their highly **endothermic** vaporization. This peculiarity is both an obstacle and an attraction to using alcohols as fuels. In effect, obtaining an atomized mixture is difficult, but it causes a cooling of the intake charge that leads to better cylinder filling and more power. This question is subsequently explored in more depth (see 6.10). Within the domain of conventional fuels, small variations in the heat of vaporization between fuels are imperceptibly small and they have no effect on fuel systems or combustion processes.

3.6 Octane rating

The octane rating is the characteristic that describes a fuel's resistance to autoignition in an experimental laboratory engine, which has been specifically designed and installed for that purpose. The measurement is **comparative** and the fuel's performance is expressed as a **dimensionless number** between 0 and 100, with the possibility of extrapolating the number up to 120.

Table 3.10 *Heat of vaporization of some organic compounds.*

Product	Heat of vaporisation (kJ/kg)	Product	Heat of vaporisation (kJ/kg)
Methane	510	Benzene	394
Ethane	489	Toluene	363
Propane	426	*o*-Xylene	347
Butane	385	Cumene	312
n-Pentane	357	Methanol	1 100
Isopentane	339	Ethanol	855
n-Hexane	337	Propanol	695
n-Heptane	320	Isopropanol	667
n-Octane	306	*n*-Butanol	591
Ethylene	483	Acetaldehyde	570
Propylene	438	Acetone	521
But-1-ene	391	Dimethylether	377
Isobutene	394	Gasoline (Eurosuper)	335
Cyclopentane	390	Water	2 256
Cyclohexane	358		
Acetylene	829	Ammonia	1 391

The reference scale is based on two hydrocarbons, one that is susceptible to knock (see 3.4.3.2) and another one that is highly resistant to it. The octane rating reflects the fuel's behavior under the particular thermodynamic conditions of the test engine. It is therefore desirable to correlate this precise but unrealistic measure with the real knock tendency of the fuel in an actual vehicle, which varies with the type of engine, the type of control system, the operation conditions, ...

Following a review of the various historical steps that led to the definition of octane rating, this section describes the test equipment, the test conditions, and the recommended operating procedures. The methods used to determine the knock resistance of fuels in production engines and vehicles is also described. A collection of data that relates the knock resistance of pure hydrocarbons to conventional fuels is also provided. The section concludes by describing why, and under what conditions, additives like lead alkyls and other organometallics have been used for a long time to improve the octane rating of fuels.

3.6.1 Some historical notes

During the first world war, the need to establish gasoline quality standards became very obvious to avoid destructive results with high-performance aircraft engines. At that time, two grades of product differentiated by their volatility characteristics were already available.

In the 1920s, knock rose to great practical importance and several attempts were made to determine the auto-ignition resistance of fuels. A first attempt was proposed by Ricardo in 1921; this attempt consisted of using a characteristic value for each hydrocarbon or fuel, which was based on the maximum possible compression ratio without the occurrence of knock. This method was known as Highest Useful Compression Ratio (HUCR) and it was found to be imprecise and difficult to use.

Around 1928, the Cooperative Fuel Research Committee (CFR) in the United States established a working group that was charged with the task of developing a standard procedure for classifying fuels. The committee sought the help of the Waukesha Engine Co. and requested that they design, develop, and produce an experimental engine that was especially adapted to determine knock resistance. This engine, known as the CFR engine, is currently in use throughout the world and it still produced by the same company.

At that time, there was also an agreement that knock resistance would not be expressed as an absolute number, but as a comparative rating with respect to two reference hydrocarbons with very distinct behaviors. Graham Edgar of the Ethyl Gasoline Corporation proposed the adoption of the following reference products in 1931:

- *n*-heptane
- 2,2,4-trimethylpentane, which is more commonly known as **isooctane**

The first of these products is very susceptible to knock and the second one is very resistant to knock. Edgar's scale enabled the introduction of the definition of octane rating, which is still in use today.

A first draft test method appeared in 1931; it led to a characteristic like the present "Research" octane number. However, the members of the CFR committee wanted to develop an experimental procedure that was closer to the conditions observed in vehicles and proceeded to develop a parallel procedure that became the basis for the "Motor" octane number.

Since the beginning, gasoline specifications in different countries have used either the Research Octane Number (RON) or the Motor Octane Number (MON). Currently, both ratings are taken into consideration as separate entities (Europe, Japan) or jointly as an average [(RON + MON)/2], as in the United States.

From about 1950 to 1970, there was a rapid increase in octane levels worldwide. The RON for premium gasoline was an average of 80-85 in 1950, 85-90 in 1955, 95 in 1960, and 99 in 1970. Since then, the reduction in the use of additives containing lead alkyls has contributed to a stabilization of gasoline RON at about 95.

As a final note, the first test methods for the measurement of "road" octane numbers were proposed in 1946 in the United States. The variants of this method still carry the name Uniontown as a reminder of the city in Virginia where they were developed.

3.6.2 Measurement techniques

3.6.2.1 General methodology

By established convention, the reference hydrocarbons *n*-heptane and isooctane are assigned octane numbers of 0 and 100 respectively. When a fuel is tested under tightly-controlled experimental conditions in a CFR engine, that fuel is given an octane number of X when its knock behavior is the same as a blend of $X\%$ by volume of isooctane and $(100-X)\%$ of *n*-heptane. The principle of the methodology is to raise the compression ratio until **knock occurs** with the test fuel and assign this point the value ε_0; this test fuel value is then bracketed by the values ε_1 and ε_2 for two neighboring heptane-isooctane blends. The octane number for the test fuel is determined by using linear interpolation to interpret the reference blend that provides the same behavior as the test fuel.

When road octane numbers are measured on production engines, knock is induced by advancing ignition timing; the advance point at which knock occurs is compared to a calibration curve based on *n*-heptane and isooctane blends.

3.6.2.2 Reference fuels

Reference fuels are blends of *n*-heptane and isooctane that meet the strict quality requirements detailed in Table 3.11. These blends are referred to as **Primary Reference Fuels (PRF)**. This acronym is used frequently and is often followed by the octane number. For example, a PRF 95 is by definition a blend of 95% (vol) isooctane and 5% *n*-heptane.

Questions have often been asked as to the reasons why the early researchers chose a C_7 hydrocarbon and a C_8 hydrocarbon as reference fuels, when it would appear to have been more logical to have chosen two compounds with the same number of carbon atoms per molecule. The answer can be found in Table 3.11; *n*-heptane and isooctane have almost identical boiling points, which permits the exclusion of physical parameters such as evaporation from their otherwise very-different behavior in engines.

To extend Edgar's scale beyond 100, small quantities of tetraethyl lead (TEL) are added to isooctane. The relationship between the octane numbers "beyond 100" and the TEL content is established by extrapolating one curve and comparing it to another curve; the critical compression ratio versus CFR-engine octane-number curve is extrapolated and then it is compared to the compression ratio versus lead-content (in isooctane) curve.

The formula is

$$\text{Octane number} = 100 + \frac{28.28\,T}{1 + 0.736\,T + (1.0 + 1.472\,T - 0.435\,216\,T^2)^{1/2}}$$

where T is the TEL content in milliliters per gallon (1 gallon = 3.785 liters).

For example, adding 0.40 g Pb/liter to isooctane by this definition yields an octane number of 110.8.

Table 3.11 *Quality standards for reference fuels. (Source: ASTM)*

Characteristics	Isooctane	*n*-Heptane
Minimum purity (%)*	99.75	99.75
Maximum lead (mg/liter)*	0.53	0.53
d_4^{20} **	0.69193	0.68376
Freezing point (°C)**	–107.38	– 90.61
Boiling point (°C)**	99.23	98.42

* ASTM specifications.
** Supplier quality assurance (*Phillips Petroleum*).

3.6.2.3 Relevance and limitations of comparison methods

To determine octane numbers, identical behavior is established between the fuel being tested and the reference fuels. The experimental parameter that is changed to induce knock on a CFR engine is the compression ratio, which involves a combined change of temperature and pressure.

Overall, this pressure-temperature combination can be reproduced on various types of engines with known variations, but without fundamental changes. The classification of fuels established by CFR engines is thus applicable to most current engines. However, when one of the two variables, namely pressure and temperature, are changed in a very-specific way (supercharging, major or minor cooling), the correlation of the classifications can be significantly altered.

Therefore, the usefulness of "Road" octane numbers (see 3.6.5), which are established under actual use conditions, becomes apparent.

In conclusion, it is worth mentioning that for a given fuel, an infinity of possible octane numbers can exist, depending on the engine used and the experimental conditions of the test.

3.6.2.4 CFR engines

The CFR engine is a single-cylinder experimental engine, which is very ruggedly built to resist prolonged knock. CFR engines have the following geometry:

Bore: 82.55 mm; stroke: 114.30 mm; displacement: 611 cm^3

The compression ratio can be adjusted **while running** from a ratio of 4 to 1 to a ratio of 18 to 1 by the vertical movement of the cylinder with respect to the slider-crank assembly. The exact position of the cylinder, and hence the compression ratio, is established using a micrometer. The CFR engine is equipped with a single carburetor fitted with a multiport valve that is linked to several small reservoirs of about 400 cm^3 each, which contain the fuel to be tested and the reference fuels (see Fig. 3.46). Mixture equivalence ratio is varied by changing the level of the fuel in one of the reservoirs. There is no choke

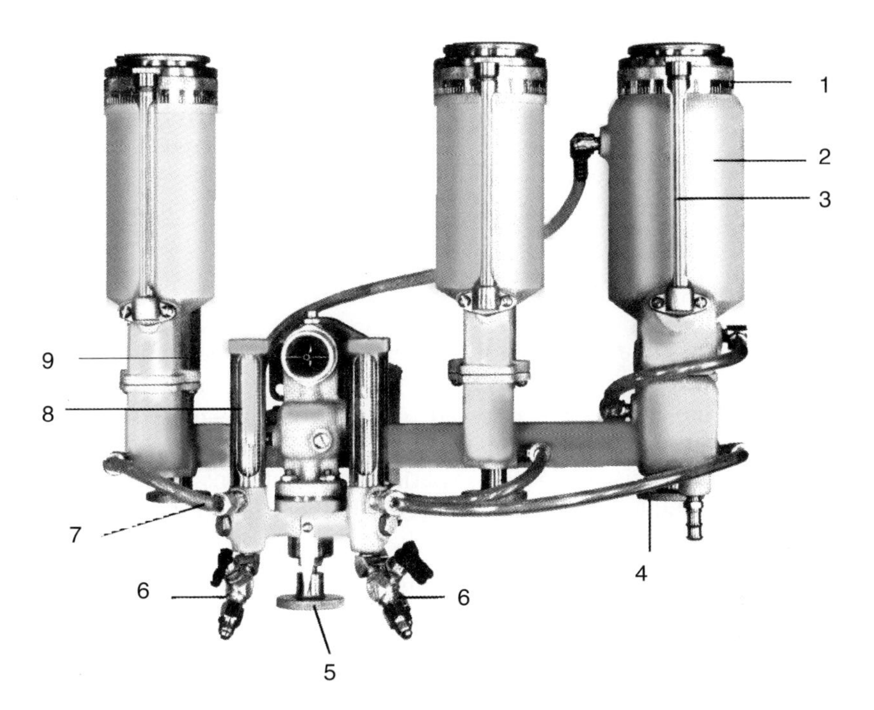

1. Rotateable collar used to indicate the octane number of the reference fuel
2. Carburetor tank
3. Reservoir fuel level indicator
4. Reservoir height control knob
5. Multiport valve
6. Drain valve
7. Connecting lines
8. Fuel level and mixture equivalence ratio indicator
9. Air intake

Figure 3.46 *CFR engine intake and fuel system. (Source: ASTM)*

or throttle plate to control the flow of the intake charge, therefore, the CFR engine runs at **full load**.

The engine is coupled to an asynchronous electric motor that ensures starting, strictly controls engine speed, and returns energy to the power grid.

Knock is detected using a **magnetostriction sensor** that is mounted in the cylinder head. The sensor provides a voltage signal that is proportional to the derivative of the cylinder pressure with respect to time. These impulses are processed and displayed on a Detonation Meter (DM). The DM can be adjusted for threshold (Meter Reading) and sensitivity (Spread) readings.

The output signal from the DM is read from a **knockmeter** that indicates an arbitrary value of knock intensity between 0 and 100.

To obtain reliable results, it is essential to periodically check the mechanical condition of the CFR engine and the knock detection system. This long, delicate, and often fastidious operation is called **calibration**. The calibration must then be checked using standard fuel blends, referred to as calibration blends containing heptane, isooctane, and toluene. This calibration process should be done daily.

The laboratory apparatus described above is a relatively complex and expensive assembly, given the precision required for the engine components and measuring apparatus, and it is supplied exclusively by the Waukesha company or its representatives.

3.6.3 Research and Motor octane numbers

There are two standard octane rating procedures: the Research or F1 (ASTM D 2699) method and the Motor or F2 (ASTM D 2700) method. The corresponding octane ratings are referred to by the acronyms RON (Research Octane Number) and MON (Motor Octane Number), which are used in this text. Each method makes use of specific experimental conditions, whereas the technique used to establish the ratings is precisely the same for both methods.

3.6.3.1 Test conditions for determining RON and MON

The RON and MON test conditions are described in Table 3.12. The major differences are engine speed, intake temperature, and spark advance. Thus, dur-

Table 3.12 *Test conditions for the measurement of RON and MON. (Source: ASTM)*

<table>
<tr><th>Operating parameters</th><th>Research
or F1 method
(RON)</th><th>Motor
or F2 method
(MON)</th></tr>
<tr><td>Engine speed (rpm)</td><td>600 ± 6</td><td>900 ± 9</td></tr>
<tr><td>Spark advance (°CA)</td><td>13</td><td>variable (14 to 26)</td></tr>
<tr><td>Intake air temperature (°C)</td><td>28.3* at 1 bar</td><td>38.0 ± 2.8</td></tr>
<tr><td>Intake charge temperature (°C)</td><td>not stated</td><td>149.0 ± 1.1</td></tr>
<tr><td>Coolant temperature (°C)</td><td colspan="2">100.0 ± 1.5</td></tr>
<tr><td>Oil temperature (°C)</td><td colspan="2">57.0 ± 8.5</td></tr>
<tr><td>Oil pressure (mPa)</td><td colspan="2">0.17 to 0.20</td></tr>
<tr><td>Oil viscosity</td><td colspan="2">SAE 30</td></tr>
<tr><td>Spark plug gap (mm)</td><td colspan="2">0.51 ± 0.13</td></tr>
<tr><td>Ignition breaker point gap (mm)</td><td colspan="2">0.51</td></tr>
<tr><td>Valve adjustment (mm)</td><td colspan="2">0.20 ± 0.03</td></tr>
<tr><td>Humidity (g water/kg of air)</td><td colspan="2">3.6 to 7.1</td></tr>
<tr><td>Venturi diameter (mm)</td><td colspan="2">14.3</td></tr>
<tr><td>Mixture equivalence ratio**</td><td colspan="2">Adjusted to obtain maximum knock intensity</td></tr>
</table>

* Variable as a function of atmospheric pressure.
** The mixture equivalence ratio is usually between 0.7 and 1.7.

ing RON measurement, the engine runs at 600 rpm with a fixed spark advance (13°CA), and without heating of the intake charge. The MON measurement occurs at 900 rpm, with a spark advance that varies (from 14 to 26°CA) along with the compression ratio, and, of particular importance, with an intake charge temperature of **149°C**.

The mixture equivalence ratio is adjusted to obtain maximum knock. The other operating parameters for ignition, cooling, and lubrication are controlled and adjusted in the same way for both methods.

3.6.3.2 Test method

The test method consists of varying the compression ratio to obtain a standard knock intensity for the test sample. This knock intensity is then compared to the knock intensity for two PRFs whose octane numbers bracket the octane number of the sample. Thus, the measurement consist of a first estimate followed by an interpolation. The sequence of operations are described here and they are the same for both RON and MON.

A. The first-estimate octane number

A sample is placed in one of the reservoirs connected to the carburetor and the compression ratio is varied to obtain a knock intensity of 50 on the knockmeter. The mixture equivalence ratio is changed to obtain maximum knock intensity and then the compression ratio is re-adjusted to yield a knock intensity of 50 on the knockmeter. A combination of the micrometer reading and the information in the PRF performance tables provides a first estimate of the octane number, for example, 98. However, this number can be very erroneous because the *Detonation Meter* (DM) has not been calibrated.

B. Choosing references and calibrating the DM

A PRF blend is prepared, which has a RON of one number less than the number derived from the first estimate, that number being 97 for this example. Using this PRF blend, which is referred to as the first reference, the compression ratio is set to the standard indicated in the tables and the equivalence ratio is set to obtain maximum knock. The *Meter Reading* potentiometer on the DM is adjusted to indicate a knockmeter reading of 50. Another PRF blend is prepared, which has a RON two numbers greater than the first reference. This blend is referred to as the second reference.

C. Setting DM sensitivity

The second reference, whose octane number is 99 for this example, is evaluated. The mixture equivalence ratio is readjusted and the knockmeter reading is recorded. The standard requires that the spread between the two references must be between 20 and 36.

For example,
first reference 97 = knockmeter 50
second reference 99 = knockmeter 14 (deviation of 36)

If this condition cannot be met, the "Spread" potentiometer of the DM is adjusted to place the reading within the sensitivity zone. When such an adjustment is required, evidently the first reference must be retested and the knockmeter reset to 50 by adjusting the "Meter Reading" potentiometer.

Another check consists of verifying that the knockmeter reading for the test sample is between the readings obtained for the two references.

D. Bracketing procedure

Using the test fuel, the compression ratio is set to obtain the standard knock intensity corresponding to 50 on the knockmeter (the mixture equivalence ratio having been set during the first test). The engine is then fueled alternately with the first and second references and, without changing any settings, the knockmeter readings are recorded. This process is repeated and the readings, which are evidently very close, are averaged. The octane number is obtained by interpolation as shown in the following example:

Fuel	Readings		Average
1st reference: PRF 97	60	58	59
Sample	50	51	50.5
2nd reference: PRF 99	40	39	39.5

$$\text{Octane number} = 97 + \frac{2 \times 8.5}{19.5} = 97.9$$

E. Validity checking

The compression ratio used during the preceding tests should be checked against the standards for a 97.9 PRF blend. Therefore, the micrometer reading used during the tests is compared to the theoretical standard listed in the tables. The deviation must not be above a given threshold, with the tolerance for MON being slightly higher than RON.

In addition, each time the tables are used, the readings must be corrected for the ambient atmospheric pressure.

This detailed description of the test method shows that measuring octane numbers involves a **series of detailed operations** that can take 30 to 45 min. The volume of the test samples must be 600 to 800 cm^3. However, there is a micro-method that only uses 200 cm^3 of fuel. This test method variant is not approved by ASTM and it does not provide exact readings unless an order-of-magnitude octane number for the test fuel is known ahead of time.

3.6.3.3 Thermodynamic parameters Definition and significance of sensitivity

This section describes the thermodynamic conditions that occur during the operation of the CFR engine when RON and MON measurements are derived. Fig. 3.47 shows that the compression ratios are low, for example 7 to 8 for RONs and MONs between 90 and 100. The mixture equivalence ratio is about 1.00 to 1.10 and its adjustment has a pronounced effect on the results.

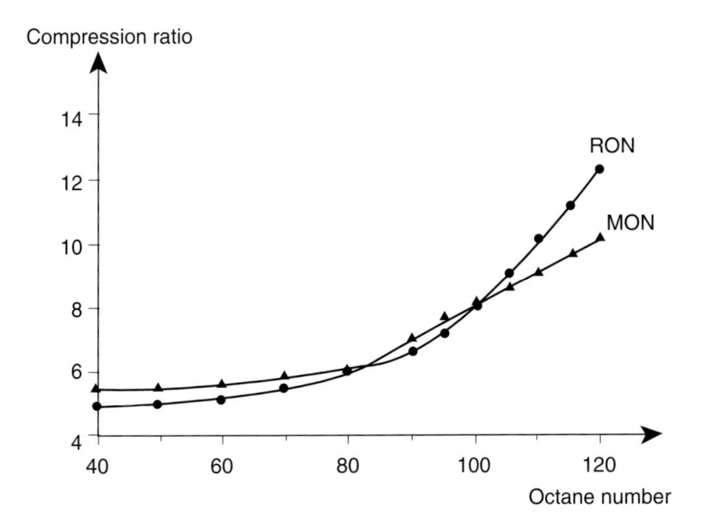

Figure 3.47 *Octane number measurement using a CFR engine. Compression ratio variation versus standard octane numbers. (Source ASTM)*

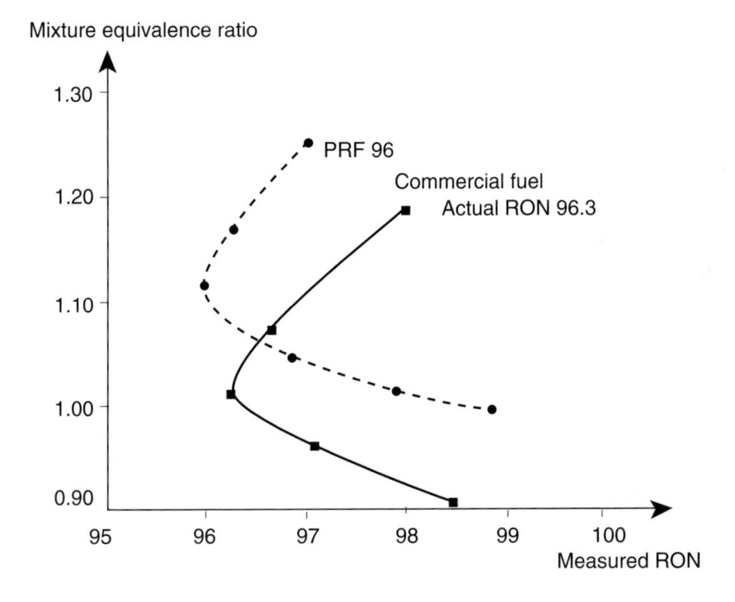

Figure 3.48 *The effect of mixture equivalence ratio on RON measurement.*

Figure 3.48 shows that deviations of 2 to 3 points in RON can be obtained using mixture equivalence-ratio values other than the one that produced maximum knock.

Consequently, the specific operating conditions of the CFR engine—compression ratio, mixture equivalence ratio, speed, spark advance—cause a change in the pressure and the temperature of the gases as a function of time, which is very different from those found in a production engine undergoing knock. As indicated previously, octane numbers are only **benchmarks**—they are not absolute indications.

However, this measurement scale provides a relatively precise discriminator. For example, a correlation can be established between the auto-ignition delay of PRFs and their RON (or MON).

The correlation is expressed in the following equation,

$$\theta = 0.01931 \left(\frac{\text{RON}}{100}\right)^{3,4017} P^{-1,7} \exp\left(\frac{3800}{T}\right)$$

where

θ = delay (s)

P and T = pressure (bar) and temperature (K)

Based on the previous equation, when P and T are constant

$$\frac{d\theta}{\theta} = 3.4017 \frac{d(\text{RON})}{\text{RON}}$$

Therefore, a one number change in the octane rating of a reference fuel that is rated around 100 results in only a 3.4% change in auto-ignition delay.

Additional information is also provided by comparing the **thermodynamic conditions** present during the RON and MON measurements. For example, the 98 PRF data in Table 3.13 shows that a change from research to motor methods results in an increase of about 130K in the temperature of the gases undergoing knock—from 900 to 1030 K—whereas the pressure is slightly reduced from 26 to 24 bar.

The MON of commercial fuels is usually lower than the RON. This is a qualitative indication that these fuels are **more sensitive** than PRFs to changes in experimental conditions like those applied to the CFR engine when switching from Research to Motor methods.

The difference between RON and MON is referred to as **sensitivity** (S):

$$S = \text{RON} - \text{MON}$$

The sensitivity of a conventional fuel is usually between 6 and 13.

The **high-speed** knock performance of production engines is often better represented by the MON—or the sensitivity—instead of the RON. This is prob-

Table 3.13 *A comparison of thermodynamic conditions during RON and MON measurement (PRF 98)*

Indicator	RON	MON
Compression ratio	7.45	7.75
Pressure* (bar)	26.3	24.5
Temperature* (K)	902	1 031

* Pressure (measured) and temperature (derived) of the fresh charge at the point when knock occurs.

ably due to a degree of similitude between the temperature-pressure history that occurs during high-speed operation and the conditions applied to the CFR engine during MON measurement.

3.6.3.4 Accuracy of measurements

As in all matters related to regulation, it is important to know the degree of precision that can be attributed to octane numbers.

This section covers the reproducibility, repeatability, and conformity that are essential determinants of measurement precision.

A. Reproducibility

Reproducibility represents the spread in the measurements taken at **two different laboratories**. It is the maximum permissible difference, within a given level of probability, between the results from two different laboratories for the same sample. For example, for a probability threshold of 95% this difference can only be exceeded in one case in 20.

Reproducibility (R) is defined by the following relationship,

$$R = a\sigma$$

where

a = **numeric factor** based on the probability threshold chosen (2.327 for 90%, 2.772 for 95%, and 3.645 for 99%)

σ = a **standard deviation** derived from ASTM standards for average octane numbers, which is based on a large experimental database

Figs. 3.49 and 3.50 show the change in σ as a function of the range of RON and MON under consideration. Maximum precision occurs in the 90 to 95 octane range. The values for RON are more precise than those for MON, since their minimum values of σ are 0.22 and 0.36 respectively. Reproducibility becomes very poor for octane numbers higher than 105.

B. Repeatability

Repeatability is the maximum difference (Z) between two measurements taken in the same **laboratory** by the same operator as represented by this relationship,

$$Z = a\sigma'$$

where a is the probability threshold defined earlier and σ' is the standard deviation for the laboratory, which is based on a sufficiently large number of experiments. In any case, σ' must be less than σ.

C. Conformity

Since the RON and the MON, are used as part of very-precise fuel specifications, it is essential to have criteria for determining if a sample conforms.

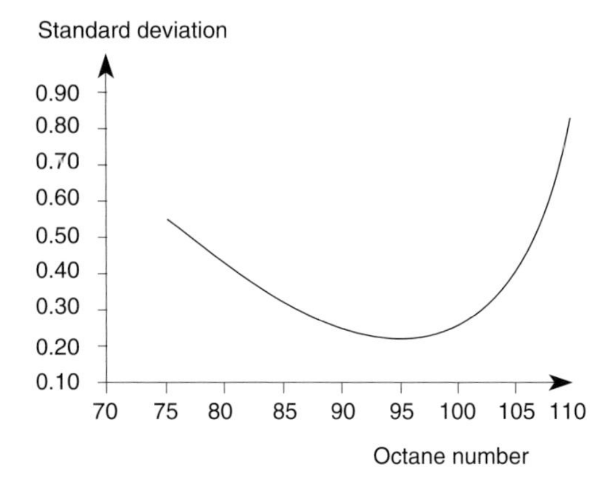

Figure 3.49 *Change in the standard deviation for reproducibility as function of octane number. Research method. (Source: ASTM)*

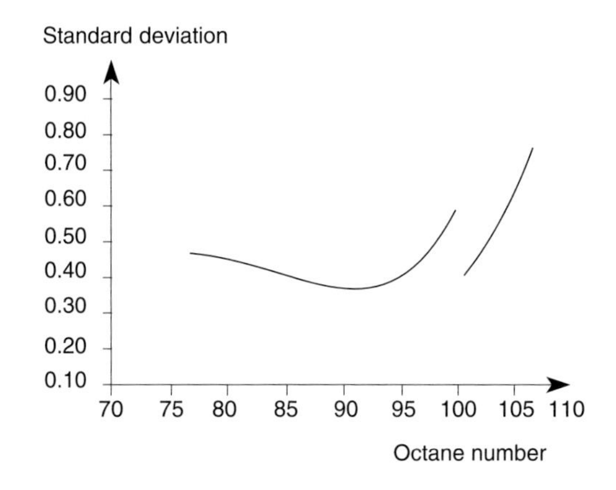

Figure 3.50 *Change in the standard deviation for reproducibility as a function of octane number. Motor method. (Source: ASTM)*

This need led to the definition of an **Acceptance Limit** (AL), which is described in ASTM D 3244 or ISO 4259 and provided by the following equation,

$$\text{AL} = S \pm \frac{\sigma D}{\sqrt{N}}$$

where

S = specified octane number

σ = standard deviation (previously defined)

D = factor related to probability (1.282 for 90%; 1.645 for 95%; 3.090 for 99%)

N = number of laboratories testing the sample

A numerical example of the application of this formula follows:

Four laboratories test a sample and the average of all their results is an average RON of 96.75 for a specified RON of 97 (minimum). The difference is 0.25.

The acceptance limit with a 95% probability is

$$AL = 97 - \frac{0.22 \times 1.645}{2} = 96.82$$

In this particular case, the sample does not conform to the specification with a probability of 95%.

3.6.3.5 Updates to ASTM standards D 2699-86 and D 2700-86

ASTM standards D 2699-86 and D 2700-86, which are the respective standards used to determine RON and MON, contain a procedure to standardize CFR engines. This procedure consists of adjusting, if required, certain engine parameters to obtain the octane numbers indicated in the standard tables, while using standard fuel blends of **toluene, isooctane, and *n*-heptane**. The adjustments permitted are the air intake temperature for RON measurements and the mixture temperature for MON measurements.

In 1969, a round robin test conducted by the National Exchange Group (NEG) in the United States found a small systematic shift among the CFR engines in the test. The exact cause of this change has not been determined. The change could be related to the cleanliness of the combustion chambers of CFR engines as a result of the widespread use of unleaded gasoline. The use of new knock measuring equipment on the CFR engines could also be a possible explanation.

Whatever the reason, the NEG adopted new RON and MON values for the ternary fuels (toluene/isooctane/*n*-heptane) and published them in the standards in 1989.

A **global round robin test** involving 150 CFR engines, which was conducted in 1990, confirmed the NEG findings. In 1996, the ASTM standards D 2699 and D 2700 were revised to adjust the RON and MON values of the three standard blends. As shown in Table 3.14, the differences are as high as 0.4.

Table 3.14 *Revised ASTM standards D 2699-86 and D 2700-86.*

Composition of the toluene/ isooctane/*n*-heptane blends (vol %)	RON			MON		
	ASTM table			ASTM table		
	D 2699-86	D 2699-89	D 2699-96	D 2700-86	D 2700-89	D 2700-96
74/0/26	93.4	93.4	93.4	81.1	81.6	81.5
74/5/21	96.7	96.9	96.9	84.9	85.3	85.2
74/10/16	99.6	99.9	99.8	88.5	88.8	88.7

Such a change may appear small or insignificant to the uninformed reader. However, as indicated in previous sections, refiners adjust the octane numbers of gasolines as **precisely as possible** to meet the required specifications. The new ASTM standard permits a MON rating of 0.3 points higher without changing the composition of the gasoline!

3.6.4 Octane numbers for volatile fractions

The occurrence of knock in some older vehicles during acceleration demonstrates the complete concealment of some phenomena by the standard CFR engine procedures. During the acceleration of a carburetor equipped vehicle, what occurs is a **segregation of the carburetted mixture** in both time and space, due to the inertia of the fuel droplets with respect to the air. The engine readily ingests the more volatile fuel fractions, which means that they must have a high degree of knock resistance.

To evaluate the real behavior of gasolines with respect to segregation, the knock performance of the fuel components must be determined as a function of their **distillation range**.

This leads to the definition of new classifications, of which the most common are "**delta R100**" (ΔR100) and the Distribution Octane Number (DON). Both are often referred to as Front-End Octane Number.

ΔR100 is the difference between the RON of the fuel and the RON of the fraction distilled at 100°C. This is measured using three steps:

- conventional RON measurement of the gasoline
- collection of the distillate up to 100°C using ASTM distillation
- measurement of the RON of the volatile fraction

The repeatability of the ΔR100 measurement is about one octane number.

The ΔR100 for commercial gasolines is usually between 5 and 15, with a value of less than 10 being considered satisfactory. Fig. 3.51 shows two examples of RON distribution based on the various distilled fractions of gasoline. This distribution indicates the importance of ΔR as a parameter (Owen et al. 1989).

The need for low ΔR values has been, until recently, a severe constraint on fuel formulation. The change in refinery schemas, especially the increased production of **cracked gasolines** that have low ΔR values, has resolved this problem almost completely. The problem persists only in simple refineries without cracking facilities, whose average gasoline constituent is a reformate that has a high ΔR value. Also, the increasing use of **fuel-injection systems** will eliminate any possibility of segregation.

Another classification similar to ΔR100 is the Distribution Octane Number (DON) described in ASTM D 2886, which was proposed by the *Mobil* company. The measurement principle is based on the **continuous separation** of the heaviest fractions at the intake of the CFR engine. For this purpose, the engine is equipped with a chilled separation chamber between the carburetor and the intake system. Some of the less volatile constituents are trapped in the cham-

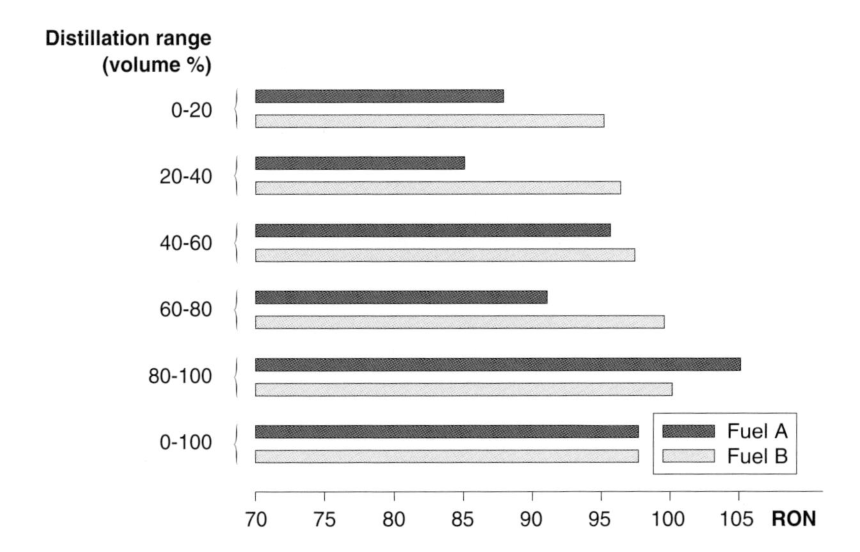

Figure 3.51 *RON distribution based on the distilled fractions of two different gasolines that have the same overall RON. (Source: Owen 1989)*

ber and separated from the rest of the intake mixture. This procedure is probably more realistic than the ΔR100 procedure, but it is less discriminating. However, both procedures are mostly of historical interest.

3.6.5 Road Octane number

The Road Octane number is obtained using production engines to compare the knock performance of a fuel to a well defined blend of isooctane and *n*-heptane. The Road Octane number is defined by this blend. The word "road" recalls the origin of this procedure, which was originally done on roads or test tracks. Today, the procedure is carried out on chassis dynamometers or test benches.

3.6.5.1 Test procedures

Fig. 3.52 shows the principle used to determine Road Octane numbers using an engine running under full load at a given speed. Knock is induced by increasing ignition advance. First, a calibration curve that includes the knock points for various PRF blends is established, then the ignition advance required to produce trace knock is recorded for the fuel being tested. Locating this point on the Road Octane curve (for example, 94) provides the Road Octane number. This technique is often referred to as **Knock Limited Spark Advance (KLSA).**

Other methods consist of accelerating the engine instead of operating it at constant speed. One of these methods is referred to as the Modified Uniontown (UTM) procedure, which is named after the town in the United States where the procedure was developed in 1946.

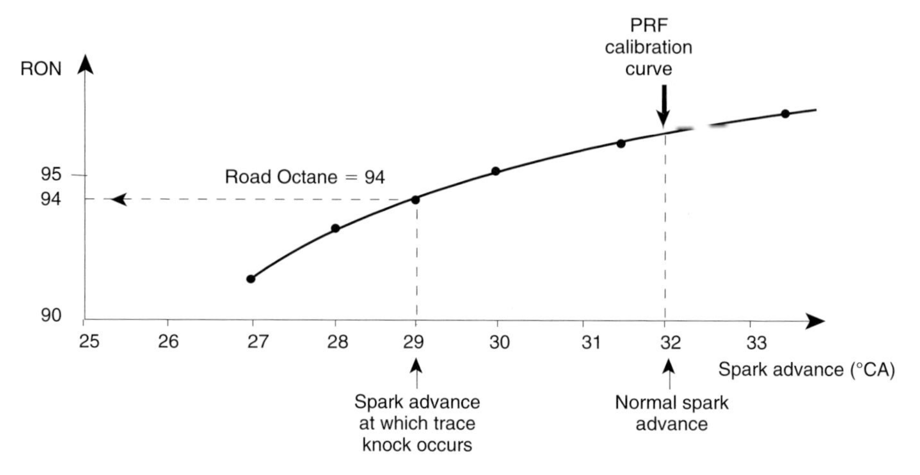

Figure 3.52 *Measuring Road Octane number.*
Diagram representing the principle and an example (full throttle at 4500 rpm).

The UTM follows the same procedures as the KLSA method. The only difference is the conditions under which the spark advance is used to induce trace knock. The measurement is not taken at a constant load, but when the engine is undergoing full-throttle acceleration.

Another technique referred to as Modified Borderline is described schematically in Fig. 3.53. The spark advance is adjusted **continuously** during the acceleration in such a way that trace knock is maintained throughout the acceleration. A series of curves is constructed using each of the PRF blends (80, 85, 90, 95, …).

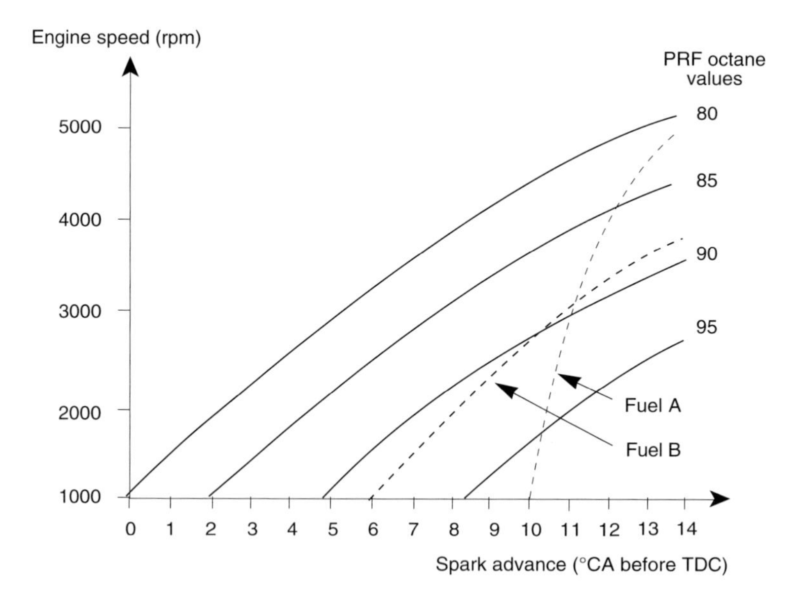

Figure 3.53 *Modified Borderline method for determining Road Octane number.*

A similar curve is constructed for each fuel to be tested. The Road Octane number at different speeds can be determined from the curves. For the examples in Fig. 3.53, fuel A has a higher octane number than fuel B at speeds less than 3000 rpm, but a lower number at higher speeds.

It is very difficult to change the ignition timing that is programmed into the engine map of newer vehicles. Using the KLSA, UTM, and Modified Borderline procedures requires a key to "**unlock**" the ignition timing, which must be obtained from the ignition system manufacturer.

3.6.5.2 Correlation with conventional octane numbers

It only takes a few Road Octane number ratings of a fuel to confirm that the correlation with conventional octane numbers varies according to the following criteria:

- the type of vehicle tested
- the experimental conditions

To illustrate this, Table 3.15 lists the Road Octane numbers derived, on the one part, from the same fuel used with several vehicles, and on the other part, from the same vehicle using several fuels with the same RON under identical conditions.

Table 3.15 *Examples of Road Octane number ratings.*

1. *Behavior of the same fuel with different vehicles*
Fuel: RON = 98.8; MON = 88.9; S = 9.9.

Vehicle	Engine speed (rpm)	Road Octane number	Depreciation
A	5800	86.8	12.0
B	4500	94.3	4.5
C	4500	97.4	1.4
D	5800	98.2	0.6
D	2000	99.2	−0.4

2. *Behavior of different fuels with the same engine*
Engine speed = 6 000 rpm.

RON	MON	Road Octane number	Depreciation
98.7	90.2	101.0	−2.3
98.2	87.7	98.2	0
98.5	87.3	97.3	1.2
98.7	87.7	96.6	2.1
98.6	85.8	94.2	4.4

It is useful to compare the Road Octane number with the RON; the difference between them is referred to as the **depreciation** (D):

$$D = \text{RON} - \text{Road Octane number}$$

This definition is derived from the fact that a fuel is depreciated in comparison to a PRF blend when its behavior is worse with a production engine than it is with a CFR engine using the Research method.

Usually, the Road Octane number is between the RON and MON, but it can be outside this range. Negative depreciation occurs when a production engine demonstrates better knock behavior with the test sample than it does with a PRF blend with the same RON.

When attempting to establish a relationship between Road Octane number and other conventional knock resistance characteristics, two contrasting situations appear if the engine is operating at low or high speed:

- At **low speeds** (about 2000 rpm), ΔR becomes important, especially with carburetor equipped vehicles, and the Road Octane number is expressed by the following relationship,

  $$\text{Road Octane number} = a + b\ \text{RON} - c\ \Delta\text{R}$$

  where a, b, and c are positive constants.

  In extreme cases the term c can be 0.5, which means that a change of 10 in the value of ΔR results in a change of 5 in the value of the Road Octane number, for the same RON.
- At **high speeds** (over 3000 or 4000 rpm), the MON is often more representative of a vehicle's resistance to knock than the RON.

  This results in the following relationship,

  $$\text{Road Octane number} = a\ \text{RON} + b\ \text{MON}$$

  where: $a + b = 1$ and $b > a$ (for example: $a = 0.20$; $b = 0.80$).

An example of this is shown in Fig. 3.54, which shows the high-speed behavior of various Eurosuper gasolines. The knock resistance of each fuel, which is expressed as the ignition timing at the appearance of trace knock, appears to be strongly correlated with the MON.

In the United States, equal weight is given to the RON and MON values ($a = b = 0.50$), which is justified by the lower level of high-speed operation. In Europe, on the contrary, the maximum permissible speeds on freeways are much higher, and therefore it is more logical to consider the MON as the **best indicator** of overall knock resistance.

3.6.6 Octane numbers of various organic compounds

3.6.6.1 Numeric values

Table 3.16 lists the RON and MON values for a number of organic compounds likely to be used in the blending of gasoline. For values greater than 100—for example, aromatic hydrocarbons—whole numbers are used because the decimals are not significant, given the degree of precision in octane rating. This also applies to alcohols and ethers with short hydrocarbon chains.

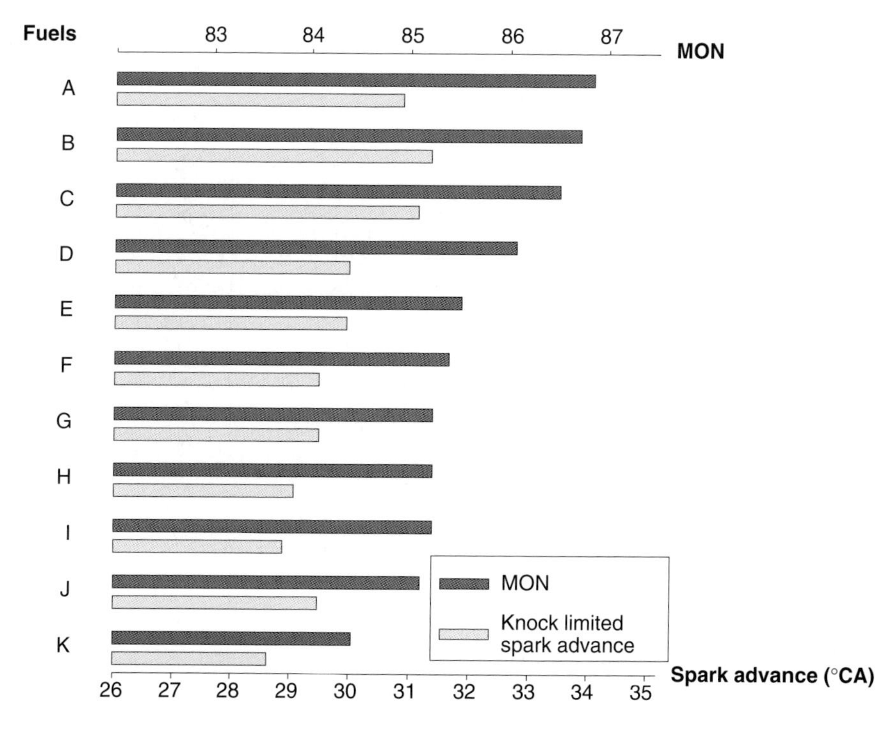

Figure 3.54 *Correlation between MON and knock resistance at high engine speeds (4000 rpm).*
(Fuels with a constant RON [95])

Table 3.16 *RON and MON values for pure hydrocarbons.*

Hydrocarbons	RON	MON
Paraffins		
Methane	> 100	110.0
Ethane	> 100	104.0
Propane	> 100	100.0
n-Butane	95.0	92.0
2-Methylpropane	> 100	99.0
n-Pentane	61.7	61.9
2-Methylbutane	92.3	90.3
2,2-Dimethylpropane	85.5	80.2
n-Hexane	24.8	26.0
2-Methylpentane	73.4	73.5
3-Methylpentane	74.5	74.3
2,2-Dimethylbutane	91.8	93.4
2,3-Dimethylbutane	103.5	94.3
n-Heptane	0.0	0.0
2-Methylhexane	42.4	46.4

Table 3.16 *RON and MON values for pure hydrocarbons (continued).*

Hydrocarbons	RON	MON
3-Methylhexane	52.0	55.0
3-Ethylpentane	65.0	69.3
2,2-Dimethylpentane	92.8	95.6
2,3-Dimethylpentane	91.1	88.5
2,4-Dimethylpentane	83.1	83.8
3,3-Dimethylpentane	80.8	86.6
2,2,3-Trimethylbutane	112.1	101.3
n-Octane	< 0	< 0
2-Methylheptane	21.7	23.8
3-Methylheptane	26.8	35.0
4-Methylheptane	26.7	39.0
3-Ethylhexane	33.5	52.4
2,2-Dimethylhexane	72.5	77.4
2,3-Dimethylhexane	71.3	78.9
2,4-Dimethylhexane	65.2	69.9
2,5-Dimethylhexane	55.2	55.7
3,3-Dimethylhexane	75.5	83.4
3,4-Dimethylhexane	76.3	81.7
2,2,3-Trimethylpentane	108.7	99.9
2,2,4-Trimethylpentane	100.0	100.0
2,3,3-Trimethylpentane	106.1	99.4
2,3,4-Trimethylpentane	102.7	95.9
2-Methyl-3-ethylpentane	87.3	88.1
3-Methyl-3-ethylpentane	80.8	88.7
n-Nonane and higher *n*-alkanes	< 0	< 0
Olefins		
Ethylene	100.0	81.0
Propylene	102.0	85.0
But-1-ene	–	80.0
But-2-ene	100.0	83.0
Pent-1-ene	90.9	77.1
Pent-2-ene	98.0	80.0
2-Methylbut-1-ene	102.5	81.9
2-Methylbut-2-ene	97.3	84.7
Hex-1-ene	76.4	63.4
Hex-2-ene	92.7	80.8
Hex-3-ene	94.0	80.1
2-Methylpent-1-ene	95.1	78.9
3-Methylpent-1-ene	96.0	81.2
4-Methylpent-1-ene	95.7	80.9
2-Methylpent-2-ene	97.8	83.0
3-Methylpent-2-ene	97.2	81.0
4-Methylpent-2-ene	99.3	84.3
2-Ethylpent-1-ene	98.3	79.4
3,3-Dimetylbut-1-ene	111.7	93.5

Table 3.16 *RON and MON values for pure hydrocarbons (continued).*

Hydrocarbons	RON	MON
2,3-Dimetylbut-2-ene	97.4	80.5
2,3-Dimetylbut-1-ene	101.3	82.8
Hept-1-ene	54.5	50.7
Hept-2-ene	73.4	68.8
Hept-3-ene	89.8	79.3
Methylhex-1-ene	90.7	78.8
3-Methylhex-1-ene	82.2	71.5
4-Methylhex-1-ene	86.4	74.0
5-Methylhex-1-ene	75.5	64.0
2-Methylhex-2-ene	90.4	78.9
3-Methylhex-2-ene	91.5	79.6
4-Methylhex-2-ene	96.8	83.0
5-Methylhex-2-ene	94.5	81.0
2-Methylhex-3-ene	96.4	81.4
3-Ethylpent-1-ene	95.6	81.6
3-Ethylpent-2-ene	93.7	80.6
2,4-Dimethylpent-1-ene	99.2	84.6
4,4-Dimethylpent-1-ene	104.4	85.4
2,3-Dimethylpent-1-ene	99.3	84.2
2,3-Dimethylpent-2-ene	97.5	80.0
2,4-Dimethylpent-2-ene	100.0	86.0
Oct-1-ene	28.7	34.7
Oct-2-ene	56.3	56.5
Oct-3-ene	72.5	68.1
Oct-4-ene	73.3	74.3
Diisobutylene	105.3	88.6
2-Methylhept-1-ene	70.2	66.3
2-Methylhept-3-ene	94.4	80.4
2,3-Dimethylhex-1-ene	96.3	83.6
2,2-Dimethylhex-3-ene	103.5	89.0
2,4,4-Trimethylpent1-ene	106.0	86.5
Naphthenes		
Ethylcyclopropane	102.5	83.9
Ethylcyclobutane	41.1	63.9
Methylcyclopentane	91.3	80.0
Ethylcyclopentane	67.2	61.2
Propylcyclopentane	31.2	28.1
1,1-Dimethylcyclopentane	92.3	89.3
(Z)-1,3-Dimethylcyclopentane	79.2	73.1
(E)-1,3-Dimethylcyclopentane	80.6	72.6
1,1,3-Trimethylcyclopentane	87.7	83.5
Cyclohexane	83.0	77.2
Methylcyclohexane	74.8	71.1
(Z)-1,3-Dimethylcyclohexane	71.7	71.0
(E)-1,3-Dimethylcyclohexane	66.9	64.2

Table 3.16 *RON and MON values for pure hydrocarbons (end).*

Hydrocarbons	RON	MON
Diolefins and cyclenes		
Cyclopentene	93.3	69.7
Cyclohexene	83.9	63.0
Methylcyclopentene	93.6	72.9
3-Methylbuta-1,2-diene	61.0	42.4
2-Methylbuta-1,3-diene	99.1	81.0
Cyclopenta-1,3-diene	103.5	86.1
Cyclohexa-1,3-diene	74.8	53.0
Aromatics		
Benzene	–	114.8
Toluene	120.0	103.5
Ethylbenzene	107.4	97.9
1,2-Dimethylbenzene (*o*-xylene)	–	100.0
1,3-Dimethylbenzene (*m*-xylene)	117.5	115.0
1,4-Dimethylbenzene (*p*-xylene)	116.4	109.6
n-Propylbenzene	111.0	98.7
Isopropylbenzene (cumene)	113.1	99.3
1-Methyl-2-ethylbenzene	102.5	92.1
1-Methyl-3-ethylbenzene	112.1	100.0
1-Methyl-4-ethylbenzene	–	97.0
1,2,3-Trimethylbenzene	105.3	100.8
1,3,4-Trimethylbenzene	110.5	106.0
1,3,5-Trimethylbenzene	> 120	120.0
n-butylbenzene	104.4	95.3
Isobutylbenzene	111.4	98.0
1,3-Diethylbenzene	> 115.5	97.0
1,4-Diethylbenzene	106.0	96.4
1-Methyl-2-*n*-propylbenzene	103.5	92.2
1-Methyl-3- *n*-propylbenzene	112.1	100.5
1-Methyl-2- isopropylbenzene	106.0	96.0
1-Methyl-4- isopropylbenzene	110.5	97.7
1,2-Dimethyl-3-ethylbenzene	104.4	91.9
1,3-Dimethyl-4-ethylbenzene	106.0	95.9
1,3-Dimethyl-5-ethylbenzene	114.8	102.5
1,4-Dimethyl-2-ethylbenzene	106.0	96.0
1,2,3,4-Tetramethylbenzene	105.3	100.3
1,2,3,5-Tetramethylbenzene	–	102.5
Indane	103.5	89.8

3.6.6.2 Property-structure relationships

The RON and MON values of hydrocarbons are narrowly dependent on their **chemical structures**. A general view of the RON characteristics of each group of hydrocarbons is shown in Fig. 3.55.

The RON of ***n*-paraffins**, which is very high for the lighter constituents, decreases progressively as the length of the molecular chain increases and ends by definition with *n*-heptane. **Branching** (the number and complexity of lateral molecular chains) always increases RON. The MON of paraffins is sometimes higher and sometimes lower than the RON, with the differences being generally insignificant. However, there are some exceptions; for example, the MON (52.4) of 3-ethylhexane is much higher than the RON (33.5). As it is with RON, branching is an important factor in increasing MON.

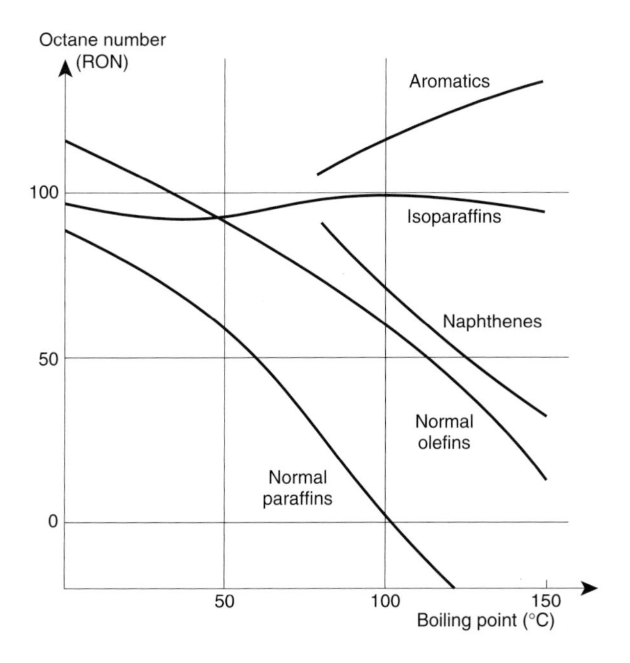

Figure 3.55 *Octane numbers of various hydrocarbon families.*

The octane numbers of **olefins**, like paraffins, are narrowly dependent on the length and the branching of molecular chains. The RONs of olefins are generally higher than those of paraffins with the same molecular structure. The occurrence of double bonds near the center of the molecules tends to increase the RON, at least for the first items in the molecular series. The MON of olefins is almost always lower than the RON (*n*-octene being one of the rare exceptions).

The difference can be as high as 10 or 15 octane numbers, which means increasing the **activation energy** of the kinetic chemical processes that lead to auto-ignition. In practice, the **MON is always closely correlated to the olefin content** of commercial fuels.

Naphthenes always have higher octane numbers—both RON and MON—than their open-chain counterparts. Thus, the RON of *n*-hexane is 24.8 as compared to 83 for cyclohexane.

Aromatics all have RONs higher than 100 (from 115 to 120). The MON of aromatics is equally high, greater than 100, but about 10 numbers lower than the RON. It is almost impossible to discern significant differences in the octane ratings of the major aromatic constituents of gasoline: benzene, toluene, xylene, ethylbenzene, cumene, ... All aromatics have excellent knock resistance.

Alcohols and **ethers** also have high octane numbers, which are usually greater than 100. For the same product group, alcohols for example, the octane numbers vary with the length and the branching of the hydrocarbon radicals, as previously discussed. Thus, the RON and MON values decrease continuously from methanol through *n*-butanol to *n*-hexanol ... . In the same manner, the octane values of isopropanol are higher than those of *n*-propanol; also, a large difference can be observed in the behaviors of diethylether and methyl tertiary butyl ether (MTBE). Lastly, the MON is always greater than the RON for all oxygenated products and the sensitivity (RON-MON) can be 15 to 20 octane numbers.

3.6.7 The use of additives to improve octane ratings

The use of additives to improve octane ratings is one area of the engine-fuel interface that has seen the largest amount of research work. Today, there are few prospects of finding new additives that are both effective and completely safe from an environmental point of view.

3.6.7.1 Lead alkyls

Lead alkyls have been added to gasolines for the last 50 years because they were the most flexible and economic way to achieve high octane ratings. Environmental problems (the toxicity of lead itself and the problem of catalytic converter poisoning) prompted many countries to adopt legislation banning the use of lead alkyls beginning in the 1970s. This will inevitably evolve to a conclusion around the year 2000 when a total ban on lead additives in gasoline will be in effect.

This section continues with a short retrospective on the use of lead alkyls followed by a description of the mechanics of lead's effects and the improvements in octane ratings that result from their use.

A. The introduction, peak usage, and disappearance of lead alkyls

During the First World War, three researchers at General Motors (Thomas A. Boyd, Thomas Midgley, and Charles Kettering) were actively seeking additives to reduce the knock tendency of gasolines. At first they proposed iodine and aniline, but later they devoted their attention to **organometalic** compounds, which were discovered a few years earlier by the Frenchman Victor Grignard.[1] December 09, 1941 was the first time they observed the remarkable properties of tetraethyl lead (TEL), which has the chemical formula Pb $(C_2H_5)_4$. In 1922, the American air force began to use the additive, but it caused several incidents linked to deposits on valves and spark plugs.

Further testing was discontinued until the development of scavengers. Two products were identified for this use: **dibromoethane** $(C_2H_4Br_2)$ and **dichloroethane** $(C_2H_4Cl_2)$. Thus, the first delivery of leaded gasoline occurred on February 2, 1923 at Dayton, Ohio. It took five years (1928) before lead alkyls appeared in England and then they gradually began to appear in the other European countries (Goodacre 1958). A few years later, tetramethyl lead (TML) with the formula Pb $(CH_3)_4$ was produced, but it was not widely accepted because it was expensive and only marginally better than TEL.

The discovery of the knock suppression properties of lead alkyls had widespread effects. About 1920, the RON of gasoline was barely over 60; with a lead content of about 0.8 g Pb/liter, RON increases by about 15, which permits a two- to three-number increase in compression ratios (for example, from 4 to 7) and a resulting increase in efficiency of 15 to 20%. Even if these capabilities were not immediately put to direct use, due to mistrust or ignorance, they deserve, even today, a degree of admiration.

From 1935 to about 1970, countries used gasolines containing lead alkyls at allowable maximums of 0.50 to 0.60 g Pb/liter (up to 0.80 g Pb/liter in some exceptional cases).

From 1935 through to about 1970, the addition of lead to gasolines with a RON of about 90 increased the octane rating by 6 or 7 numbers at very-low cost—1 to 2% of the cost of refining. Obtaining the same effect by process improvements would have been either technically impossible or very costly (energy penalties of 15 to 20%).

Beginning in the 1970s, United States manufacturers began to install **catalytic converters** on new vehicles as an efficient way to reduce tailpipe emissions. This decision implied the reduction of lead alkyls, because their presence in gasoline rapidly **poisoned catalytic converters**.

In 1974, unleaded gasolines began to appear in the United States. This meant that only traces of lead—less than 0.013 g Pb/liter—could be used to fuel new vehicles equipped with catalytic converters. Japan and Canada followed the same route in 1975 and 1977 respectively.

1. Victor Grignard (1871-1935) discovered organo-metalic compounds in 1901. He received the Nobel prize in Chemistry in 1912.

Beginning in the 1980s, Europe, and especially England (Yule 1983), began to reduce and even replace lead alkyls in gasolines due to their toxicity. It took the appearance of catalytic converters in 1989 for the first unleaded gasolines to appear on the European market.

The use of leaded gasolines is now decreasing throughout the world with the introduction of catalytic converters. The maximum permitted lead content for leaded gasoline is generally 0.15 g Pb/liter. The effects of this on the formulations of gasoline are described in subsequent sections (see 3.10.2).

B. How lead alkyls act to reduce knock

Lead alkyls have an **inhibiting** effect on the oxidation reactions of organic compounds and thus help to delay auto-ignition. The normal flame propagation process does not change, at least within the limits of observational precision that are applicable to engines.

The active agent is one of the following **lead oxides**: PbO, PbO_2, or Pb_3O_4. It is most likely PbO, which results from the chemical decomposition of the organo-metallic compound. PbO deactivates the free radicals, primarily the $OH^•$ types that are directly involved in the propagation and branching of molecular chains during the oxidation that precedes auto-ignition. It has not been possible to precisely determine if the inhibition takes place during the heterogeneous phase, when the fine PbO solids are dispersed in the intake charge, or during the homogeneous phase when PbO is vaporized. Regardless, the general deactivation reaction can be described by the following equations:

$$\mathrm{PbO} + \mathrm{OH}^\bullet \longrightarrow (\mathrm{PbO} - \mathrm{OH})$$

$$(\mathrm{PbO} - \mathrm{OH}) + \mathrm{OH}^\bullet \longrightarrow \mathrm{PbO_2} + \mathrm{H_2O}$$

According to this simple but comprehensible model, each PbO molecule captures two $OH^•$ radicals, whereas, during a conventional branching, one $OH^•$ radical contributes to the generation of three others. Various other metallic oxides other than lead oxide could also have the same inhibiting effect.

At first, TEL was the only lead alkyl added to gasoline. About 1950, it became possible to easily produce TML, which is more volatile and more toxic, but in some cases more effective. Table 3.17 shows a comparison of some of the characteristics of the TEL and TML products.

TML has a boiling point of 110°C instead of the 198°C of TEL. It can therefore accompany the more volatile gasoline fractions and act effectively on the ΔR rating. In addition, TML is more thermodynamically stable than TEL and it decomposes **later** in the combustion cycle, which is a more opportune time for the appearance of knock. Therefore, under severe conditions, for example at high rpm, TML could be more effective than TEL.

The foregoing explains why, depending on the circumstances, past practice consisted of the use of pure TEL, pure TML, physical blends of the two, or mixed chemical compounds made up of various possible combinations of CH_3 and C_2H_5:

$$\mathrm{Pb\,(C_2H_5)_2\,(CH_3)_2,\quad Pb\,(C_2H_5)_3\,CH_3,\quad Pb\,(C_2H_5)\,(CH_3)_3}$$

Table 3.17 *Physical properties of TEL and TML.*

Characteristics	TEL	TML
Formula	$Pb(C_2H_5)_4$	$Pb(CH_3)_4$
Density at 20°C (kg/liter)	1.650	1.995
Boiling point (°C)	198.9	110.0
Vapor pressure (mbar) at 15°C	0.22	23.3
at 50°C	2.80	133.0
Freezing point (°C)	–130.2	–30.3
Refractive index n_D^{20}	1.520	1.512
Molecular mass (g/mol)	323.5	267.4
Metal content (% mass)	64.06	77.51
Metal content per ml (g of Pb)	1.057	1.546

The lead oxides formed by the use of TEL and TML can be deposited on the combustion chamber walls and on various other engine components (valves, spark plugs, pistons). **Scavengers** are used to clear away the deposits by changing the oxides and other refractory compounds of lead into more **volatile** combinations that are expelled with the exhaust gases.

The only truly effective scavengers are **dibromoethane** (DBE: $C_2H_4Br_2$) and **dichloroethane** (DCE: $C_2H_4Cl_2$). These compounds are easily converted in the combustion chamber to either halogens (Br_2 and Cl_2) or acids (HBr and HCl), which react in turn with PbO to produce volatile halides according to the following equations,

$$2\,PbO + 2\,X_2 \rightleftarrows 2\,PbX_2 + O_2$$

$$PbO + 2\,HX \rightleftarrows PbX_2 + H_2O$$

where X = Br or Cl.

To appreciate the effectiveness of the scavengers, consider the melting points of PbO (888°C), $PbBr_2$ (378°C), and $PbCl_2$ (501°C).

DBE is usually used alone, which is the most effective way, or it is used in combination with DCE; the former is used in semi-stoichiometric doses and the latter is used in stoichiometric doses.

The use of scavengers does have its disadvantages, because they promote **corrosion** of exhaust systems and the bromine and chlorine contribute, independently of lead, to the poisoning of catalysts.

C. Improvements in octane ratings that are attributable to lead-alkyl additives

The improvements in octane ratings from the addition of lead alkyls are greatest when the octane ratings of the recipient products are lowest. The term **sus-**

ceptibility is used to describe the effectiveness of lead-alkyl additives. A base stock is considered more susceptible if it demonstrates a greater improvement in octane rating for a given lead content. For example, the susceptibility of various refinery stocks are shown for illustrative purposes in Fig. 3.56.

A basestock with a RON of almost 70, which is derived from the direct distillation of crude oil, can have its rating improved by 10 with the addition of 0.6 g/liter of lead. Therefore, countries that do not have elaborate refinery capability can use the addition of lead as a **very effective** means of achieving satisfactory octane levels.

For basestocks whose octane ratings are closer to conventional gasolines (RON 90 to 95), the addition of lead alkyls results in rating improvements of much less than 10, but still appreciable given the difficulty of progressing from this level. Thus, from a RON of about 92, the rating improvement is 2 to 3 for 0.15 Pb/liter and 5 to 6 for 0.4 Pb/liter. For higher concentrations, a **saturation** effect begins to appear and additional improvements are more modest. The ratings in the preceding discussion apply to MON as well RON. However, slightly greater improvements are usually observed for the RON. In other words, adding lead tends to provide a **slight increase in sensitivity** (about one number for 0.4 Pb/liter).

The presence of **sulfur** in refinery basestocks tends to reduce their susceptibility. How this secondary effect occurs is unknown, but its extent depends on both the concentration and the nature of the sulfur compounds (disulfides, thiophenes, thiols, ...). Thus, in the presence of 1000 ppm of one of these sulfur compounds, the decrease in RON or MON with the addition of 0.15 g Pb/liter is one octane number, which is a one third reduction in the effectiveness of the lead. Henceforth, the required desulfurization of gasolines will minimize this type of problem.

D. Lead content determination in gasolines

Modern methods for testing lead content use a flame atomic-absorption spectrometer. Procedures EN 237 and ASTM D 3237 can be used in principle to detect lead levels of 2.5 to 26 mg/liter. In fact, it is possible to obtain acceptable readings to a precision of 1 mg Pb/liter.

For leaded gasolines, the DIN 51 769 standard is used, which works well for standard levels of 0.15 Pb/liter.

In both cases, the measurement is preceded by a short conditioning process: the sample is diluted with methyl isobutyl ketone (MIBK) and then the lead alkyls are reacted with iodine. The lead iodide ions are then stabilized by the addition of quaternary ammonium salt and measured by atomic absorption.

3.6.7.2 Other additives that can improve octane ratings

A. MMT

Many products have been tested over the last 80 years to find a product to match, exceed, or replace lead alkyls. The one that has, without question,

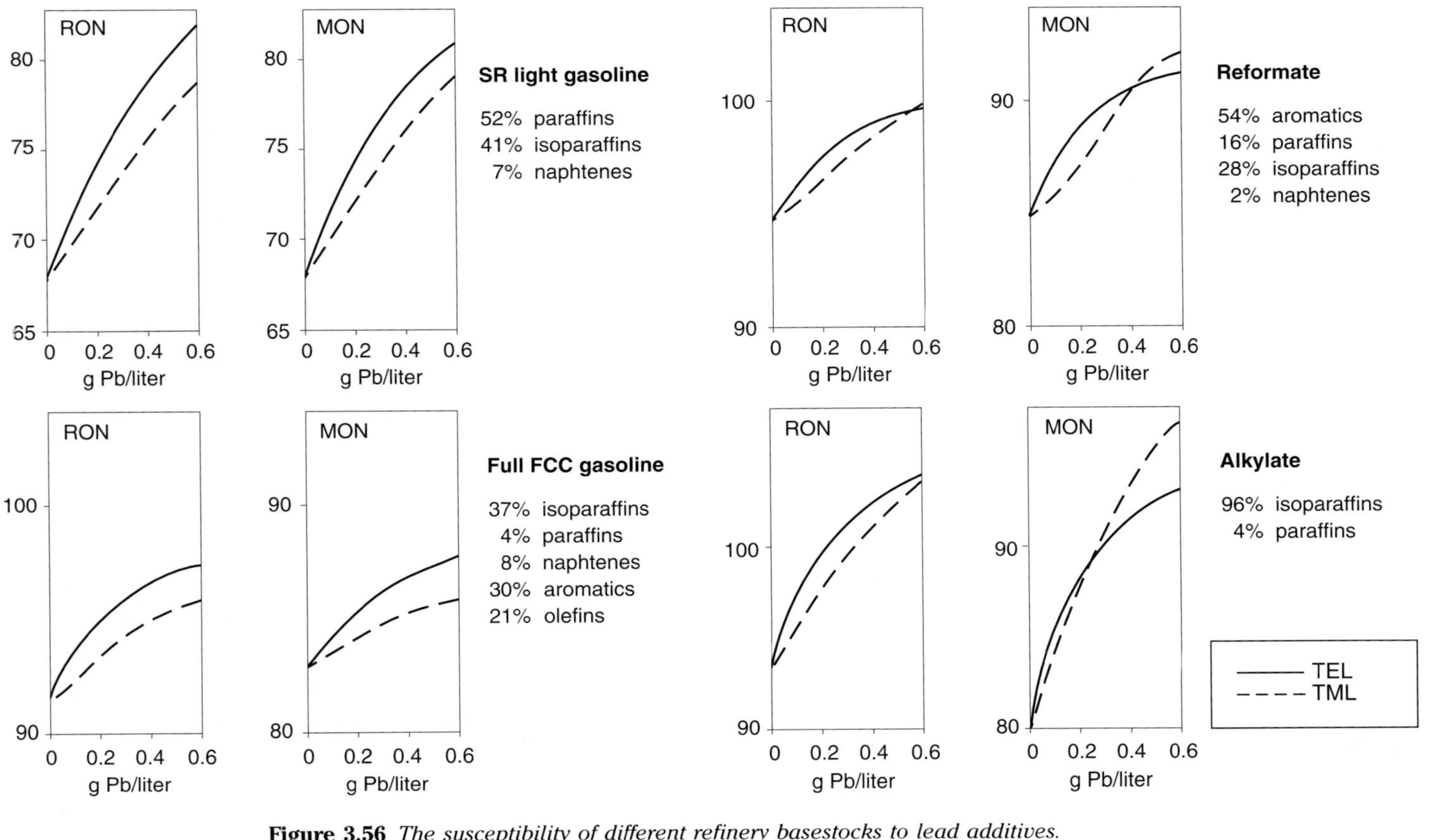

Figure 3.56 *The susceptibility of different refinery basestocks to lead additives. (Source: Elf)*

raised the most interest is methyl cyclopentadienyl manganese tricarbonyl (**MMT**), which has the chemical formula CH_3–C_5H_4–$Mn(CO)_3$. It is produced by the Ethyl Corporation who commercialized it first under the name AK33X and used it in conjunction with lead alkyls (Ethyl 1983).

With the introduction of unleaded gasoline in the United States and Canada, MMT was proposed as an octane rating improver for vehicles equipped with catalytic converters. However, even though it is less toxic to catalytic converters than lead, MMT did not receive an approval for general use from the Environmental Protection Agency (EPA) in the United States. In 1996, MMT was still used in Canada at levels of 0.0185 g Mn/liter.

From the standpoint of intrinsic effectiveness, **manganese** is more effective than lead—for equal concentrations of metal. At a level of 0.0185 g Mn/liter (1/16 g per gallon), MMT provides a two-number octane-rating improvement for gasolines with a RON of 88 to 90 and a one-number improvement for gasolines with a MON of 81 to 83. MMT also has the following characteristic: octane rating improvements are lowest when the aromatic content is highest; thus, the improvements quoted are valid for aromatic levels of 25 to 30%, but they can be cut in half if the aromatic level exceeds 40%.

There are no scavengers that can remove MMT with the exhaust gases ($MnCl_2$ boils at 1190°C). Also, at levels of 0.0165 g Mn/liter, rust colored deposits of manganese oxide are found on spark plugs and the interior of the combustion chamber, but these do not appear to be harmful to engine behavior or performance.

B. *Other additives*

Aside from MMT, there are several other types of organometallic compounds than have antiknock properties similar to lead alkyls (Ungelman et al. 1972). Consider **iron pentacarbonyl** [$Fe\ (CO)_5$], **ferrocene** (dicyclopentadienyliron) [$Fe(C_5H_4)_2$], **nickel carbonyl** [$Ni(CO)$], and **tetraethyl tin** [$Sn\ (C_2H_5)_4$]. None of these products have reached industrial production for reasons of toxicity, cost, or secondary effects, related primarily to wear. Rare earth compounds (ytterbium, praseodymium, neodymium) have also been proposed.

Of the non-metallic compounds, **iodine**, **aniline**, and ***n*-methylaniline** have beneficial effects, which have been known for over 50 years. However, these compounds are insufficiently effective for use in current gasolines. The addition of almost 1% methylaniline was required to provide an 8 RON improvement for a direct distillate gasoline with an initial octane rating of 60 RON. The effectiveness would certainly be much lower with the present refinery basestock levels closer to 90 RON.

3.6.8 The octane ratings of commercial fuels

At many service stations, motorists are confronted with two to four types of gasolines with different lead levels and octane ratings.

3.6.8.1 Mandated octane ratings

Table 3.18 lists the minimum RON and MON ratings for the various quality grades of gasoline sold in some of the major world markets. The most widely used product in Europe is unleaded **Eurosuper**, which is defined by the European directive of December 16, 1985 (EN 228). Eurosuper must have a minimum RON of 95 and a minimum MON of 85. A product that is also sold in Europe, and especially in France, is a **super premium** unleaded gasoline with a minimum RON of 98 and a minimum MON of 88.

Leaded gasolines will persist in Europe for a few more years (until 2000). Their octane ratings are 97 RON and 86 MON.

"Regular" type gasolines, with and without lead, have practically disappeared in Europe, except in Germany where an unleaded regular grade is available that has minimum 91 RON and 82.5 MON.

The same classifications are used worldwide for gasolines: super and regular, with and without lead. Octane ratings can vary from one country to another depending on the level of modern equipment in the vehicle fleet and the local refinery capabilities.

The suppression of lead additives generally occurs in those countries that have high automobile densities and very-stringent emissions regulations. As a result, only unleaded fuels are available in the United States, Canada, and Japan.

In the United States, gasoline quality grades are not defined by minimum values of RON and MON. Instead, a combination of the two ratings is used,

Table 3.18 *Octane-rating specifications for some countries.*

<table>
<tr><th rowspan="2">Unleaded gasoline grades</th><th rowspan="2">Properties</th><th colspan="4">Europe (standard EN 228)</th><th rowspan="2">Japan (JIS standard K2202)</th><th rowspan="2">United States (ASTM standard D 4814)</th></tr>
<tr><th>Germany</th><th>France</th><th>Italy</th><th>UK</th></tr>
<tr><td rowspan="3">Regular</td><td>RON min</td><td>91</td><td colspan="3" rowspan="3"></td><td>89</td><td></td></tr>
<tr><td>MON min</td><td>82.5</td><td></td><td>82</td></tr>
<tr><td>AKI min</td><td></td><td></td><td>87***</td></tr>
<tr><td>Intermediate</td><td>AKI min</td><td></td><td colspan="3"></td><td></td><td>89****</td></tr>
<tr><td rowspan="3">Premium</td><td>RON min</td><td rowspan="3"></td><td colspan="3">95*</td><td>96</td><td></td></tr>
<tr><td>MON min</td><td colspan="3">85*</td><td></td><td></td></tr>
<tr><td>AKI min</td><td colspan="3"></td><td></td><td>91*****</td></tr>
<tr><td rowspan="2">Super premium 98</td><td>RON min</td><td>98</td><td>98**</td><td></td><td>98</td><td></td><td></td></tr>
<tr><td>MON min</td><td>88</td><td>88**</td><td></td><td>87</td><td></td><td></td></tr>
</table>

* Eurosuper
** Octane level required by the French automobile manufacturers *(Cahier des charges "Qualité")*
*** Must meet the requirements of vehicle models older than 1971
**** For vehicles with higher octane requirements
***** For vehicles with very high octane requirements
The antiknock index (AKI) is equal to (RON + MON)/2; its value can be adjusted to match altitude and seasonal requirements (ASTM standard D 4814).

which is the sum of (RON + MON)/2 and it is referred to as the **Anti-Knock Index** (AKI). There are three grades of gasoline (Regular, Intermediate, and Premium) with (RON + MON) /2 octane ratings of 87, 89, and 91 respectively. The fact that the sensitivity is often close to 10 enables the individual RON and MON values to be determined.

3.6.8.2 Identifying products

Coloring additives are used in small amounts to distinguish between the various types of gasolines, to control their use, and to prevent fraud.

Hence, unleaded gasoline is colored **green** in Europe by using the following two additives:

- blue (1,4-di-n-butylaminoanthraquinone)
- yellow (diethylaminoazobenzene)

Both additives are used in quantities of 2 mg/liter.

Leaded regular gasoline, where it is still in use, is colored using a red dye (*o*-toluene-azo-*o*-toluene-azo—βnaphthol) at the rate of 4 mg/liter.

Super premium gasoline is not colored because mixing it with other fuels is of no interest to tax evaders because it carries a higher tax.

In addition, to avoid the risk of inadvertently using the wrong fuel (catalyst poisoning from the use of leaded fuels), fuel filler ports in vehicles and service station nozzles are carefully matched.

3.7 Behavior during storage and distribution

Gasoline can undergo changes during the time between blending in the refinery and engine combustion, which can have serious consequences. This section describes the quality control methods used during storage and the methods used to prevent deterioration.

3.7.1 Oxidation tendencies of gasolines

In the presence of oxygen, even at ambient temperatures, hydrocarbons are subject to an oxidation process that leads to the formation of viscous products commonly referred to as **gums**. Gums can cause a variety of problems: blockage of fuel pump membranes, blockage of nozzles and injectors, and piston ring sticking.

Olefins, which are derived primarily from conversion processes (catalytic and thermal cracking), are the most likely gasoline constituents to form gums. Other compounds like aromatics (substituted or not), heterocyclics, and sulfides (even in trace amounts) also lead to gum formation.

There are many quality control methods that can be used to control gum formation in the final product. There is a clear distinction in the behavior of gasoline at a given point in time (existent gum content) and the prediction of how it will change during prolonged storage (storage stability).

The **existent gum content** (EN 5 and ASTM D 381) is obtained by evaporating a 50 cm^3 fuel sample, which is placed in a controlled temperature bath at 160°C under a jet of air for 30 min. After evaporation, the weight of the residue provides the **unwashed** gum content. The residue is then extracted with heptane, which leaves the **existent** gum. Specifications apply to the existent gum after extraction. The gum content allowed in various countries depends on the local storage conditions (temperature and duration) and an understanding of the risks.

An upper limit of **5 mg/100 ml** for existent gum content is normally sufficient protection.

The gum content test procedure can also provide information about the use of **detergent additives** in a gasoline. When detergent additives are used, there is usually a large difference in weight between the unwashed and existent gum residue.

The method used to evaluate the **potential gum** content (NF M07-013) consists of artificially stimulating oxidation to simulate prolonged storage conditions. A fuel sample is retained for 16 hours at 100°C under an oxygen pressure of 7 bar and the gum is recovered by filtration. The gum content derived by this method is usually not subject to official specifications and it is simply considered an indicator; as well, its significance becomes very questionable when oxygenated compounds (alcohols, ethers, ...) are included in the gasoline.

Lastly, determining the **induction time** (ISO 7536 and ASTM D 525) also enables an assessment of gum formation potential during storage. A gasoline sample is placed in a bomb filled with oxygen at a temperature of 100°C and a pressure of 7 bar. The change of oxygen pressure with time is recorded using a manometer. The time at which the pressure begins to drop signals the onset of oxidation. If a pressure drop has not occurred after 960 minutes, the test is considered complete because this duration corresponds to the maximum attainable induction period. The European specification (EN 228) for unleaded gasoline (Eurosuper) requires a minimum induction period of **360 min**. In the United States, the requirement is 240 min (ASTM D 4814).

The simplest way to meet these quality criteria would be to completely eliminate the undesirable constituents and impurities. This is not a realistic alternative and **antioxidant additives** are preferred.

3.7.2 Use of antioxidants

The antioxidant additives used in gasolines belong to one of the following chemical families:

- alkyl-*p*-phenylenediamines

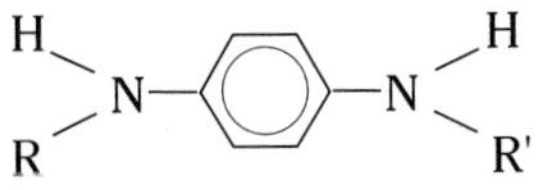

H(R)N–C_6H_4–N(H)R'

- alkyl-*p*-aminophenols

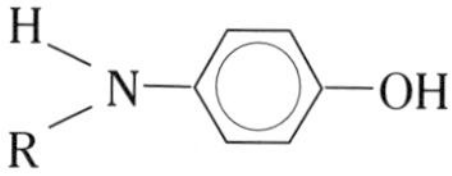

$$\begin{matrix} H \\ R \end{matrix} > N - \bigcirc - OH$$

- alkylphenols

$$\text{(R)}_2\text{-}\bigcirc\text{-OH}$$

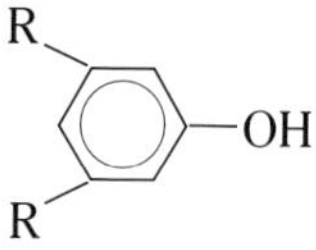

All of these additives either stabilize of decompose **peroxide radicals** of the ROO• type, which are usually very active in the oxidation process. The reaction mechanism, which is most readily accepted, is the **transfer of a hydrogen atom** from the antioxidant (AH) to the free radical:

$$AH + ROO^{\bullet} \rightarrow A^{\bullet} + ROOH$$

The peroxide group is deactivated and the A• radical is stabilized by resonance and it cannot become involved in propagating chain reactions.

Other processes can also occur:

- the formation of a **complex substance** between the antioxidant and the ROO• radical according to the following schema

$$ROO^{\bullet} + AH \rightarrow (ROO \leftarrow AH)$$
$$(ROO \leftarrow AH) + ROO^{\bullet} \rightarrow R-R + 2O_2 + AH \text{ (inactive product)}$$

- deactivation by **transfer of electrons**, which is represented by the following reaction

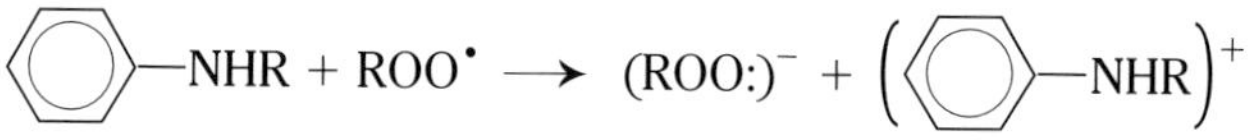

$$\bigcirc\text{-NHR} + ROO^{\bullet} \rightarrow (ROO{:})^{-} + (\bigcirc\text{-NHR})^{+}$$

This mechanism helps to explain the actions of some additives that do not have unstable hydrogen. For example, this is the case with N,N,N',N'-tetramethyl-*p*-phenylenediamine:

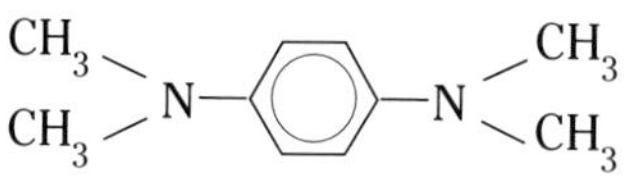

$$\begin{matrix} CH_3 \\ CH_3 \end{matrix} > N - \bigcirc - N < \begin{matrix} CH_3 \\ CH_3 \end{matrix}$$

In practice, the products used most often are blends **of *p*-phenylenediamines** and **alkylphenols**, which are adjusted to suit the olefin content of the fuel being treated (Tupa et al. 1986). The most common dose is 10 to 20 ppm. These additives should be mixed in a soon as possible after blending, preferably when the basestocks are mixed.

The one additional advantage to the use of antioxydants is that they stabilize lead alkyls. The latter can be attacked by peroxides and they will decompose during storage. Obviously, avoiding such a degradation is essential.

3.7.3 Use of metal deactivators

Traces of **metals** that derive from crude oil or refinery processes can, even in very-small quantities (0.01 mg/liter), promote oxidation reactions in gasoline. **Copper** is the most active of these metals. It becomes involved in either the formation of free radicals that propagate chain reactions or it inhibits the action of conventional antioxidant additives.

Metal-deactivator additives deactivate copper by forming **chelates** that inhibit its reactivity.

The product most commonly used is N,N'-disalicylidene-1,2-propanediamine. The chelate structure is

```
 (C6H4)—O         O—(C6H4)
   |      \      /     |
   |       Cu         |
   |     ..  ..       |
 CH = N        N = CH
      |        |
 CH3 - CH  -  CH2
```

Metal deactivators are used in the range of 2 to 10 ppm.

Gasoline blending

Gasoline blending is a combination of operations that consists of mixing various refinery basestocks to obtain sufficient quantities of a final product that meets all specified requirements. The most difficult requirements to meet are volatility (vapor pressure and distillation curve) and octane rating (RON and MON).

Following a description of the major characteristics of the available refinery basestocks, this section describes how to predict the behavior of component blends, especially the octane rating. The increased production of unleaded gasoline is also discussed, which is the present situation in some countries and a new situation for others.

The various requirements for improved environmental protection that now apply in the United States will soon come into effect in other countries. These requirements have led to the production of **reformulated** gasoline that must meet stringent requirements for performance (RVP) and chemical composition (levels of benzene, aromatics, olefins, sulfur, ...). Reformulated gasolines are discussed in complete detail in Chapter 5.

3.8 Characteristics of gasoline basestocks

An analysis of some of the basestocks used for gasoline production is provided in Table 3.19. This data is provided for illustrative purposes only. Factual data derives from the type of crude oil used, the separation methods chosen, the production technology, and the severity of the processes. The characteristics to be considered are mainly the following ones: density, vapor pressure, distillation range, octane rating, and chemical composition by family.

To be more precise, a refiner is primarily interested in the following two requirements:

- **vapor pressure**
- **MON**

In fact, experience has shown that among the various specification requirements, those that are related to MON and RVP are the most difficult to meet from both an economic and a technical perspective.

This leads to minimizing the use of refinery basestocks that are too volatile or too low in MON, or both.

With respect to RVP, the most vulnerable products are mainly the C_4 and C_5 fractions, which are the direct distillates and the isomerates. Their addition to the gasoline pool must be **closely controlled**. This is particularly true for C_4 fractions, which should not be more than 8 to 10% of gasoline-pool content.

Table 3.19 *Properties of refinery basestocks used in gasoline blending.*

Properties	Refinery basestock							
	Butane	**Iso-pentane**	**SR light gasoline**	**Low-pressure reformate**	**Medium-pressure reformate**	**High-pressure reformate**	**Heavy reformate**	**Straight FCC gasoline**
Density (kg/dm³)	0.579	0.610	0.655	0.819	0.778	0.790	0.867	0.752
RVP (kPa)		15.0	70.0	40.0	41.4	41.9	6.0	42.7
ASTM distillation (°C)								
IBP		20	30	50	34	41	117	39
10%		22	38	75	63	56	121	56
30%		24	45	113	92	85	126	73
50%		27	52	138	112	108	132	105
70%		29	61	146	128	126	140	146
90%		31	77	170	146	145	153	179
FBP		36	101	204	180	184	194	204
Analysis (% vol)								
Paraffins	100	100	91	27	44	41	2	41
Naphthenes	–	–	7	1	2	–	–	8
Olefins	–	–	–	–	–	–	–	21
Aromatics	–	–	2	72	54	59	98	30
RON	95	92	68	102	94	99	113	91
MON	92	89	67	91	85	88	102	83
S	3	3	1	11	9	11	11	8
ΔR100	–	–	–	16	15	19	–	–1

Table 3.19 *Properties of refinery basestocks used in gasoline blending (end)*

Properties	Refinery basestock						
	FCC light gasoline	FCC heavy gasoline	Alkylate	Isomerate	Dimersol	ETBE	MTBE
Density (kg/dm³)	0.687	0.830	0.710	0.643	0.686	0.750	0.746
RVP (kPa)	67.9	5.0	40.0	78.0	49.0	40.0*	55.0*
ASTM distillation (°C)							
IBP	37	115	32	35	56		
10 %	47	134	72	42	58		
30%	53	146	100	44	59		
50%	63	158	106	47	60	72.8	55.3
70%	79	172	110	52	70		
90%	106	192	126	61	130		
FBP	134	208	198	75	188		
Analysis (% vol)							
Paraffins	57	22	100	95	–	–	–
Naphthenes	9	6	–	5	–	–	–
Olefins	28	10	–	–	100	–	–
Aromatics	6	62	–	–	–	–	–
RON	91	95	97	83	97	117*	118*
MON	82	84	94	82	82	101*	101*
S	9	11	3	1	15	16	17
ΔR100	– 1	–	–	–	0	–	–

* Blending index.

The basestocks with the lowest MON ratings are those that contain olefins (cracked gasoline, dimersol, ...).These compounds should only be used **in moderation**, even if their other properties (RVP, RON, ...) are worthwhile.

Fig. 3.57 provides a panoramic view of the RON and MON ratings of the various basestocks suitable for gasoline blending. The most interesting ones are those high in aromatics (reformate) or isoparaffins (isomerate, alkylate). However, as described elsewhere (see 5.12), aromatic levels are precisely controlled in **reformulated** gasolines.

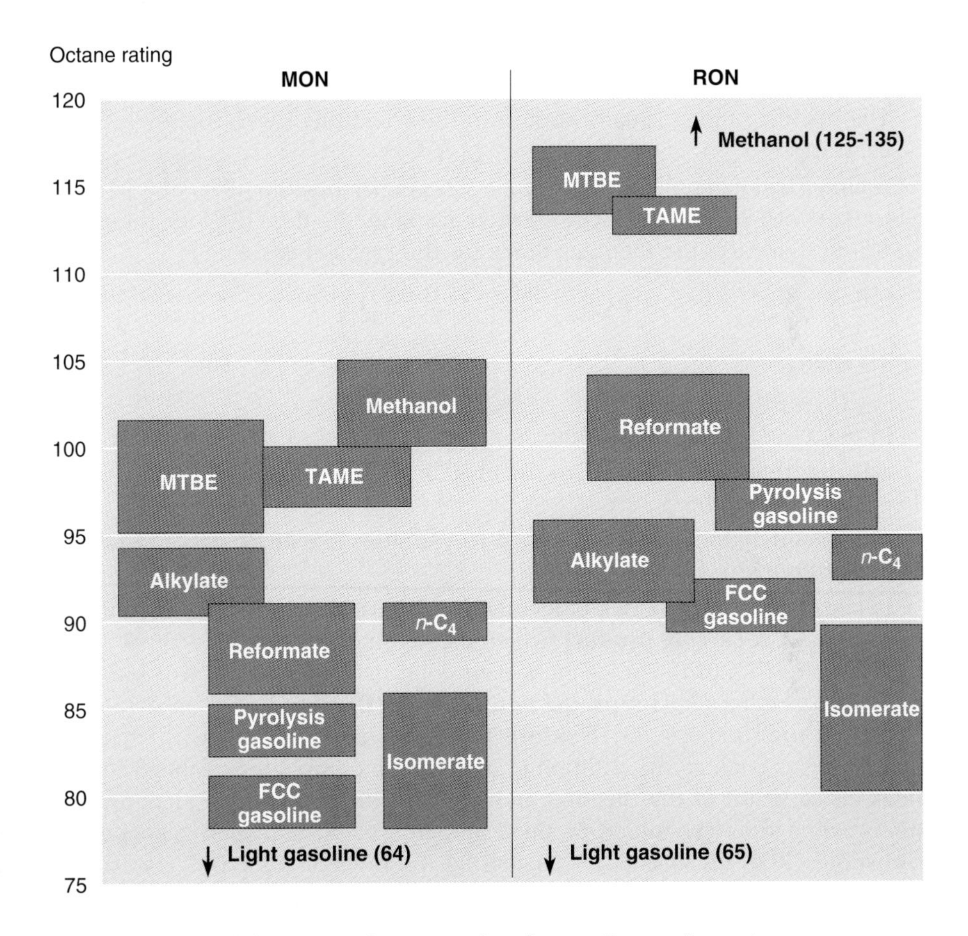

Figure 3.57 *Range of octane ratings for gasoline pool constituents.*

3.9 Behavior of blends

It is obvious that the behavior of gasoline blends, with respect to their physical and chemical properties (volatility, octane rating, ...), does not follow the rule of linear addition as might be expected. To account for these variances with ideal behavior, the concept of **blending indexes** is often used.

3.9.1 Blending indexes

Assume that the RON of a blend (AB) of two basestocks (A and B) must be predicted.

The blending index (M_A) of one of the two constituents, A for example, can be calculated from the following equation,

$$RON_{AB} = x\, M_A + (1 - x)\, RON_B$$

where

RON_B and RON_{AB} respectively represent the RON of constituent B and of the blend AB

x is the volume fraction of constituent A in the blend

The blending index usually applies to the constituent with the lowest concentration.

A numerical example of the formula follows: by blending 80% of a 97 RON reformate with 20% of a 92 RON cracked gasoline, the final RON for the blend is 96.5. In this case, the blending index for the cracked gasoline is

$$M = \left(\frac{96.5 - 97 \times 0.8}{0.2} \right)$$

which yields:

$$M = 94.5$$

In this case, the blending-index is 2.5 numbers higher the RON rating.

The difference with respect to the ideal is 0.5 RON (measured RON = 96.5; theoretical RON = 96.0).

This blending index can be applied to variables like MON, vapor pressure, and volatility (E70, E100, ...).

In fact, the blending index of a constituent varies with the **content** and the **nature of the receiving product**, which means that it is not an intrinsic property. Despite this problem, refiners have long used the concept of blending indexes to predict and set **production schemas** that are based on data gathered from their experience. This practice is tending to disappear, except in specific cases such as the addition of oxygenated compounds. Table 3.20 provides an estimate of the blending indexes for various types of alcohols and ethers when they are added in small amounts to an unleaded gasoline like Eurosuper (RON 95/MON 85). Considering the diverse situations involved in the composition of the receiving basestock, instead of a single value, a blending index is given as a range of possible values.

3.9.2 Predicting octane ratings of gasolines based on their composition

Modern gasoline blending techniques make use of chemical analysis and octane-composition rating correlations. A technique proposed by the BP (Descales et al. 1989) uses **infrared spectrometry** to predict gasoline octane ratings. The technique has the advantage of being very fast (one minute). It

Table 3.20 *Blending indexes for some alcohols and ethers.*

Type of product	Blending index (M)	
	RON	MON
Methanol	125-135	100-105
MTBE	113-117	95-101
Ethanol	120-130	98-103
ETBE	118-122	100-102
Terbutyl alcohol (TBA)	105-110	95-100
TAME	110-114	96-100

has also proven itself to be very useful for adjusting blends by making small changes to the standard composition, and it is therefore most useful in fixed refining situations at a specific site.

Other more generally applicable and widespread techniques use **gas chromatography** (Durand et al. 1987). These techniques can be used to identify and quantify almost 200 constituents whose octane ratings are known.

Using this data, a primary method consists of developing linear models like this one

$$\mathrm{OR} = \sum_{i}^{n} (\mathrm{OR}_{ppi} + K_i) C_i$$

where

OR = octane rating of the gasoline (RON or MON)

OR_{ppi} = octane rating of the pure product i

C_i = content (% mass) of constituent i

K_i = coefficient that represents the divergence from the ideal—negative or positive—while accounting for the product's behavior in the blend

The K_i term in this equation is the same for all constituents that belong to the same hydrocarbon family, that have the same level of branching, and that have the same number of carbon atoms. Computerized chromatographic analyzers for octane-rating analysis have now been commercialized. The Chromoctane® is manufactured under license from IFP and distributed by the *Vinci-Technologie* company. This device enables the automatic online octane rating of reformates.

Non-linear methods are required to predict the octane ratings of complex blends in which the blending behavior of an i constituent depends on the hydrocarbon environment in which it is found. Therefore, the octane rating (OR) for gasoline is expressed by the following equation,

$$\mathrm{OR} = \sum_{i}^{n} (\mathrm{OR}_{ppi} + K_i^{\bullet}) C_i$$

where the preceding notations apply except for K_i^*, which is a function of the content of various hydrocarbon groups as follows:

$$K_i^* = f\,(C_{f1}, C_{f2}, C_{f1} \cdot C_{f2} \ldots)$$

There are rather significant **interactions** between the constituent being added and the receiving system, for example:

- in an **aromatic-olefinic** environment: the addition of paraffins and isoparaffins accompanied by a methyl group is usually benefic
- in an **aromatic** environment: the addition of olefins and isoolefins accompanied by a methyl group is also benefic
- in an **olefinic** environment: the addition of light aromatics (C_6 to C_8) is detrimental and the addition of heavy aromatics (C_{9+}) is benefic
- in a **paraffinic** environment: the addition of any aromatic is detrimental

These mathematical models help considerably in the fine tuning of octane-rating prediction. Fig. 3.58 shows the difference in measured and calculated RON and MON for one hundred samples of a wide spectrum of gasoline blends. The differences are in the range of ± 3 numbers for 70% of the population evaluated.

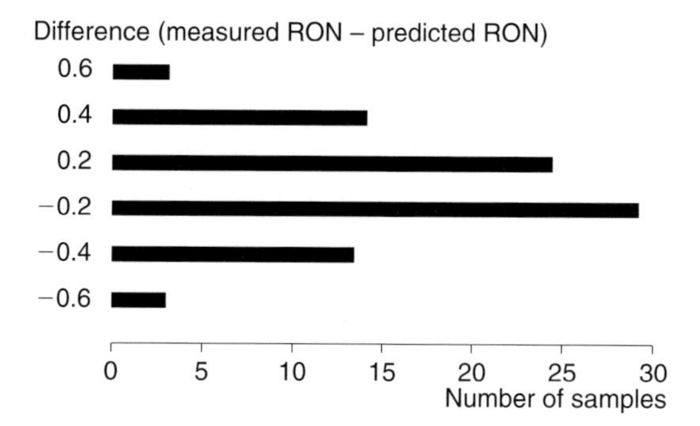

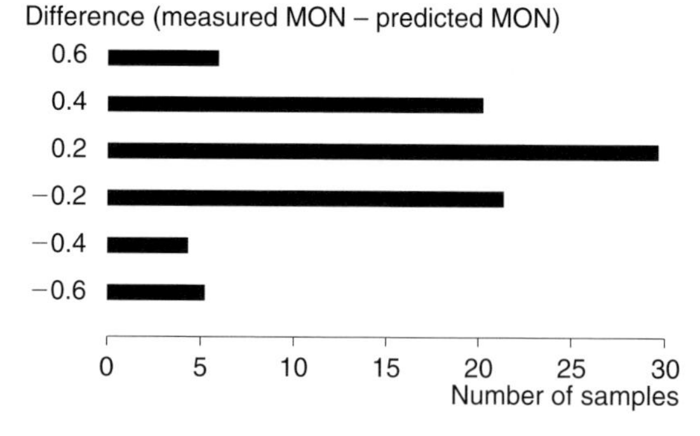

Figure 3.58 *Difference between measured and predicted octane ratings.*

Measurement type	*RON*	*MON*
Standard deviation of all results	*0.34*	*0.42*
Standard deviation of results with RON > 95	*0.30*	*0.40*

The preceding analysis is obviously just an example, but it reveals the constraints that will be placed on refiners in the 2000s. The composition of the fuel pool will become more and more complex and more-precise modeling tools will be required.

Even today, current gasoline blends are being formulated by adjusting the most critical parameter (the MON) very close the required specifications. In fact, any excess octane rating beyond a vehicle's requirements does not result in better performance and it incurs additional production costs.

3.10 Composition and characteristics of finished products

The production methods used to provide basestocks and blends in adequate proportions have undergone significant changes since 1990. These changes occurred in Europe with the rapid introduction of unleaded gasolines and in the United States with the introduction of reformulated gasoline. This section primarily describes the blending of conventional gasolines, while the specific problems involved in the production of reformulated gasolines are covered in Chapter 5 (see 5.12).

3.10.1 Leaded and unleaded European gasolines

The most probable scenario for the penetration of unleaded gasolines into the European market, through to the year 2000, is shown in Fig. 3.59. Gasolines containing 0.4 g Pb/liter have been unavailable since 1992 and those with 0.15 g Pb/liter will also disappear by the year 2000.

At that point, unleaded gasolines will be the only ones available to the **entire vehicle fleet**.

Even though Eurosuper grade (RON 95 and MON 85) is sufficient, some countries in Europe, notably France, produce and distribute higher-octane unleaded gasolines (referred to as Super 98) with ratings of 98 and 88 respectively for RON and MON. These fuels can also be used by some older vehicles unequipped with catalytic converters (see 3.13).

In Europe, the market **proportions of Eurosuper and Super 98** will remain ambiguous for the next few years. This question is of fundamental importance for the refiner because it affects the development and installation, or the retirement, of various processes. For example, Table 3.21 lists the typical composition, by basestock and content level, of the main gasoline grades currently sold in Europe: conventional premium at 0.15 g Pb/liter, Eurosuper, and Super 98.

The following comments apply to Table 3.21:

- There is no fixed composition for a given type of gasoline. Instead, there is a **range** of possible concentrations of each type of basestock. Eurosuper can contain between 15 and 35% catalytically-cracked gasoline and between 35 and 60% reformate.

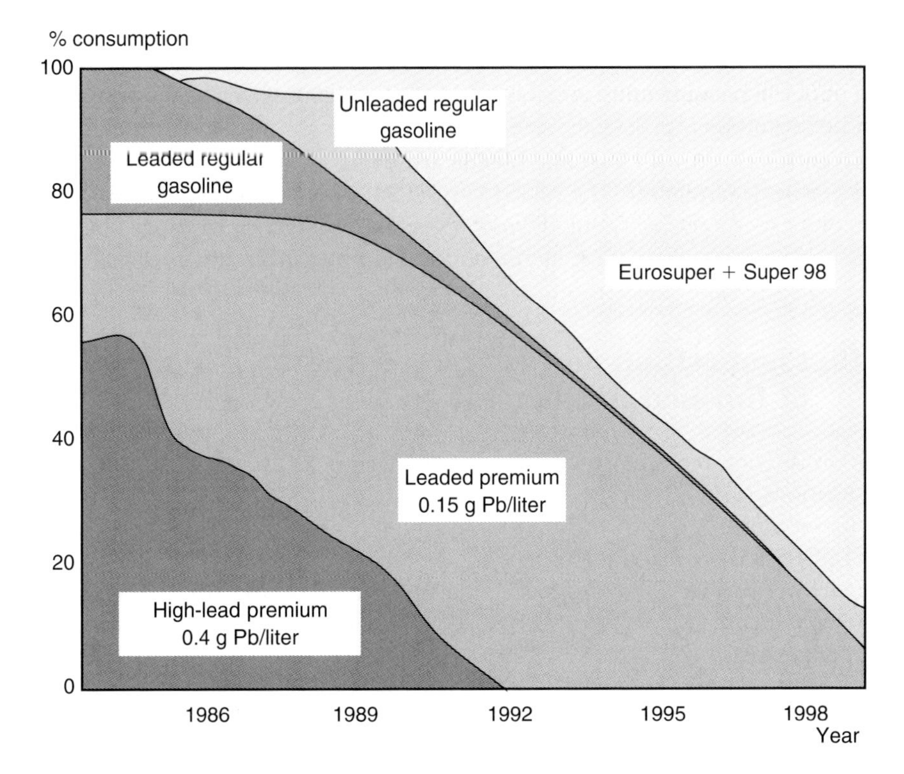

Figure 3.59 *European consumption predictions for various gasolines.*

Table 3.21 *Typical gasoline composition by basestock and content level.*

Basestock	Composition range (% vol)		
	Premium at 0,15 g Pb/l	**Eurosuper**	**Super 98**
C_4 fractions	2-4	2-4	2-4
Straight-run gasoline	5-10	0-8	0-5
FCC gasoline	20-40	15-35	10-25
Reformate	30-60	35-60	45-80
Isomerate	0-5	0-5	0-8
Alkylate	0-10	0-15	0-20
MTBE	0-3	0-5	0-9
RON	95*	96.5	99.5
MON	84*	85.5	88

* Before adding lead.

- The octane ratings of leaded premium gasoline, before the addition of lead, are only **slightly lower** than Eurosuper (about 1.5 numbers). Therefore, the reduction of lead in Europe does not cause serious refining problems as long as this product is replaced by Eurosuper (that will not be the case, however, if Super 98 is chosen).
- Although the official specifications for unleaded gasolines require a **sensitivity** (RON - MON) of 10 numbers, this is usually exceeded in practice leaving the MON as the refiner's real constraint. In other words, once the MON of 85 is reached for Eurosuper, the RON is usually greater than 95.

3.10.2 Improving the quality of gasolines: refining constraints and flexibility

In all countries, gasoline marketing faces changes in both **quantity** and **quality**.

In Europe for example, the gasoline share of the market for all petroleum products is rising slightly (about 35% in 1999). The major quality problem is increasing the octane rating of the gasoline pool. This is required by the progressive introduction of unleaded gasolines and the requirements of a vehicle fleet that is vulnerable to knock (vehicles with small displacement engines that operate at high loads).

Increasing octane levels is also an essential objective for those countries that have recently decided to reduce or eliminate lead (regions of South-East Asia).

In the United States, instead of increased octane ratings, the reformulation of gasoline is being sought for environmental reasons.

The next section briefly describes the effect that these large scale changes will have on refining.

3.10.2.1 Improving octane ratings

To improve octane ratings, the contribution of specific processes like catalytic reforming and MTBE synthesis must be expanded. Also, increases in alkylation and isomerization are needed, but these processes cannot contribute large quantities to the gasoline pool because of limited feedstocks.

More sophisticated processes are conceivable: for example, gasoline from the middle of the catalytic-cracking distillation range (the fraction from 130 to 160°C), which has a very-low octane rating, could be separated and sent to catalytic reforming after hydrotreatment.

These techniques require large investments and result in high energy consumption. This is shown in Fig. 3.60, where the refinery **process-consumption** is plotted against the average MON of the gasoline pool. The beneficial effect of MTBE is apparent in this example.

3.10.2.2 Reducing the aromatic and olefin content

The reduction of aromatic and olefin content is a trend that has begun to appear throughout the world, especially in the United States with the produc-

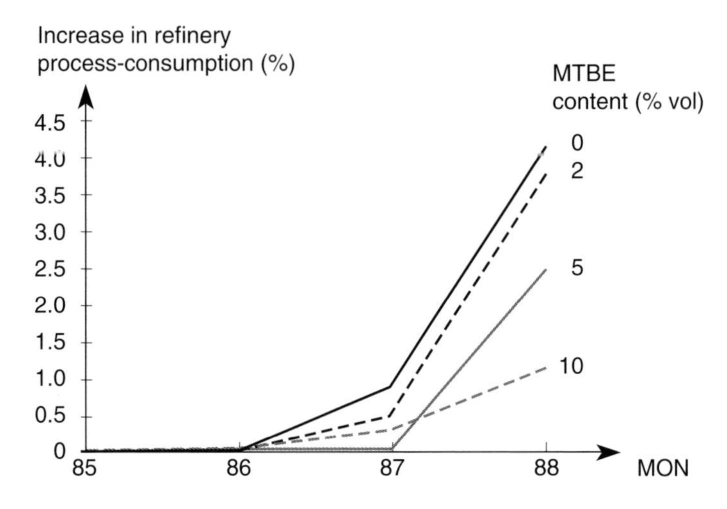

Figure 3.60 *Increases in refinery process-consumption for the production of high-octane unleaded gasoline.*

tion of reformulated gasoline. Reformulated gasoline is dealt with subsequently (see 5.12); this section provides only general information on the topic.

In a low-aromatic-content production scenario, a dependence on catalytic reforming would be considerably reduced while the contributions of alkylation, isomerization, and etherification would be increased.

Exclusively olefinic blendstocks of the dimersol type would be of little interest in this context. The best course would be to **etherify** them using methanol (Nocca et al. 1993). Table 3.22 lists the comparative properties of **dimate** (a sidestream of the dimersol process) and an etherified product referred to as **dimatol**. Note that the olefin content is considerably reduced as well as the vapor pressure—a favorable result. In addition, the RON rating and especially the MON rating, are noticeably improved.

Table 3.22 *Improvements with the etherification of dimersol.*

Properties	Dimate	Dimatol
Product volume (relative quantity)	100	107.5
RON	96	98
MON	81	86
RVP (bar)	0.5	0.3
Olefin content (% vol)	100	57

3.10.3 The gasoline pool

Fig. 3.61 provides a comparative view of the gasoline-pool composition in two of the world's largest consumption areas, the United States and Western Europe. The figure provides an **order-of-magnitude** view only, since changes will occur from the introduction of reformulated gasolines in the United States, from the elimination of lead in Europe, and from the introduction of new specifications for gasoline.

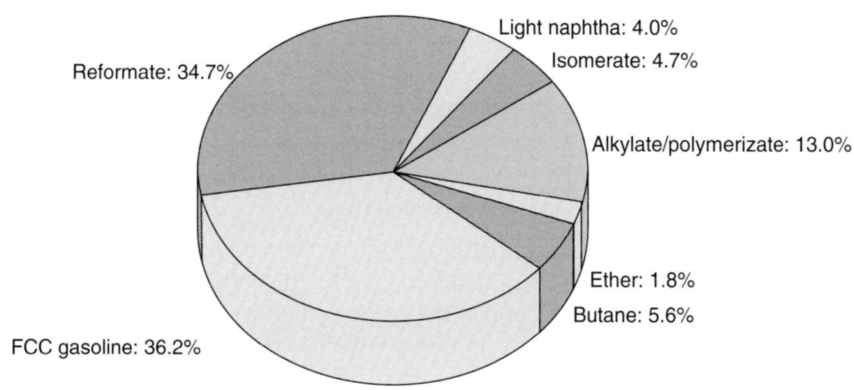

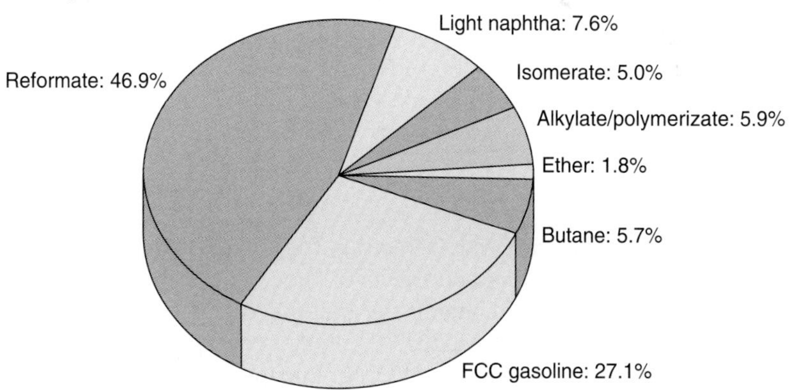

Figure 3.61 *Average composition of the gasoline pools in the United States and Europe.*

Engine-fuel matching

When designing new engines and vehicles, automobile manufacturers must account for the limitations resulting from the characteristics of available fuels. Likewise, refiners seek to obtain the best possible products at the lowest cost, within the constraints imposed by fuel specifications. These complimentary efforts are referred to as engine-fuel matching.

This optimization brings several criteria into play. The most important ones are

- obtaining maximum performance (efficiency, torque, power) without knock
- minimizing emissions (this is covered in Chapter 5)
- providing the best possible driveability (starting, moving off, smooth acceleration)
- maintaining the preceding items over time

The first item considered in this section is octane requirement and the methods used by manufacturers to avoid knock. This is followed by a description of the methods used to control deposits in fuel systems and combustion chambers, and how these deposits affect vehicle maintenance and longevity.

3.11 Octane requirements

Octane requirements initially apply to each new vehicle introduced to service and then to the whole vehicle fleet without neglecting the **extreme diversity** of conditions that can be encountered (vehicle age, mileage, driving conditions, mechanical condition, ...).

3.11.1 Definition

A gasoline vehicle's octane requirement is the octane rating that is both **necessary and sufficient** to avoid all traces of knock under the most-severe operating conditions, without changing the manufacturers recommended adjustments–ignition timing and mixture equivalence ratio. Octane ratings are normally expressed in terms of the RON in Europe and the (RON + MON)/2 in the United States.

It is important to distinguish between an octane requirement, which applies to an engine or an **engine-vehicle** system, and a Road Octane number that applies to the behavior of a gasoline in a given vehicle. From a methodology perspective, octane requirements are established at a fixed spark advance, whereas Road Octane ratings are derived by varying the spark advance.

3.11.2 Methods of establishing octane requirements

There is no universal method of measuring octane requirements. However, there are technical groups in various regions of the globe who develop procedures that are adapted to local vehicle operating conditions and who also collect data on vehicle fleets. These organizations are the following: the Coordinating Research Council (CRC) in the United States, the Cooperative Octane Requirement Committee (CORC) in Europe, the Japanese Petroleum Research Institute in Japan, and the Australian Cooperative Octane Requirement Council (ACORC) in Australia. Instead of describing the detailed activities of each of these groups, this section covers the principles of the procedures they use and the general trends that are observed.

3.11.2.1 General methodologies

For the explanation of these methodologies, assume that only PRFs are available (isooctane-heptane blends).

Octane requirements are determined either during a vehicle **acceleration phase** or at a **steady speed under full-load conditions**.

Under **acceleration**, the operating mode and the type of results obtained are shown schematically in Fig. 3.62. For example, an acceleration under full load from 50 km/h to the maximum speed of the vehicle is carried out using various PRF blends with octane ratings that increase in increments of one octane number. At lower RON values, knock is present throughout most of the acceleration. As the RON value increases, repeating the same acceleration results in knock occurring only at certain speeds, then gradually reducing, and ultimately disappearing.

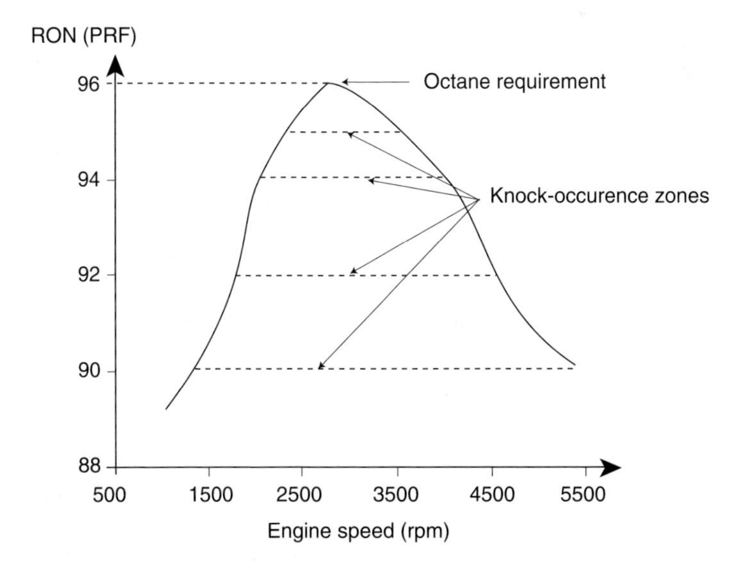

Figure 3.62 *Principles of octane-requirement measurement during acceleration.*

The RON that corresponds to the last traces of knock is the octane requirement of the vehicle. The most critical vehicle speed for knock can also be determined with this procedure.

Another approach consists of **constant speed** operation, usually at full load. With this method, vehicle behavior at various speeds (for example, 500 rpm increments from 1500 rpm) can be evaluated. The engine is fueled with PRFs of increasing RON until knock-free operation is obtained. The octane requirement for the engine is the RON needed to avoid knock at the most-critical speed. This method can be used with engines installed in chassis or mounted on test benches.

To increase the precision of the results, the Knock Limited Spark Advance (KLSA) technique can be used. This technique uses changes in spark advance to initiate knock for each gasoline undergoing evaluation. From this, a series of spark-advance curves for trace knock can be plotted for various speeds (Fig. 3.63). The engine's octane requirement can be read from the series of curves, because the standard spark advance for each speed is known. This technique provides more information than the previous one because it provides the octane requirement at each speed.

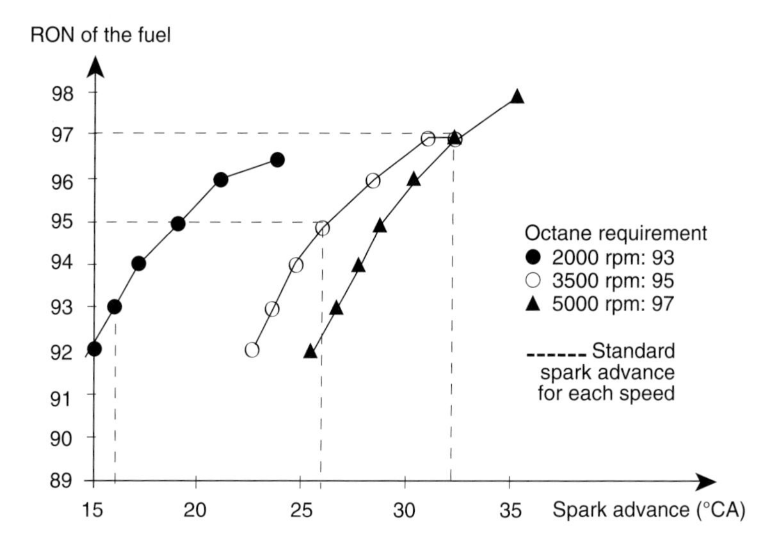

Figure 3.63 *Example of evaluating an octane requirement using the KLSA technique (schematic representation).*

3.11.2.2 Families of fuels

Octane requirements measured with PRFs indicate a vehicle's octane needs, but they do not provide information on the octane level of the fuel that must be used.

To obtain that information, **families** or ranges of commercial fuels must be considered. Each family is prepared from all the possible blends of two product bases, one of **high** octane and the other of **low** octane. The blending is done in such a way that the base products and the blends have, to the extent possible, the same sensitivity.

In Europe, families of fuels have been defined as high sensitivity (HS), medium sensitivity (MS), and low sensitivity (LS). The range of sensitivity has been defined as 12 to 13 for HS, 9 to 10 for MS, and 7 to 8 for LS, as shown in Table 3.23.

The previous procedures using PRFs can be repeated with each family of commercial fuels. The results obtained depend on the family chosen. Usually, the requirements are higher with commercial fuels than they are with PRFs. This result supports the fact that commercial fuels often demonstrate a degree of "depreciation" when tested in vehicles.

In practice, the octane levels of the MS family represent the level of octane needed by commercial fuels to meet vehicle requirements. This does in fact occur, because the refinery gasolines, as mentioned earlier, have sensitivities between 9 and 11.

Table 3.23 *Fuel families.*

Group	Designator	Octane rating			
		High		Low	
		RON	MON	RON	MON
High sensitivity	HS	101	88	88	76
Medium sensitivity	MS	101	91	88	79
Low sensitivity	LS	101	93	88	81

3.11.3 Parameters affecting octane requirements

With any given vehicle, octane requirements depend on **engine speed**; the direction and the amplitude of these variations in octane requirement are themselves a function of the type of fuel. The changes shown in Fig. 3.64 are typical. With PRF blends and low-sensitivity gasolines, octane requirements decrease as the speed increases. With high-sensitivity gasolines, octane requirements increase along with speed up to about 5000 rpm. Beyond that speed, other parameters (for example, reduced volumetric efficiency) begin to take effect and contribute to a drop in octane requirement. These trends confirm the findings that indicate that high-sensitivity gasolines have a high propensity to cause knock at high speeds.

Table 3.24 provides a general overview of the relationship between octane requirements and engine design and control parameters. These are **orders-of-magnitude** values, because, in practice, each one is separately involved. However, the following points are worth mentioning:

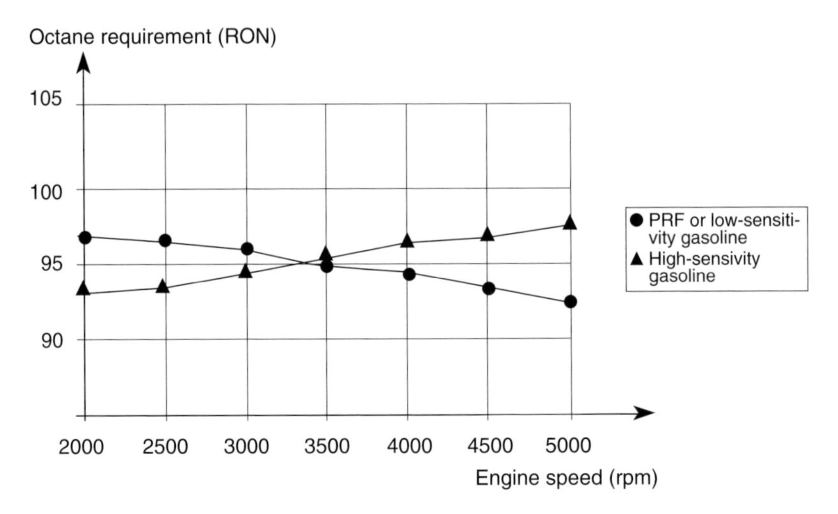

Figure 3.64 *Example of the change in octane requirement with engine speed.*

Table 3.24 *Engine parameters that affect octane requirements.*

Engine parameters	Parameter variation	Octane requirement variation (RON or MON)
Compression ratio	+1	+4 to +7
Spark advance (°CA)	+1	+0.5 to +1
Intake air temperature (°C)	+25	+1 to +4
Intake air pressure (mbar)	–10	–0.5 to –1
Equivalence ratio	+0.2*	–4
Hygrometry (g water/kg of dry air)	+4	–1
Altitude (m)	+300	–1 to –1.5 (without air-fuel ratio adjustment) –0.2 (with air-fuel ratio adjustment)

* Based on 1,10 (increasing from 1.10 to 1.30).

- A one number increase in the **compression ratio** results in a 4- to 7-number increase (average of 6) in octane requirement. In practice, some flow characteristics—chamber shape, turbulence—are changed along with the compression ratio and they modify its intrinsic effect. Most manufacturers of modern engines account for the octane ratings of available gasolines and use compression ratios of about 9.5 to 1.
- An increase in **spark advance** of 1°CA increases the octane requirement by 0.5 to 1 number (0.5 for engines with high initial requirements and about 1 number for the rest).

- A 5% increase in **volumetric efficiency** increases the octane requirement about 1 number. Conversely, operating a vehicle under conditions of low atmospheric pressure, for example, at high altitude, results in a decrease in octane requirements (Callison 1987). This situation is used to advantage in certain areas of the globe, especially South Africa, to make local use of fuels with low octane ratings.
- An increase in **intake air temperature** of 25°C (for example, going from 25 to 50°C) is solely responsible for increasing the octane requirement by 2 to 4 numbers. In reality, this parameter is very difficult to isolate from the others (volumetric efficiency, equivalence ratio, ...).
- The **hygrometry** in the ambient air also affects octane requirements. The endothermic evaporation of water vapor contributes to lower temperatures throughout the combustion cycle. Furthermore, water vapor acts as a diluant. Ultimately, the reduction in octane requirement can be up to 1 number for a hygrometry increase of 4 g of water per kg of dry air.
- Octane requirements are usually at a maximum for **stoichiometric or slightly-rich mixtures** (1.05). The octane requirement decreases for very-rich mixtures (above 1.10). This can be explained by the fact that the auto-ignition delay of a fuel increases as a function of the increasing equivalence ratio in excess of stoichiometry.
- **Combustion-chamber shape** and **flow dynamics** (turbulence, directed gas movement, ...) also have a determining effect on octane requirements. This is a very-competitive area among automobile manufacturers and numerical data is not provided. However, a broad general trend is to promote an orderly combustion progress that ensures that the last traces of unburned mixture are located in the **low-temperature areas** of the combustion chamber, which means away from the exhaust valve.

3.11.4 Octane requirement spread

Several examples of the same vehicle model, in the same mechanical condition (essentially the same mileage), and operated under the same conditions can often exhibit **differences** of several numbers in octane requirement. This situation occurs because of slight variations in design parameters and adjustments (actual compression ratio, equivalence ratio, spark advance, ...) The wide spread use of electronic fuel and ignition controls has recently resulted in a reduction of the range of dispersion to about 4 to 5 numbers, instead of the 7 to 10 numbers that occurred during the 1980s.

However, the effect of manufacturing tolerances on the actual compression ratio of each cylinder is still an unresolved issue (Betts 1979). This has important repercussions: it has been estimated that eliminating all variability in actual compression ratios would enable improved engine-fuel compatibility and reduce fuel consumption by up to 2%.

3.11.5 ORI — Octane requirement increase with mileage

As a perhaps colorful but pertinent comparison, **octane requirement increase (ORI)** is like high blood pressure in humans—it must not be allowed to become too high to ensure a lifetime of good health.

3.11.5.1 Observation and interpretation

A vehicle's octane requirement varies throughout its lifetime. It is usually low for a new vehicle and it increases steadily as a function of kilometrage until it stabilizes after 15 000 to 25 000 km. **ORI** is often about 5 to 7 numbers, but it can vary from 2 or 3 numbers to 12 or 13 numbers, depending on the vehicle and the mode of operation. After 25 000 km, the octane requirement can still vary about an average value, but the difference is usually small.

ORI is linked to combustion chamber **deposits** (Kalghatgi 1990) and, to a lesser extent, deposits on intake valves.

Deposits cause

- changes in heat transfer
- changes in combustion-chamber gas flows that can possibly affect the speed of energy release
- increases in compression ratio due to the volume they occupy
- chemical and catalytic effects on the unburned gases susceptible to knock

The **thermal insulator** role played by deposits is a deciding factor. Deposits cause an increase in the temperature of the intake charge, a reduction in wall temperature, and reduced heat transfer to the coolant. These processes alone could be responsible for 80 to 90% of the total ORI.

The increase in compression ratio—about 0.2 (Diedrichs 1988)—would have a less pronounced effect and it would only account for 10 to 20% of the ORI.

ORI is narrowly dependent on engine **operating conditions**. ORI is most pronounced after a long period of operation at low loads, which leads to heavy deposit formation. On the other hand, frequent high-load operation leads to low ORI. It is also possible to reduce the octane requirement of an engine by operating it continuously at full load, with or without trace knock.

3.11.5.2 Classification procedures

The procedures used in Europe make use of Renault F2N carburetor-equipped engines or Renault F3N fuel-injected engines, both of which are considered sensitive to ORI. The test duration is 200 to 400 hours and it includes a repetition of the operational sequences described in Table 3.25. The full-load octane requirement at various speeds is recorded at regular intervals. The KLSA method is used to determine **trace knock**, which can be detected by ear or by using a stethoscope. The change in KLSA (referred to as ΔKLSA) is measured between the beginning of the test and a given point in the 400-hour cycle.

Overall knock performance is normally expressed as the maximum value of ΔKLSA for all engine speeds combined, after the ORI has stabilized.

Table 3.25 *Test procedures used to measure ORI.*

1. Renault procedure 22710 (Renault F3N engine) (duration: 200 h).

Phase	Duration (min)	Engine speed (rpm)	Torque (N.m)	Equivalent road speed (km/h)
1	7	800	0	0 (idle)
2	35	1 455	17.3	50
3	5	800	0	0 (idle)
4	35	1 455	17.3	50
5	4	800	0	0 (idle)
6	30	0	0	0 (stopped)
7	2	800	0	0 (idle)
8	41.3	2 620	35.6	90
9	35.7	3 500	56.2	120
10	10	1 450	17.3	50
11	5	800	0	0 (idle)
12	30	0	0	0 (stopped)

2. Renault procedure 22700 (Renault F2N engine) (duration: 400 h).

Phase	Duration (hours)	Engine speed (rpm)	Torque (N.m)
1	1	idle	0
2	4	2 500	60
3	3	3 500	0
4	4	2 500	60

In their *cahier des charges "qualité"*, French automobile manufacturers (Advenier 1994) require that ΔKLSA be less than

- 8°CA for unleaded gasolines
- 12°CA for leaded gasolines (0.15 g Pb/liter)

Three examples of the results obtained with the Renault F2N procedure are shown in Fig. 3.65. As the data indicate, the above requirements are not easily achieved and conventional detergent additives can aggravate the ORI problem (see the next section).

The above procedures are not exactly ideal because they are lengthy and are therefore costly, and they have insufficient discrimination and poor repeatability. A shorter procedure (35 hours) that **uses a CFR engine** (Leduc et al. 1995; Faure 1997) has been also proposed. This procedure uses an oscilloscope to precisely determine the KLSA point by viewing the cylinder-pressure signal. In addition, the ability to vary the compression ratio on the CFR engine provides considerable flexibility to modify the deposit accumulation conditions, if necessary.

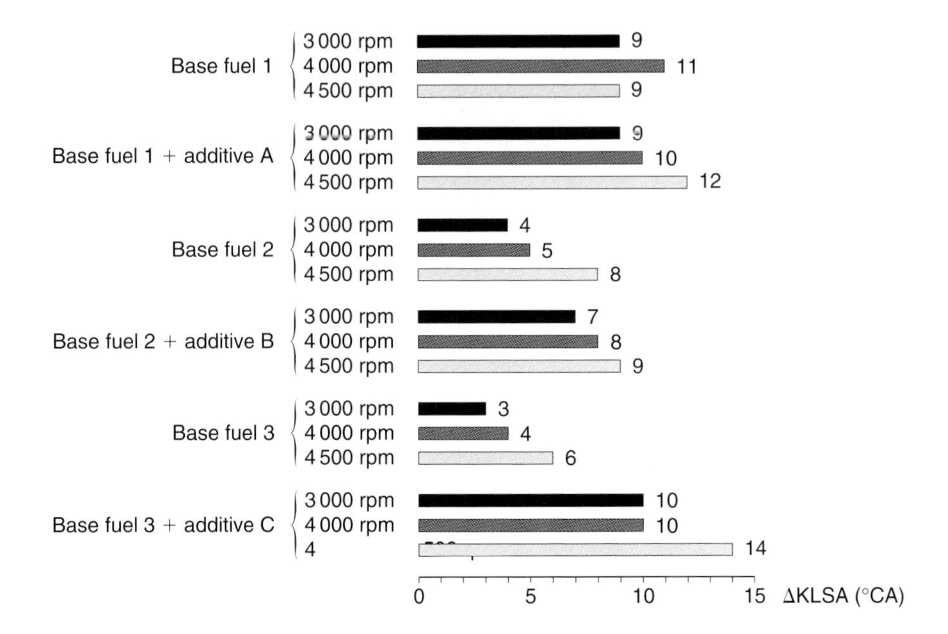

Figure 3.65 *ORI test results. Behaviour of various fuels ans additives (400 h test on a Renault F2N engine).*

3.11.5.3 Factors affecting ORI

As stated earlier, the longer an engine operates, the more ORI deposits it accumulates, especially under low-load conditions that favor deposit formation.

In a general way, and no doubt due to the nonlinearity of the octane rating scale, ORI varies inversely with the initial requirement (IR) at 0 kilometers (Nakamura et al. 1985).

As a result, the following equation has been proposed:

$$ORI = 43.4 - 0.44\ IR$$

This explains the problem with the use of the ΔKLSA criteria detailed earlier. In fact, the ΔKLSA point varies inversely with the initial octane requirement. Also, the ΔKLSA can vary noticeably between engines of the same production series and between cylinders in the same engine.

ORI is a function of the mass, composition, and thermal characteristics of the combustion chamber deposits. Several studies have been done on this subject (Ebert 1985; Anderson et al. 1982; Hayes et al. 1992).

For illustrative purposes, two examples are shown in Tables 3.26 and 3.27, which include the chemical composition and the properties of the deposits found on the head, piston, and inner face of the intake valves.

Table 3.26 *Examples of the chemical analysis of deposits.*

Type of analysis	Intake-valve deposits	Head deposits	Piston deposits
Element analysis (% mass)*			
C	73.3	63.3	53.3
H	5.1	3.8	2.4
N	2.5	2.4	2.5
O	14.6	26.5	30.1
Infrared and thermogravimetric procedures**			
Aromatics	••••	•	•
Alkyl chains	•••	••	••
Acids	••	••••	••••
Sulfates	•	•••	•••
Spectrography (semi-quantitative content in mg/kg)			
Al	100	300	1 650
Ca	3 000	5 400	4 500
P	1 650	5 000	4 200
Sn	< 30	300	1 150
Zn	3 600	3 600	3 600

* The other constituents are mineral based.
** The number of dots (•) indicates the relative importance of the chemical compounds.

Table 3.27 *Examples of the thermal characteristics of combustion chamber deposits.*

Thermal analysis of the deposits	Leaded gasoline	Unleaded gasoline*		
Heat capacity C (in $kJ/m^3 \cdot K$)	250	70 to 120	1 000 to 1 800	250 to 1 000
Thermal conductivity λ ($W/m \cdot K$)	0.7	0.4 to 0.7	0.3 to 0.7	0.2 to 0.3
Thermal diffusivity a (m^2/s)	$2.7 \cdot 10^{-6}$	$5.6 \cdot 10^{-6}$	0.3 to $0.4 \cdot 10^{-6}$	0.2 to $0.9 \cdot 10^{-6}$
Density ρ (kg/m^3)			900 to 1 500	1 530
Specifc heat C_p ($J/m^3 \cdot K$)			700 to 2 000	160 to 700

* Some values are slightly different depending on the source.

With respect to the composition of the deposits, what is important is the **presence of oxygen** in the following chemical structures,

$$-\mathrm{OH}, \qquad >\mathrm{C}=\mathrm{O}, \qquad -\overset{|}{\underset{|}{\mathrm{C}}}-\mathrm{O}-$$

which is evidence of partial oxidation reactions of the fuel and the lubricant. Trace metals are also found, which come mainly from lubricant additives.

The thermal properties of the deposits likely to affect knock, and hence ORI, are heat capacity (C), thermal conductivity (λ), and thermal diffusivity (a). The term $C = \lambda/a$, which is an expression of a deposit's ability to retain heat, is an important parameter.

These analyses indicate that ORI depends on the **properties of the fuel and the lubricant**. With respect to the lubricant, the most effective course of action is to reduce oil consumption and, if possible, the content of mineral ash in the oil.

With respect to the fuel, the initial efforts during the 1970s focused on the effects of reduced lead content on ORI. It was finally established that there were no important differences between leaded and unleaded gasoline with respect to ORI or its stabilization point.

Of greater interest is the intrinsic effects of some gasoline properties. In a general sense, the mass of combustion chamber deposits, and subsequently the ORI, have a tendency to increase along with the **aromatic content** (Choate et al. 1993), the **final boiling point**, and the concentration of **heavy constituents** that are derived from catalytic cracking (Peyla 1994).

Fig. 3.66 shows a comparative behavior of five fuels, one of which has a high proportion of catalytic-cracked components combined with unsatisfactory behavior.

Some years ago, when a detergent additive was mixed with gasoline to clean injectors and the intake system up to the intake valves (see 3.13), the mass of combustion chamber deposits increased along with the ORI (Schreyer et al. 1993). Presently one of the objectives in formulating additives is ensure that their use results in essentially the **same level** of ORI. The data shown in Fig. 3.67 demonstrates that it is possible to formulate additives which have no deleterious effect on ORI.

A more ambitious development would be an **anti-ORI additive** that would alter the mass, volume, or characteristics of combustion chamber deposits. Despite several development efforts (Papachristos 1992), these products are not widely used.

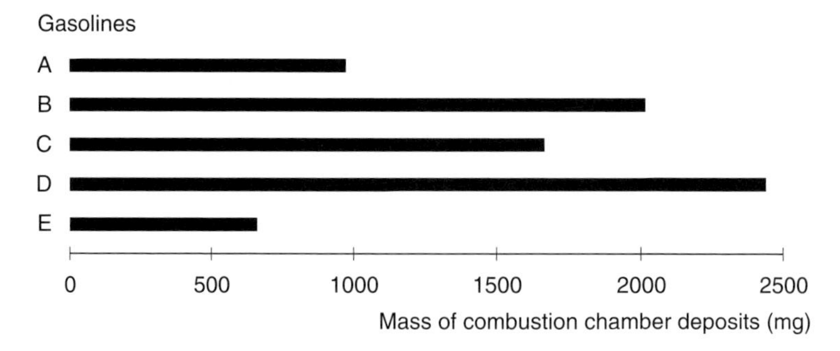

Figure 3.66 *Effect of fuel type on combustion chamber deposits.*
A, B, and C are conventional European or American gasolines
D is a fuel with a high proportion of FCC products
E is a reformulated gasoline

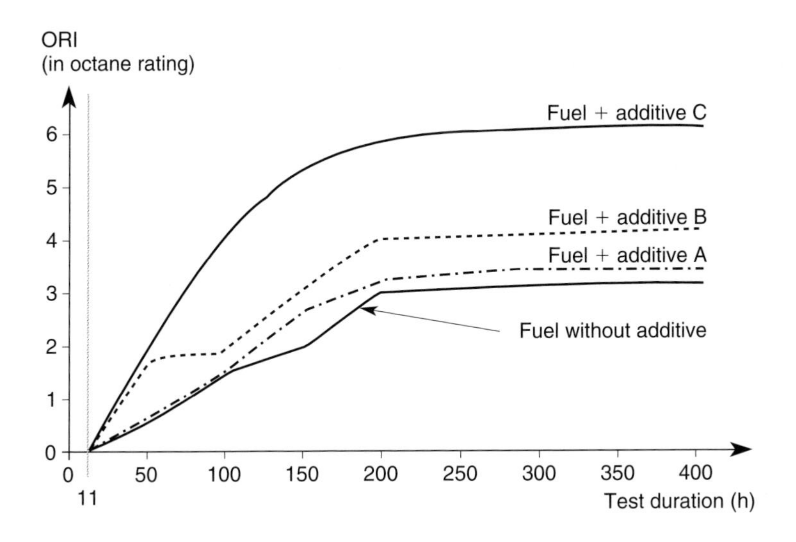

Figure 3.67 *Comparative effects of various detergent additives on ORI. (Mercedes M 102E engine fueled with Eurosuper)*

3.12 Controlling knock

This section describes the theoretical tools and the technological means that can be used to control knock at the engine-design stage and throughout the life of the vehicle.

3.12.1 Modeling

The following equation, which was developed to express auto-ignition delay, is a qualitative expression of knock tendency:

$$\theta = AP^{-n} \exp (B/T) \tag{1}$$

Knock occurs at a time t_c expressed by the following equation:

$$\int_{t_0}^{t_c} \frac{dt}{AP^{-n}(t) \exp\left(\frac{B}{T(t)}\right)} = 1 \tag{2}$$

The following items are represented:

t = time

t_0 = point of origin (for example, BDC)

A, n, and B = constants related to the fuel

$P(t)$ and $T(t)$ = pressure and temperature respectively as a function of time during the combustion cycle

Usually, constants A, n, and B are unknown beforehand. They can be derived by processing experimental data from the **pressure-temperature-time** changes of the fresh gases in the combustion cycle, when knock occurs at various times t_c. The mathematical resolution consists of determining the numerical values of A, n, and B that satisfy equation (2) as closely as possible for a given gasoline. When this technique is applied to PRF blends, it shows that their auto-ignition delay has the following unique formula,

$$\theta = 1{,}931 \cdot 10^{-2} \left(\frac{\text{RON}}{100}\right)^{3.4017} P^{-1{,}7} \exp\left(\frac{3800}{T}\right)$$

where the pressure P is in bar and the temperature T is in K.

The following important information can be derived from the formula:

- The respective effects of **pressure and temperature** parameters can be calculated and compared.
- The formula shows that the sensitivity to pressure and temperature stays constant within the PRF family. Variations in auto-ignition delay as a function of composition are accounted for by the **pre-exponent term A**.
- The octane rating is also shown to be a very-fine measure of knock tendency since a 1 RON change corresponds to a 3.4% change in delay.

This model can be applied to all types of gasoline. Table 3.28 lists the values for A, n, and B that have been derived for the following products: 95 PRF, high-sensitivity (HS) conventional gasoline, and low-sensitivity (LS) conventional gasoline. The numerical values for A, n, and B were subsequently used to determine the delay in milliseconds for a pressure-temperature combination (60 bar and 875 K) that is typical of the unburned gases in a production engine.

Note that term B, which could be considered as **activation energy**, is greater for a high-sensitivity gasoline that is rich in olefins. In fact, these types of hydrocarbons exhibit very temperature-dependent auto-ignition behavior.

Determining the values for parameters A, n, and B as described above involves considerable acquisition and processing of experimental data. A procedure referred to as **"the four octane ratings"** has also been proposed (Douaud 1978), which would use the CFR engine itself to determine the values

Table 3.28 *Ignition delay constants for various fuels.*
The values A, n, and B are represented in the following equation:
$\theta = AP^{-n} \exp(B/T)$

Fuels	A	n	B	θ* (ms)
95 PRF	$1.47 \cdot 10^{-2}$	1.70	3 800	1.07
LS gasoline	$3.86 \cdot 10^{-4}$	1.07	4 850	1.23
HS gasoline	$2.28 \cdot 10^{-4}$	1.68	7 525	1.28

* Derived for T = 875 K and P = 60 bar.

for *A, n,* and *B*. The approach consists of **modifying** the temperature and the speed used in the standard methods to define two new ratings:

- R'ON = 600 rpm with an intake mixture temperature of 149°C
- M'ON = 900 rpm with an unheated intake mixture

After establishing the RON, MON, R'ON, and M'ON, the numerical values for *A, n,* and *B* are determined and subsequently used in various techniques to classify the knock behavior of all vehicle-fuel combinations.

Fig. 3.68 shows an example of the knock model being used to define the **optimal adjustments** for a supercharged engine operating under fixed temperature and pressure conditions (Écomard 1982). The effects of ignition advance and volumetric efficiency are plotted against the values for IMEP and PRF-octane requirements. Because the quantity of energy introduced is constant, the IMEP represents the work done and it is proportional to efficiency. The best compromise is obtained with a high compression ratio and reduced spark advance to avoid knock. At part load this problem does not occur and optimal advance can be re-established.

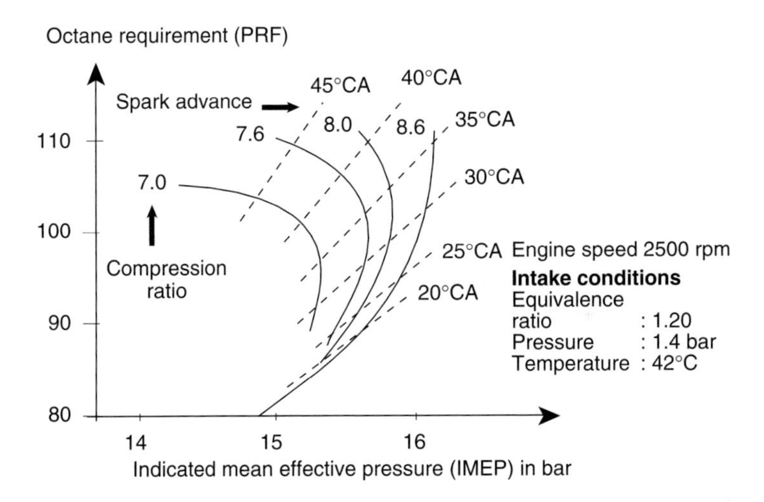

Figure 3.68 *Example of an IMEP-octane requirement compromise for a supercharged engine.*

3.12.2 Technological achievements

During engine design and development, manufacturers are always obliged to use a **safety factor** when dealing with knock and its destructive consequences. This implies that a number of events likely to affect a vehicle during its lifetime must be accounted for (ORI, out-of-adjustment mixture or spark advance settings, loss of cooling system efficiency, ...). The prospect of such serious problems, which are ill-defined and uncontrolled, encourages manufacturers to be cautious and to design for **longevity and reliability** over maximum performance. This context has recently changed however, due to a better understanding of the risks involved and the priority placed on **energy efficiency**.

It is now conceivably possible to produce engines and vehicles that can operate at the **knock limit**.

The following section describes the various methods that manufacturers can use to match engines and fuels for knock resistance. The available methods are mainly a choice of compression ratio and spark advance control, as well as the eventual use of knock detectors. Techniques like water injection, which were discontinued long ago despite their potential, are also described.

3.12.2.1 Ignition advance and compression ratio compromise

If the constraint imposed by fuel octane ratings did not exist, high compression ratios and an optimal spark advance could be used to achieve the best possible efficiency and specific power output. The constraints imposed by the limited knock resistance of fuels force reductions in compression ratios and retarded ignition timing, or simultaneous adjustment of both parameters.

Fig. 3.69 is a schematic that shows the various possible strategies and their impact on efficiency. Accordingly, to operate a vehicle that is initially designed to run on 98 RON gasoline on 95 RON gasoline, implies a compression ratio reduction (by 0.5) or reduced ignition timing (4 to 6°CA). Reduced compression would result in a 2% loss of efficiency throughout the operating range. Retarded ignition results in a smaller loss (less than 1%) and it only affects high-load use.

In Europe, adapting new vehicles to the Eurosuper RON 95 unleaded gasoline was done by applying new **ignition maps** and maintaining the high compression ratios of almost 10 to 1.

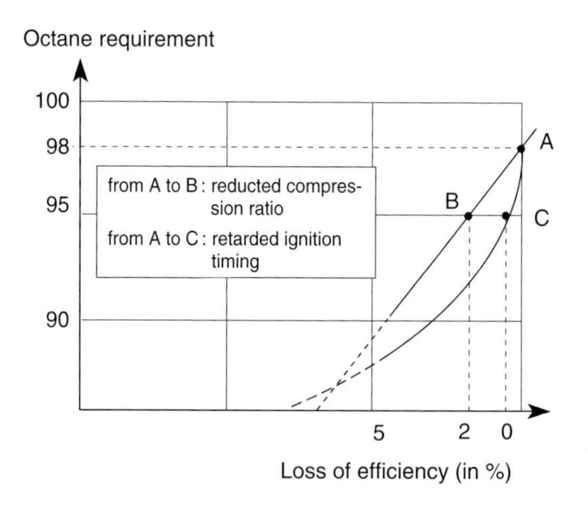

Figure 3.69 *Optimization strategies for engine-fuel matching.*

3.12.2.2 Knock detectors

Knocks detectors are only available on a large number of high-performance vehicles, which have a high propensity to knock.

The equipment includes a **knock sensor** that usually consists of an accelerometer to detect vibrations of the head and a subsystem that provides an **ignition-timing alert** (see Fig. 3.70). Knock vibrations can only occur within a narrow band of the combustion cycle—about 60°CA on either side of TDC. This fact is used to advantage to select a "window of time" in the cycle, which has a fixed reference point and a known width where the accelerometer signal occurs (Douaud 1978). Fig. 3.71 shows a combustion cycle with knock and the corresponding readout provided by the sensor.

The servo control parameters for ignition timing are

- **response time** (number of cycles with trace knock)
- **degree of correction** (It should be sufficient to suppress knock without significantly affecting performance.)
- **strategy for returning to nominal spark advance** when knock has disappeared

Knock sensors are an interesting development in the engine-fuel matching area. They are currently used as **safety, maintenance, and longevity devices** on knock-prone vehicles. Their use as **optimization devices** has yet to mature, but it looks promising.

It is understood in Europe that all recent production vehicles can use 95 RON Eurosuper without knock. However, several automobile manufacturers have requested the continued availability of high-octane unleaded gasoline (RON 98, MON 88) to enable high-performance vehicles to achieve their full potential. These vehicles could be fueled with Eurosuper on occasion without endangering their engines due to the use of knock sensors.

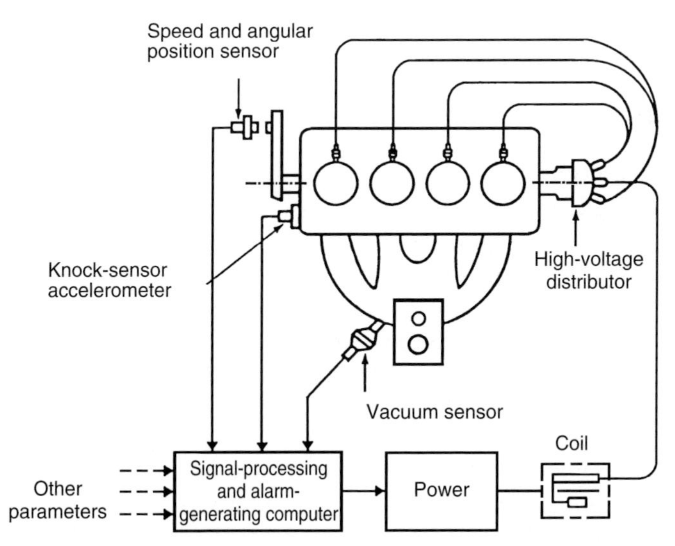

Figure 3.70 *Principle of the knock-sensing ignition system.*

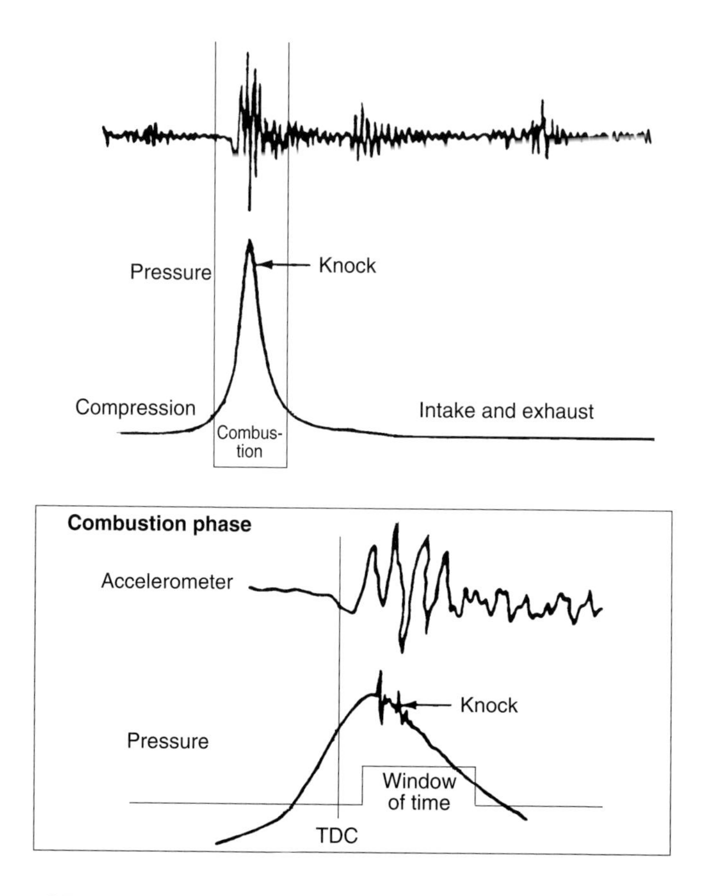

Figure 3.71 *Combustion diagram and accelerometer signal for knock detection.*

3.12.2.3 Other anti-knock devices

A. Water injection

In 1913, Professor Hopkinson of Cambridge University declared: "the idea of introducing water into internal combustion engines is not a new one!". Since that date, many "inventors" have acclaimed the idea and rediscovered its virtues. The advantages that are sought after and often demonstrated are numerous and varied (Harrington 1982; Peter et al. 1976): the possibility of increasing compression ratio and specific power, removing engine deposits, reducing nitrogen oxide emissions, engine cooling, ... Profitable use was made of these benefits in practice, initially with aircraft piston engines before World War II and later in Formula 1 automobile racing during the 1980s.

Water can be introduced directly into the intake system as a **liquid** or a vapor, or as an **emulsion** with gasoline. It acts by diluting the mixture and cooling the gases subject to knock. The latter effect is very pronounced when evaporation—a highly endothermic process—occurs within the cylinder. It is possible to add a significant flow of water—equivalent to the fuel flow—without having a significant effect on engine performance. However, if the water vaporizes in the intake manifold, a reduction in volumetric efficiency and power output occurs. Efficiency usually remains unchanged.

Fig. 3.72 and Fig. 3.73 show two examples of the effects of water injection on knock resistance. The criteria shown are the limiting compression ratio and the ignition timing that produces trace knock. Note that the amount of water necessary to provide significant benefits—a compression ratio gain of 1 or a 5°CA spark advance—is considerable because it represents a mass almost iden-

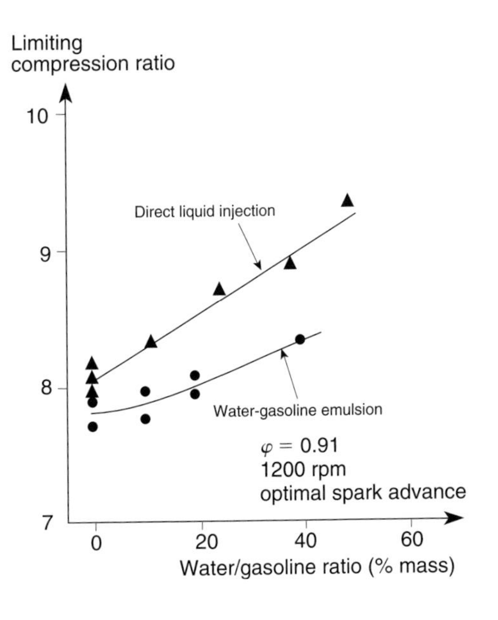

Figure 3.72
Effect of water injection on the compression ratio that results in trace knock. (Source: Peters et al. 1976)

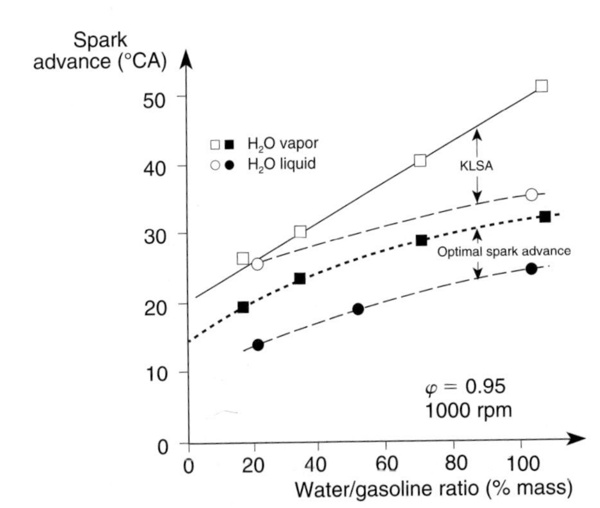

Figure 3.73
Effect of water injection on spark advance at trace knock (KLSA) and on optimal spark-advance settings. (Source: Harrington 1982)

tical to that of the fuel. Introducing water as a vapor provides better knock-protection efficiency but engine performance decreases rapidly.

Water-gasoline emulsions provide only modest potential improvements in compression ratio. It has been shown that the emulsifying agents (a type of polyoxyethylene) have a measurable tendency to cause auto-ignition. Despite this negative effect, the RON and MON of emulsions increase consistently with the water content (see Fig. 3.74).

The beneficial effect of water injection on knock resistance has been clearly proven. However, this satisfying result has **several drawbacks**: the problem of providing a second tank and supply system, problematic emulsion stability, poor combustion at low loads that requires a complex control system, problems with wear and lubricant incompatibility, ... This partial list of problems, and an evaluation of the efforts expended versus the rewards gained, explains why water injection is not a feasible solution for production vehicles.

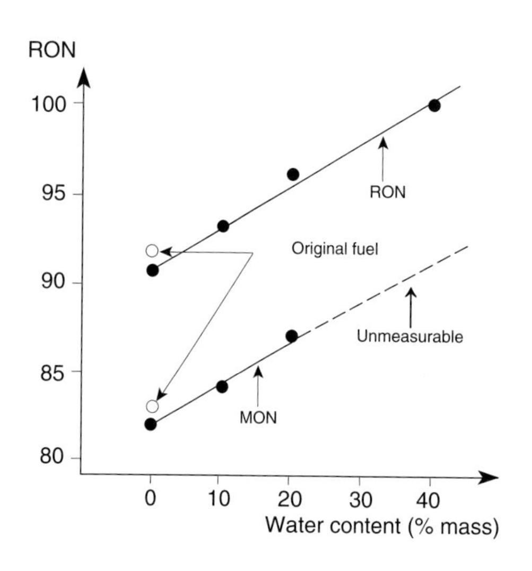

Figure 3.74
Octane rating of gasoline-water emulsions. (Emulsifier content: 3% by mass)

B. Exhaust gas recirculation

The primary objective of exhaust gas recirculation is to decrease the **emissions of oxides of nitrogen** (see 5.7), but it can have an indirect effect on octane requirements because it dilutes and cools the gases subject to knock. Therefore, high recirculation rates tend to reduce octane requirements as long as other parameters are not affected (increased intake-air temperature and increased spark advance due to reduced combustion speed). It can thus be seen that exhaust gas recirculation is a variable that can be considered, among others, in engine-fuel matching.

C. Dual-fuel operation

This technique has been used in the past on aircraft engines. It consists of using two fuel tanks and **two fuel systems**, one for conventional gasoline and one for a high-octane product like methanol. The engine operates at a high compression ratio. It is fueled with gasoline for starting and at low loads. The high-octane product is only used during acceleration and high-power cruising, when the risk of knock is high.

The current cost and complexity makes these devices unprofitable for a production vehicle.

3.12.3 Optimal octane rating

When a vehicle is placed in service, it must be supplied with a fuel that has a sufficiently-high octane rating to avoid knock. Beyond this point, further increases in octane rating do not provide any performance benefits (efficiency, power, better acceleration, ...), ... contrary to the belief of many drivers!

However, that is not the case when the octane rating is accounted for during engine **design**.

As previously indicated, increasing the compression ratio by one (for example, from 9 to 10) results in an improvement of approximately 6% in efficiency and a 6-number increase in the octane requirement. This means an increase in efficiency of 1% for each octane number increase; the relationship is referred to as Car Efficiency Parameter (CEP). The CEP represents the change in fuel consumption (% mass) that results from a change of one octane number, based on an engine with a compression ratio exactly matched to the fuel being used. In the preceding example, the CEP is 1. If an automobile manufacturer chooses to match an engine-fuel combination by changing the ignition advance instead of the compression ratio, the preceding trends are still qualitatively valid, but they point to a lower CEP (between 0.5 and 1).

Hence, it can be assumed that on average, each increase of one octane number reduces fuel consumption by 1% for a properly matched engine. Of course this change is accompanied by **supplementary energy costs** at the refinery, which are relatively small initially. However, the situation ultimately reaches a point when the refinery process consumption also consumes 1% per octane number (see Fig. 3.75). At that point, the optimal octane number represents a **state of energy equilibrium**.

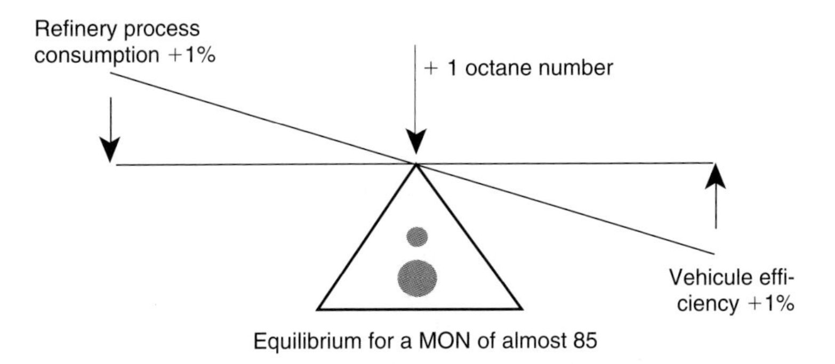

Figure 3.75 *Search for the optimal octane rating.*

The European RUFIT (Rational Utilization of Fuels in Private Transport) working group prepared techno-economic estimates when the conversion to unleaded gasoline was beginning. The estimate led the group to propose optimal values of 95 and 85 for RON and MON respectively (CONCAWE 1983). These values represent a compromise of refinery energy costs and vehicle fuel consumption. In reality, the MON (85) is the real constraint and the RON achieved in industrial production is actually higher than 95 for Eurosupers (currently 96 to 97).

Logically, the optimal octane-rating equilibrium should be under permanent review, given the constant change in the automobile fleet and in refinery processes. Hence, the increasing availability of higher octane basestocks in 1999 suggests an optimal MON higher than 85. In fact, it has been shown that variations about the optimal value have a small effect overall and do not justify frequent changes to an otherwise satisfactory situation. Furthermore, criteria that are purely energy-related must be considered separately from technical and economic ones. Consequently, the route to much-higher octane numbers implies **considerable economic expense**.

From an **economic** standpoint, estimates indicate that the octane level corresponding to the lowest gasoline production cost is about 1 number lower than the optimal energy consumption rating.

3.13 Operating older vehicles on unleaded fuels

This section describes the effects of operating older vehicles on unleaded fuels, which causes mechanical problems completely unrelated to knock.

3.13.1 Phenomenon of valve recession

To enhance the market penetration of unleaded gasolines in global regions where it is not widely used (Europe, Middle East, and Asia), it can be expected that these gasolines will be used in the older vehicles in the automobile fleet, which are not equipped with catalytic converters.

This scenario does not appear to present any major problems as long as the octane ratings of the fuels are satisfactory. In reality however, this is not an acceptable solution for all vehicles. Some vehicles are prone to **wear** and exhaust-valve **recession** that is completely at odds with engine durability. Fig. 3.76 presents a schematic view of the wear mechanism that occurs with some valve and seat materials. Small quantities of deposits consisting of lead compounds (oxides, sulfates, halides, ...) act as **solid lubricants** and they provide a beneficial effect by preventing premature wear.

Some older engines are not susceptible to valve recession because they are equipped with the appropriate metallurgy (aluminum heads with hardened steel inserts). However, for engines known to be susceptible, the problem remains.

Figure 3.76 *Exhaust-valve-seat wear mechanics.*

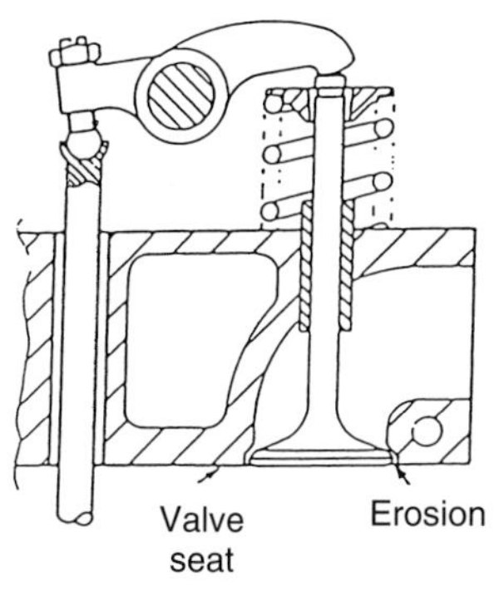

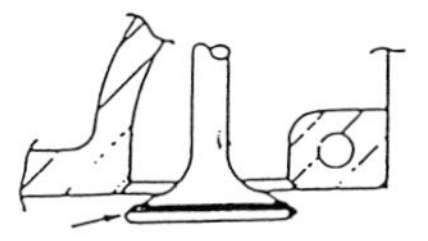

The valve seats are no longer lubricated by lead oxides and they erode, causing loss of compression.

Engine not adapted for unleaded gasoline

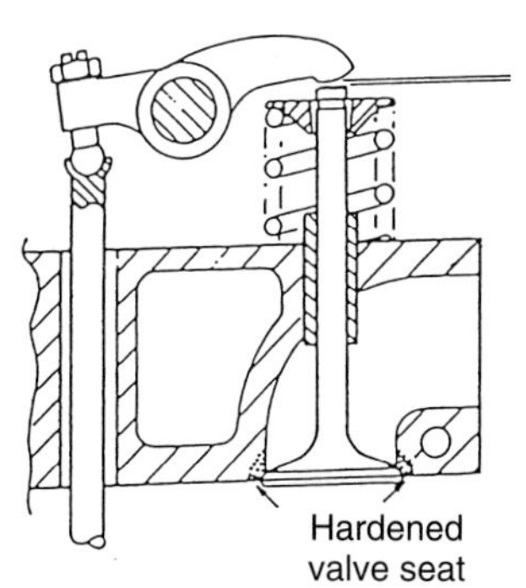

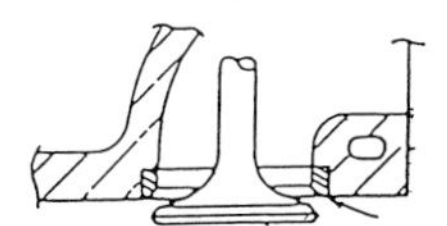

The aluminium head contains a hardened steel valve-seat insert.

Engine adapted for unleaded gasoline

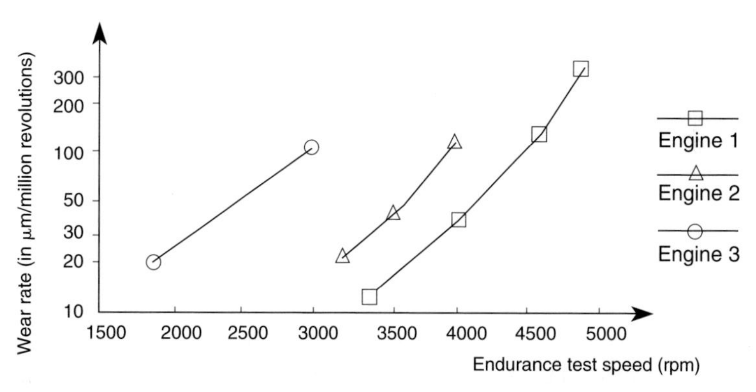

Figure 3.77 *Examples of valve-seat recession during unleaded gasoline tests on unadapted vehicles.*

There is a distinct difference between acceptable wear, which can be compensated for by valve adjustment, and **destructive deterioration** that results in loss of compression and even valve failure. The critical threshold occurs at the rate of about 4 micrometers per hour (Schoonveld et al. 1986). It has been demonstrated that valve-seat recession increases rapidly with engine load and speed (see Fig. 3.77). The most critical operating condition is continuous high-speed freeway driving.

3.13.2 Possible solutions

Sufficient protection can usually be achieved with a **low quantity** of lead (from 0.05 to 0.08 g Pb/liter) in the gasoline. This is the reason why automobile manufacturers in France have requested that conventional leaded gasoline, which will continue to be distributed until 2000, contain at least 0.08 g Pb/liter.

Otherwise, **occasional** fueling with leaded gasoline can provide temporary protection from valve wear. This protection can last for several thousand kilometers (up to 20 000 km) under low-load conditions, but that protective period becomes very short or non-existent at high loads and speeds.

Additives have been proposed as substitutes for the solid-lubricant role played by lead. Some phosphor-based additives are effective but they can cause potential toxicity problems. The most promising additives are those containing **alkaline metals** (sodium and potassium), which are available commercially in many countries.

The problem of exhaust-valve wear is still an important preoccupation in countries that have decided to stop the sale of leaded gasolines on very-short notice for environmental reasons.

With such a scenario, it is essential to identify the older vehicles in the fleet that are vulnerable. These vehicles should be fueled with gasolines containing anti-wear additives or used under low-load and low-speed conditions (city or suburban driving that excludes freeway use).

3.14 Controlling intake-system fouling

The effect of combustion chamber deposits on octane requirement increase (ORI) was described in previous sections. This section covers the fouling of the intake system (injectors, carburetor, intake system, valves, ...).

3.14.1 Causes and consequences of the fouling phenomenon

The deposits found in various engine intake-system components arise from

- recycled **crankcase or exhaust gases** that contain organic particles in suspension
- **dust** contained in the intake air, which may be poorly filtered
- the consumption of small quantities of **lubricant**
- **products of oxidation or cracking** that are derived from the fuel itself (Haycock et al. 1994)

Deposits on carburetor venturis and throttle plates cause changes in idling and low-speed settings that have a number of negative consequences: increased fuel consumption, increased exhaust emissions, rough running, stalling, ...

Deposits on the injectors of gasoline engines most often result from the thermal degradation of hydrocarbons. The deposits are usually concentrated in high-tolerance areas (0.5 mm) where they can block the passage and spray pattern of the liquid (see Fig. 3.78).

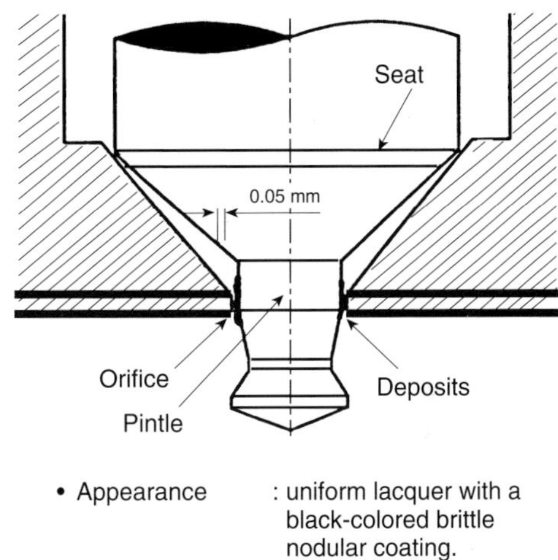

Figure 3.78 *Deposit formation in the pintle area of an injector.*

Deposits on intake valves are found mainly on the stem and the tulip. In some cases, deposits are considerable, with masses of 1g or more per valve. This level of deposits results in serious **problems**. The deposits can absorb and release fuel in an untimely manner, which results in high fuel consumption, high emissions, and driveability problems when starting and driving off (lean mixture).

Deposits on intake valves can also cause uncontrolled effects on mixture turbulence, which affects the speed of combustion. In extreme cases, the flow of the air-fuel mixture is restricted causing a loss in performance.

3.14.2 Experimental procedures

There is no single universal procedure for classifying the cleanliness of engine intake systems.

The methods vary, depending the component (injectors, carburetor, valves) under consideration. They can be further subdivided depending on the objective sought, which can be the maintenance of the system in an original state of cleanliness (keep clean) or the cleaning of deposit-laden components (clean up).

Table 3.29 provides a short description of the methods used in the United States by the CRC (Malakar et al. 1983) and in Europe by the CEC (Anon 1981) to assess the cleanliness of **carburetors**. The tests are of reasonably short duration (20 h and 12 h) and they include repetitions of the operating conditions most favorable to deposit formation (high intake temperature, exhaust gas recirculation, ...). Cleanliness is evaluated by **visual inspection** and by measuring the **CO content** of the exhaust gas while the engine is idling.

The methods used for **fuel injectors** are more difficult to carry out. The existing procedures (Table 3.30) consist of alternating phases of low-load oper-

Table 3.29 *European and American tests for evaluating the cleanliness of carburetors.*

Procedure	CRC	CEC-F-03-T-81
Engine	Ford 3.9 liter, 6 cylinder	Renault type 810-26
Test duration (h)	20	12 (18 h of soak time after 6 h of operation)
Cycle	3 min at 700 rpm 7 min at 2 000 rpm	2 min at 800 rpm 8 min at 1 800 rpm
Coolant temperature (°C)	88 to 90	80
Oil temperature (°C)	110 maximum	78
Intake air temperature (°C)	66	40 to 65
EGR	none at idle maximum at cruising speed	11%
Reference lubricant	REO-202-TI	CEC RL-051

Table 3.30 *Test used in France to evaluate gasoline fuel-injector cleanliness.*

Procedure	GFC TAE-T-87	
Engine	PSA type XU5JA	
Test duration (h)	150	
Cycle	15 min at 3 000 rpm	45 min stop
Power (kW)	18 (at the beginning of the test)	0
Equivalence ratio	1.10 (at the beginning of the test)	
Coolant temperature (°C)	96	108 to 112 after 3 min 90 minimum at the end of the cycle
Lubricant temperature (°C)	97	80 minimum at the end of the cycle
Fuel temperature (°C)	60 minimum	85 minimum at the end of the cycle
Temperature at the injector tip (°C)	-	102 maximum at the beginning 92 minimum at the end of the cycle
Exhaust temperature (°C)	approximately 650	-

ation and stops at high ambient temperatures (Shiratori et al. 1991; Tupa et al. 1986). At the end of the test, behavior can be assessed as either the loss of flow from the injectors or evidence of operating problems with the vehicle.

The methods used to assess the cleanliness of **intake valves** are described in Table 3.31. Procedures for both carbureted and fuel-injected engines are available (CRC 1979; Gairing 1986). The procedure for fuel-injected engines is the most widely used. It has a duration of 60 h. The evaluation criteria are the **mass of deposits** on each valve and a **visual inspection** of the entire intake system.

Table 3.31 *European tests used to evaluate intake-valve cleanliness.*

Procedure	CEC F04-A-87C				CEC F05-A-93			
Engine	Opel Kadett				Mercedes M 102E			
Test duration (h)	36				60			
Cycle	Time (min)	Speed (rpm)		Eff. Power (kW)	Time (min)	Speed (rpm)	Torque (Nm)	Eff. Power (kW)
	0.5	950		–	0.5	800	0.0	0.0
	1.0	3 000		11.1	1.0	1 300	29.4	4.0
	2.0	1 300		4.0	2.0	1 850	32.5	6.3
	1.0	1 850		6.3	1.0	3 000	35.0	11.0
Coolant temperature (°C)	92 maximum				85 to 95			
Lubricant temperature (°C)	94 maximum				90 to 105			
Intake air temperature (°C)	–				25 to 35			
Reference lubricant	RL 051				CEC RL-189/1			

3.14.3 Chemistry of detergent additives

The cleanliness of intake systems can only be assured by the use of detergent additives.

Detergent additives were developed in the 1960s. Many succeeding generations of products have been developed since then, leading to very-elaborate formulations that ensure effective protection for engine components, from the carburetor or injector to the combustion chamber. It is now estimated that there are 1000 to 2000 different additive formulations for gasolines.

All additive products can be represented by the general formula R-X in which

- R is an **oleophilic** hydrocarbon structure that ensures solubility in gasoline
- X is the polar part with a **hydrophilic** nature that is selectively adsorbed onto metallic surfaces in the intake system, thereby preventing the adherence of deposits

The R portion can be derived from natural fatty acids with long-chained hydrocarbon molecules (approx. C_{18}) with a number of unsaturated links. R can also be derived from dimers of fatty acids (C_{36}).

Another group of oleophilic constituents makes use of polyisobutylenes (PIB) with molecular weights between 600 and 2000. To make them reactive with the hydrophilic part, these constituents can be condensed with maleic anhydride.

A third route for obtaining an oleophilic group consists of using polypropylphenols.

The hydrophilic parts can contain oxygenated groups (glycol ethers) or amines.

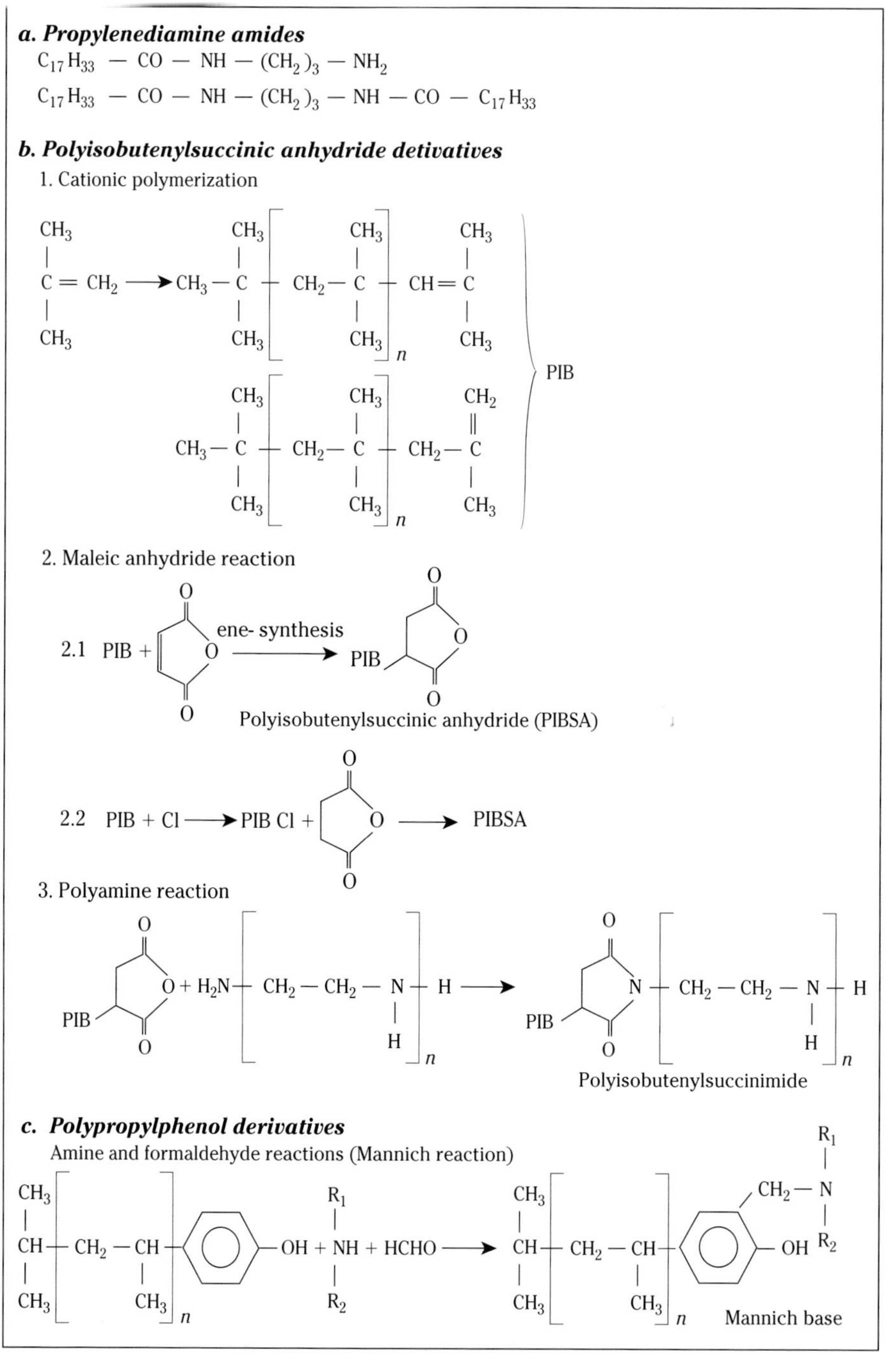

a. Propylenediamine amides

$C_{17}H_{33} - CO - NH - (CH_2)_3 - NH_2$

$C_{17}H_{33} - CO - NH - (CH_2)_3 - NH - CO - C_{17}H_{33}$

b. Polyisobutenylsuccinic anhydride detivatives

1. Cationic polymerization

$(CH_3)_2C = CH_2 \longrightarrow CH_3 - C(CH_3)_2 - [CH_2 - C(CH_3)_2]_n - CH = C(CH_3)_2$

$CH_3 - C(CH_3)_2 - [CH_2 - C(CH_3)_2]_n - CH_2 - C(=CH_2) - CH_3$

PIB

2. Maleic anhydride reaction

2.1 PIB + maleic anhydride $\xrightarrow{\text{ene- synthesis}}$ PIB-succinic anhydride

Polyisobutenylsuccinic anhydride (PIBSA)

2.2 PIB + Cl $\longrightarrow$ PIB Cl + maleic anhydride $\longrightarrow$ PIBSA

3. Polyamine reaction

PIBSA + $H_2N - [CH_2 - CH_2 - N(H)]_n - H \longrightarrow$ PIB-succinimide $- N - [CH_2 - CH_2 - N(H)]_n - H$

Polyisobutenylsuccinimide

c. Polypropylphenol derivatives

Amine and formaldehyde reactions (Mannich reaction)

$CH_3 - CH(CH_3) - [CH_2 - CH(CH_3)]_n - C_6H_4 - OH + NH(R_1)(R_2) + HCHO \longrightarrow CH_3 - CH(CH_3) - [CH_2 - CH(CH_3)]_n - C_6H_3(OH) - CH_2 - N(R_1)(R_2)$

Mannich base

Figure 3.79 *Derivation and overall chemical characteristics of detergent additives.*

Fig. 3.79 shows some of the particular aspects of detergent additive chemistry. The following points are worth noting:

- Fatty **acid amides** (group *a* in Fig. 3.79) do not offer any flexibility of proportioning between the oleophilic and the hydrophilic parts.
- The **PIB derivatives** are interesting because they allow control of the molecular weight of the polymer and some flexibility in the characteristics of the final product. Note that mono-succinimides and disuccinimides can also be obtained.
- **Alkylphenol derivatives** also offer a means of adjusting the relative portions of the oleophilic and hydrophilic groups by controlling polymerization reactions.

The active ingredient in the additive, which has just been described, must be mixed with a **carrier oil**. This oil, which has a general structure similar to a petroleum- or synthetic-based lubricant, acts as a vehicle to transport deposit-generating compounds out of the combustion chamber.

A complete additive formula therefore consists of 100 to 500 ppm of **active material** and up to 1000 ppm (up to 2000 ppm in exceptional cases) of **solvent**.

3.14.4 Effectiveness of detergent additives

Mixing additives with gasoline usually leads to spectacular results (Papachristos 1994) in terms of component cleanliness and the accumulated mass of deposits (see Fig. 3.80).

However, effectiveness should be considered on the basis of the whole intake system and the combustion chamber, and not just on a single component (for example, intake valves). For instance, there have been additives that were very effective in keeping carburetors clean, but caused an increase in intake-valve deposits. Also, as previously indicated (see 3.11.5.3), some additive formulas can certainly maintain a clean intake system, but they contribute to combustion chamber deposits and ORI.

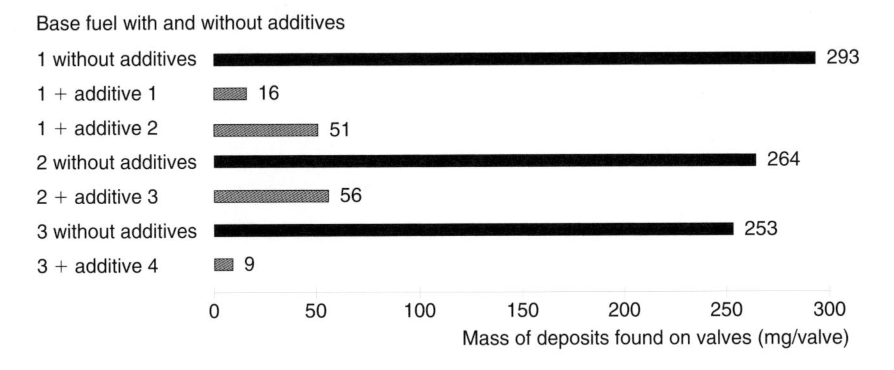

Figure 3.80 *Effects of detergent additives on intake-valve cleanliness. (60 h test on a Mercedes M102E engine).*

Therefore, the development of additives implies a careful and meticulous evaluation of their effects on all engine components. Thus, negative effects such as **valve sticking** can be found (Mikkonen et al. 1988) and alleviated.

The preceding points have explained how the overall cleanliness performance of a gasoline with additives can be estimated from a series of criteria. Table 3.32 provides an example of the criteria that must be met by a gasoline-additive package to receive the French automobile manufacturer's **quality label**.

Table 3.32 *Quality criteria required for gasoline-additive packages in France.*

Cleanliness sought	Test reference	Engine type	Test duration (h)	Criteria required
Carburetor cleanliness	CEC F-03-T-81	Renault R5	12	Visual rating ⩾ 8.0
Valve cleanliness (carburetor)	CEC F-04-A-87	Opel Kadett	36	Visual rating ⩾ 9.0
Valve cleanliness (fuel injection)	CEC F-05-A-94	Mercedes M102 E	60	Visual rating ⩾ 9.0
Injector cleanliness	CFC TAE-I-87	Peugeot 205 GTI	150	Loss of flow < 4%
Combustion chamber cleanliness	Renault 22 700 et 22 710	Renault F2N et F3N	400	ΔKLSA < 8°CA*

* Up to 12°CA is permitted for gasolines with 0.15 g Pb/liter.

Treating gasolines with detergent additives reduces emissions. In fact, with the absence of additives, emissions increase with time, but they **stabilize** after additives are used.

In conclusion, the aim of current additive strategies is to maintain intake-system cleanliness (keep clean strategy) but some of them are also used to remove existing deposits (clean up strategy). The later treatment method requires high additive content (greater than 1000 ppm) ... and patience. In fact, an accumulation of several thousand kilometers is often required before any clean-up effect becomes apparent.

Chapter 4

Diesel fuel

This chapter is mainly devoted to the diesel fuel used for on-the-road transportation in vehicles such as passenger cars and commercial vehicles. Products similar to regular diesel fuel are also used in engine applications that are closely allied to road use. For example, such applications include engines used in fishing boats, public service vehicles, agricultural tractors, ... With few exceptions, the quality criteria that apply to regular diesel fuel are also valid for these particular fuels.

There is also another category of large and powerful diesel engines that are used in locomotives, ships, and powerplants. These engines use a heavier and less-refined version of diesel fuel. The problems that are specific to this type of operation are treated in more detail in Chapter 7.

Historically, diesel fuel has been a fairly simple product to prepare. It was used mainly in older engine designs that were not very-demanding applications, especially with respect to ease of use and emissions.

Today, diesel fuel is a **premium fuel**. A fuel that must deliver high performance, stable storage qualities, and low emissions.

Furthermore, diesel fuel requires considerable care at each stage of its **distribution**. Diesel fuel is vulnerable to cold weather and biological contamination, and it must be carefully filtered to protect fuel injection systems.

Rudolph Diesel and his engine
(Patent DRP 67 207, Feb. 23, 1893)

Specific characteristics of diesel engines

This first section covers the technological and thermodynamic principles of diesel engines by describing how they function in general terms, the steps involved in diesel combustion, and diesel-engine performance characteristics. Most of the fundamentals, definitions, and conventions already used to describe spark-ignition engines (see Chapter 3) are assumed to be understood and they are not repeated in this section.

4.1 Functional principles and operating conditions

The diesel engine was first considered to be a simple variant of the spark-ignition engine, but it has since rapidly acquired first-rate status to the point where its inventor's name is spoken or written millions of times a day worldwide.

Sometimes lauded for high efficiency, sometimes criticized for its shortcomings, the diesel engine is still the subject of many debates.

4.1.1 Some historical notes

The famous German inventor Rudolph Diesel was born in Paris in 1858, where he spent most of his engineering career. He was granted his first patent in Berlin in 1892 for an engine that operated by the pneumatic injection of coal dust. This device never got beyond the development stage and it was superseded in 1897 by an engine running on crude oil, which already provided satisfactory efficiency (247 g/ch.h or 336 g/kWh). The first diesel engine developed 14.7 kW at 172 rpm from a displacement of 19.6 liters. Although Rudolph Diesel disappeared prematurely and mysteriously at sea in 1913, his invention spread quickly to boats after the First World War and to trucks between 1930 and 1939.

The first diesel-powered passenger car was introduced by Mercedes in 1936 (260 D), but very few were sold. Peugeot introduced a diesel car (402 Diesel) in 1938 with greater success (1000 were produced). After 1945, the diesel engine was in common use in utility-vehicle fleets, but its use in passenger cars remained the privilege of only a few manufacturers (Peugeot, Mercedes). From the 1970s onwards, a **rapid expansion** of diesel power occurred in both **industrial vehicles** and **passenger cars**.

In fact, the majority of diesel passenger cars use indirect injection while most industrial vehicles use direct injection. The expansion of direct-injection diesel engines in the passenger-vehicle market is very likely, primarily because of their excellent efficiency.

On a global basis, most industrial vehicles are equipped with diesel engines. However, the reverse situation applies to passenger cars. The sale of new diesel passenger cars in the United States is nil; sales in Japan are 10% and sales in "Greater Europe" (European Union plus Switzerland and Norway) are 25%; the grand prize for diesel market penetration goes to France where almost 50% of the yearly new-car sales are diesel.

4.1.2 Special characteristics of the diesel cycle

Although the diesel engine uses **compression ignition**, the four-stroke cycle of the spark-ignition engine, which has already been described, can be used as a basis for comparison. The phases of the cycle—intake, compression, combustion, exhaust—are the same for the diesel engine. The characteristics particular to the diesel engine are

- the introduction of the fuel
- combustion initiation and progression
- power control

The diesel engine aspirates **air** (undiluted or diluted by exhaust gases) and compresses it to a high level, which causes the air to increase in temperature. Just before TDC, a measured quantity of fuel from the injection pump is supplied to the combustion chamber as a high-pressure **atomized** spray; the spray vaporizes and mixes with the compressed air. Combustion occurs by **auto-ignition** in one or several areas of the combustion chamber where the conditions of temperature, pressure, and mixture concentration are sufficient to begin the process. The pressure increases rapidly and this results in the characteristic diesel combustion noise. The progress of the combustion is then controlled by a change in the quantity of fuel injected. The combustion flame is referred to as **diffused** because the mixture is formed mainly during combustion. Also, it is preferred practice that the fuel not be sprayed on the walls of the combustion chamber, to ensure that it mixes readily with the air. However, there is an exception; the MAN[1] technique sprays the fuel directly onto the walls of the combustion chamber and then progressively evaporates it using a directed flow of the intake charge.

When auto-ignition occurs, a **high-pressure gradient** results. This observation contradicts the suggestion that the diesel is a "constant pressure" cycle, in the same way that the view of the spark-ignition engine as a "constant volume" cycle is unrealistic.

Representing the diesel engine as a constant-pressure cycle was no doubt related to the fact that it is theoretically possible to consistently maintain the pressure that occurred at the beginning of the expansion by adjusting the flow of fuel injected into the combustion chamber, and thus control the speed of the energy release.

1. MAN: Maschinenfabrik Augsburg Nümberg.

Because combustion in the diesel engine is initiated by auto-ignition, it requires a **much-higher compression ratio** than the ratio used in a gasoline engine. In general, compression ratios of about 15 to 21:1 are required for direct-injection engines and ratios of about 20 to 23:1 are required for indirect-injection engines. These compression ratios are particularly useful for cold starting, even if this is assured by the use of other efficient technological devices (glow plug pre-heaters for divided combustion chambers and simply heating the intake air for direct-injection engines) (see 4.2.2).

Unlike gasoline engines, the power of diesel engines is not controlled by a simple butterfly valve that reduces the quantity of air-fuel mixture admitted. Diesel power is controlled by adjusting the flow of fuel introduced into a fixed volume of air. The pumping losses represented by the "low-pressure" loop are considerably reduced, particularly at low load. This fact combined with other thermodynamic factors (compression ratio, specific heat of the gases), is the reason why diesel engines are more efficient than gasoline engines. In addition, the reduction in thermal losses with direct injection provides a further improvement in efficiency.

By design, the diesel engine operates on a **variable overall air-fuel ratio** principle. Also, because the injected fuel has very little time to mix with the air, the mixture is **heterogeneous** and consists of rich and lean zones. The over-rich zones result in incomplete combustion and the formation of smoke. Smoke emissions are controlled by regulation because of their toxicity and as a result, the overall air-fuel ratio is also controlled. Equivalence ratios typically do not exceed 0.8 to 0.9 for divided-chamber engines and 0.6 to 0.7 for direct-injection engines. Consequently, the diesel engine runs with a **lean mixture**. This is why diesels have low specific power, which is often increased with the aid of **superchargers**. Another beneficial result of lean operation is low CO emissions. With respect to other emissions, the behavior of the diesel engine is in some cases better, and in other cases worse, than the gasoline engine (see Chapter 5).

The general architecture of diesel engines closely resembles that of gasoline engines, although its technology is more complex and the components (cylinders, head, pistons) are stronger. Heavier components are required to endure the mechanical and thermal stresses caused by the repeated auto-ignition that occurs with each cycle and the need to maintain high compression ratios at part load. Incidentally, some diesel engines that are subjected to very-heavy mechanical loads can suffer from damage analogous to that caused by knock in gasoline engines.

It may seem surprising at first that auto-ignition, which is a feared occurrence in a gasoline engine, is actually sought after in a diesel engine. This results from the fact that it occurs only in very-**localized** zones and represents only a **small portion** of the energy released. Auto-ignition in a diesel consists of one or several auto-ignition focal points instead of the massive auto-ignition of a homogeneous mixture that characterizes knock. In addition, the robust structure of the diesel engine absorbs the rapid energy release that occurs with auto-ignition. The principal manifestation of this energy release is the characteristic diesel combustion noise.

4.1.3 The fuel supply system

The diesel fuel-supply system is specific to the diesel engine. In effect, power control implies changing the flow of the fuel provided to the cycle, which cannot be accomplished by

- **a simple throttle vane** on the intake system to restrict the quantity of a homogeneous air-fuel mixture like the gasoline engine; this would lead to a small intake charge at a pressure and temperature that would be too low to obtain auto-ignition when compressed
- **a change in the equivalence ratio** of a previously prepared fuel-air mixture; in this case, combustion could not be initiated at part load because of the increased auto-ignition delay due to the lean mixture

Contrary to the above techniques, the introduction of a variable flow of liquid fuel into the combustion chamber establishes one or more areas where the temperature, pressure, and air-fuel ratio are propitious for auto-ignition under all circumstances.

The fuel supply system for a diesel engine equipped with an in-line injection pump is shown schematically in Fig. 4.1. Although the system depicted here is mechanically controlled, the principle is also valid for electronically controlled systems. The fuel circuit consists of high- and low-pressure sections.

The **low-pressure section** includes the tank and the lines that provides fuel to the injection pump. This section also includes the return lines to the tank for the fuel overflow from the pump and the filter, as well as the leak-off flow from the injector bodies.

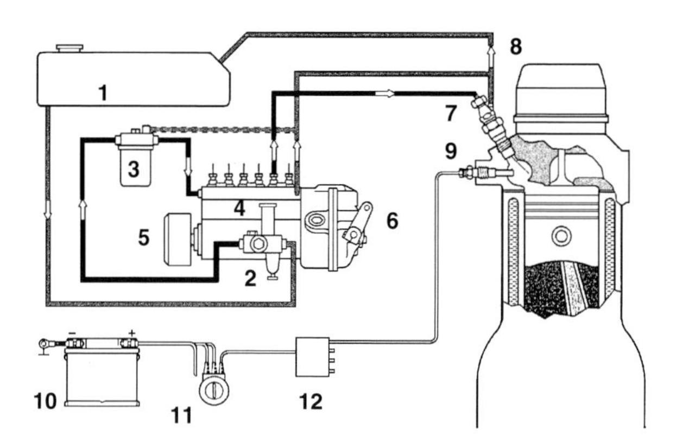

1	Fuel tank	**7**	Nozzle holder with injection nozzle
2	Supply pump	**8**	Return line
3	Fuel filter	**9**	Pencil-type glow plug
4	Inline fuel-injection pump	**10**	Battery
5	Timing device	**11**	Glow plug/starter switch
6	Speed control (governor)	**12**	Glow plug control unit

Figure 4.1 *Diesel engine fuel system with a mechanically-governed inline pump. (Source: Bosch)*

The **high-pressure section** includes the lines from the pump to the injector. The fuel flows under pressure from the pump through the delivery valve, the supply line, and the injector body before it arrives at the injector.

The fuel-supply pump, which is shown in Fig. 4.1, is effectively independent of the injection system. However, in distributor-type (rotary) fuel-injection pumps, the supply pump is a packaged subcomponent of the device.

4.1.3.1 The fuel-injection pump

The fuel-injection pump meters the quantity of fuel required to match the load imposed on the engine. At a precise crankshaft angle just before TDC, the fuel is delivered to the fuel injector under high pressure (100 to 350 bar for pre-combustion chamber engines and up to 1800 bar for direct-injection engines). The pump also controls the duration of the injection as combustion progresses.

A. The in-line fuel-injection pump

In-line fuel injection pumps are used in some automotive applications such as utility vehicles, tractors, industrial vehicles, and stationary engines. The injection pump has a **plunger** and a **barrel** for each engine cylinder. As shown in Figure 4.2, the plunger is driven by a camshaft that is connected to the engine and the plunger is repositioned by a return spring. The stroke of the pump plunger is fixed. The quantity of fuel injected is controlled by varying the effective stroke of the plunger. The effective stroke is controlled by rotating the plunger using a geared rack and a linkage attached to the accelerator pedal.

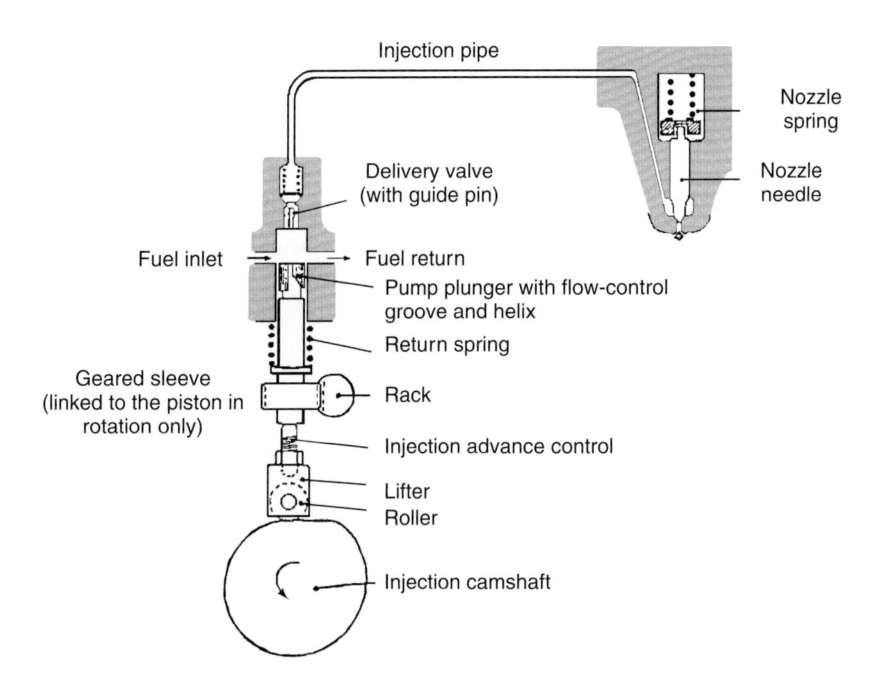

Figure 4.2 *Schematic of the control principles of an in-line injection pump.*

To achieve high fuel-injection pressures (up to 1800 bar or more) without using seals, the plunger and barrel assembly is manufactured to **very-precise clearances** on the order of 3 micrometers.

The plunger moves up and down, and with each rotation of the camshaft, it travels the full length of the stroke and completes one intake and delivery sequence. At BDC, the fuel enters the compression chamber through the intake port (see Fig. 4.3). As the plunger moves up, the intake port is covered and the fuel is gradually compressed in the high-pressure chamber until the delivery valve is forced open. This causes a **pressure wave** to proceed at the speed of sound toward the injector. When the injector-opening pressure is reached, the nozzle needle is lifted from its seat and the nozzle opens. The fuel delivery is complete as soon as the return port is uncovered. The pressure in the high-pressure pipe drops and the delivery valve closes.

The beginning of the delivery cycle is always consistent, but the end of the cycle depends on the amount of fuel being injected.

In order to reduce the exhaust emissions from commercial vehicles, a new type of in-line pump, referred to a "drawer" type, has been developed. This pump has two interesting characteristics:

- very-high injection pressures (up to 1800 bar)
- precise and easily varied fuel-delivery starting points

Also, unlike a conventional in-line pump, control of the start-of-delivery point is independent of flow control and the pump requires very-little energy. In addition, these new pumps have the added advantage of being electronically controlled (see 4.1.3.1.c).

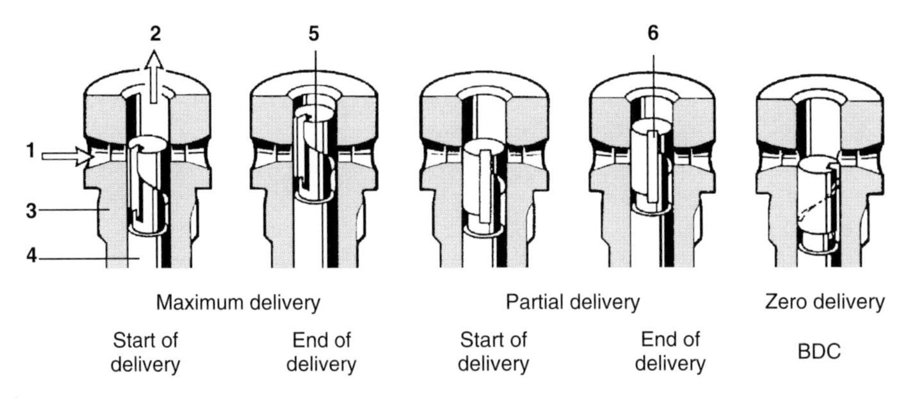

1 Inlet from the supply pump gallery
2 Fuel delivery to the injector
3 Barrel
4 Plunger
5 Lower control helix
6 Vertical groove

Figure 4.3 *Fuel delivery control in an in-line fuel-injection pump. (Source: Bosch)*

B. Distributor or rotary-type fuel-injection pumps

Distributor (or rotary) fuel-injection pumps are used mainly on high-speed small-displacement engines (passenger cars, light commercial vehicles). In contrast with the in-line pump, the distributor pump has only one pump element for all cylinders.

The Bosch pump shown in Fig. 4.4 has a unique distributor plunger assembly that reciprocates once for each cylinder during each rotation of the pump drive shaft. Fuel is delivered by a longitudinal movement of the plunger and it is distributed by a **rotary movement** to each cylinder's delivery outlet. Engine speed is governed by the position of the control collar, which controls the plunger travel and the quantity of fuel injected. In fact, only a portion of the fuel contained in the high-pressure chamber is injected and the rest is returned to the interior of the pump. Note that this type of device has a fixed start point for the delivery and the end point is variable.

The Lucas Diesel pump shown in Fig. 4.5 uses a rotor instead of a piston to distribute the fuel to the various cylinders. Depending of the model, fuel delivery takes place either before, or at, the point when it is introduced to the distributor. Thus, all the fuel provided to the inlet is delivered to the cylinders.

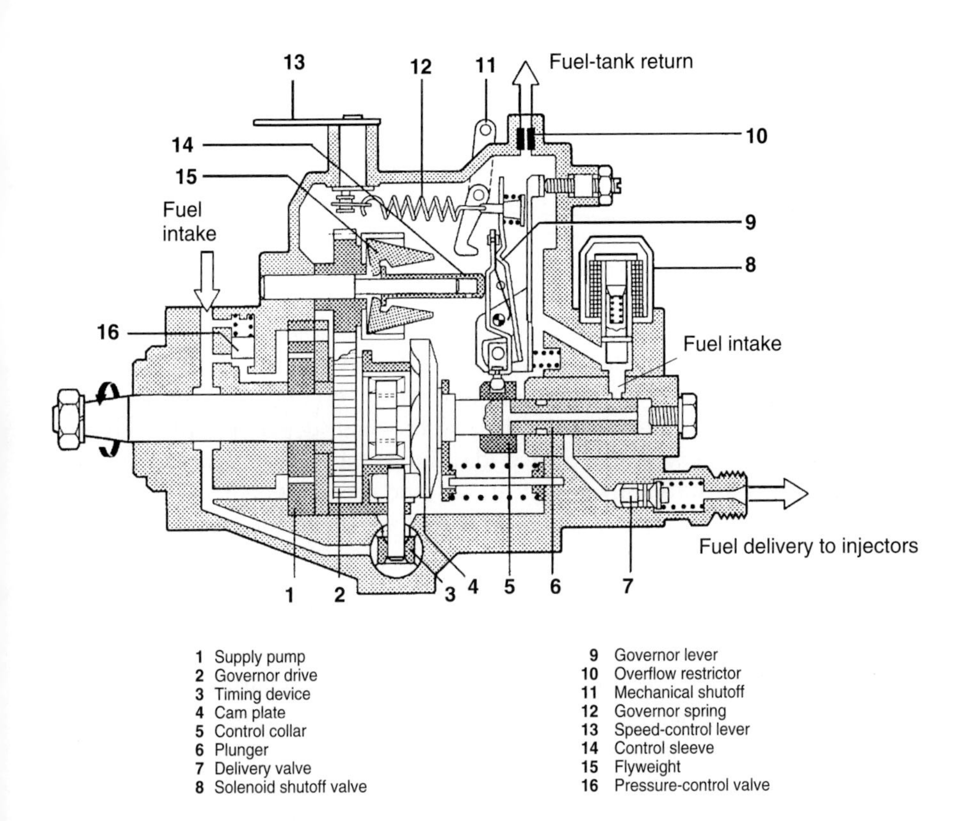

Figure 4.4 *Basic version of a Bosch distributor-type fuel-injection pump.*

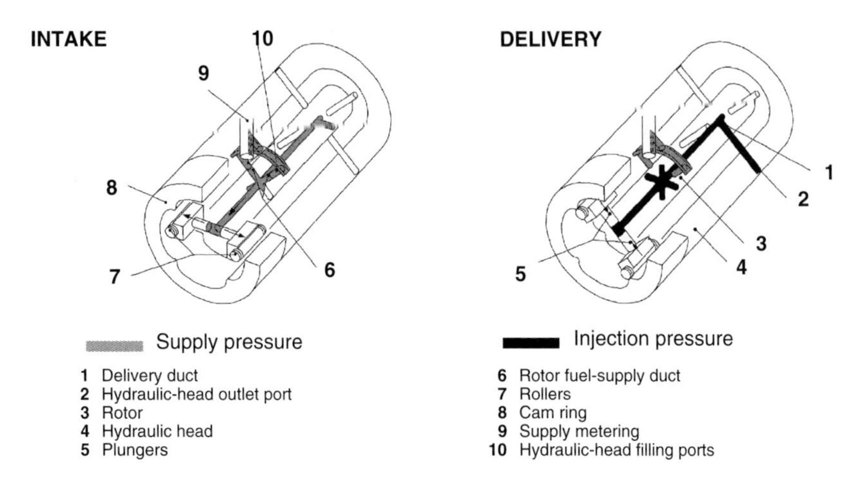

Figure 4.5 *Diagram showing the intake and injection cycles of the plunger assembly in a Lucas Diesel distributor-type pump.*

Delivery starts when the plungers are driven toward each other by the rollers riding on the internal profile of the cam ring. Delivery ends when the rollers reach the top of the cam profile; therefore, the delivery end-point is fixed.

C. Control systems

All fuel-injection pumps have control systems to improve their performance. These systems can be either mechanical or electronic.

A common system is the speed controller known as a **governor**. A governor maintains a given engine speed within a tightly defined tolerance range by adjusting the fuel flow. The governor can perform several tasks such as limiting the top speed of the engine, maintaining a stable idling speed, and, if necessary, controlling intermediate speeds, which is especially useful in stationary engine applications. The most common types of governors are "minimum-maximum-speed" and "variable-speed", the use of which depends on the chosen application. The first type acts only at idle and at maximum engine speed; they are used mainly in automotive applications. The second type is used in a variety of applications because it acts at intermediate engine speeds.

The governors now in use are based on purely mechanical technologies or mechanical components associated with electronic components. **Mechanical governors** include flyweight-based centrifugal speed-monitoring devices driven by the pump camshaft and connected to the vehicle accelerator pedal. Depending on the type of pump, governors control the quantity of fuel by acting on the control rack, the control rod, the metering valve, or the rotor. **Electronic governors** include sensors, a micro-processor, and an actuator: the micro-processor normally receives sensor information about the accelerator-pedal position, the engine speed, the coolant temperature, ... ; this information is processed using maps stored in memory and the optimum position is derived for the control rack, the control rod, the metering valve, or the rotor. The electronically-controlled device is more precise than the mechanical device and also provides new capabilities such as a choice of full-load control based on the smoke limit and the control of fuel flow based on fluid temperatures (air, fuel, coolant).

To provide the correct quantity of fuel at any operating speed, a **timing device** is used to control the injection timing. As engine speed increases, injection is delayed; the timing device compensates for this by correcting for the delay at the injection pump.

The timing devices used with in-line mechanical pumps are usually centrifugal devices and they are located between the engine and the pump. Some timing devices on distributor-type pumps are hydromechanical and they are controlled by the fuel pressure within the pump, which is proportional to the speed of the engine.

There are also electronic timing devices, which function as part of the pump electronic-control system that adjusts the fuel flow. Certain sensors, such as the one used to provide the injection start point, are obviously provided specifically for the timing device.

D. Other injection systems specific to direct-injection engines

Manufacturers like *Cummins* and *General Motors* have traditionally used injection systems that do not include high-pressure delivery pipes. These systems combine a pump and an injector into a compact pump-nozzle unit referred to as a **unit injector**. These units are mounted in the head of each cylinder and they are supplied by a low-pressure pump. The units are actuated independently by the lobes on a camshaft. Unit injectors are usually chosen for their control simplicity and the advantage of not having to pump fuel through high-pressure pipes. In most cases, control is mechanical and the start of injection cannot by readily changed.

Meeting the increasingly severe requirements of environmental protection standards for direct-injection diesel vehicles requires higher injection pressures (1800 bar). This change presents a new problem for the delivery pipes in conventional injection system, because fuel is compressible at these high pressures. In this regard, the advantages of unit injectors are such that equipment makers (Lucas Diesel, Bosch) are actively working on the development of a new type of unit injector. These new units work by the principles described in the following paragraphs and they are distinguished by high injection pressures, direct control, and a readily adjustable injection start point.

Another system intended for direct-injection diesel engines has recently become more prominent. It consists of a controllable high-pressure (1200 to 1600 bar) injection system known as **common rail fuel injection**. This system uses a pump that continuously supplies fuel to a high-pressure reservoir that is common to all injectors; the fuel is then injected into the combustion chamber in a discontinuous manner by each injector. Unlike the devices already described, common-rail injection pressure remains constant throughout the injection duration and it is completely independent of engine speed.

1 Magnet unit
2 Solenoid valve
3 Stud
4 Solenoid spring
5 Plunger cap
6 Double plunger
7 Needle travel sensor
8 Injector body
9 Injector nose
10 Connector nut
11 Lock nut
12 Spring calibration nut
13 Electromagnetic valve body
14 Fuel supply duct
15 Injector needle

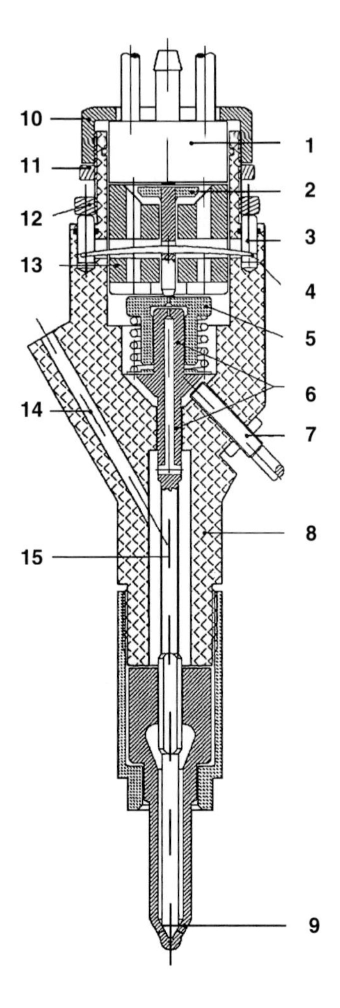

Figure 4.6 *Schematic of a Ganser-type common-rail injector. (Source: Schneider et al. 1993)*

The common rail system is monitored and controlled **electronically** (see 4.1.3.1. c). A simple micro-processor command allows the user to define the phasing, and the injection time and pressure, as well as other functions such as: fuel pre-injection (by double actuating the injector), the fuel flow at full load, the recirculation rate, ... Common-rail injection devices have several advantages, but their complex development has slowed their deployment, especially in automobiles. However, in 1999, Mercedes Benz and PSA already propose vehicles equipped with common-rail injection devices.

4.1.3.2 Fuel injectors

Fuel injectors ensure the introduction, atomization, and distribution of fuel diesel in the combustion chamber. Each engine cylinder has an injector.

A. Typical designs

In all cases, injection units consists of two sections fastened together with a nut (see Fig. 4.7) (Bosch 1991):

- The **nozzle holder** channels the fuel supply and provides for the return of overflow (leak-off) fuel. For conventional injection systems (in-line pump or distributor-pump systems), the pressure at initial delivery into the combustion chamber is established by the **calibration of the injector spring**. However, the injector used in common-rail systems has a solenoid valve in place of the spring (see 4.3.1.d).
- **Nozzle bodies** are the same for all injection systems: they consist of a long cylinder that contains a needle that is free to move. The bottom part of the needle rests on a gas-tight seat that seals the injector off from the combustion chamber when it is closed. The gap between the needle and injector body is about 2 to 3 micrometers.

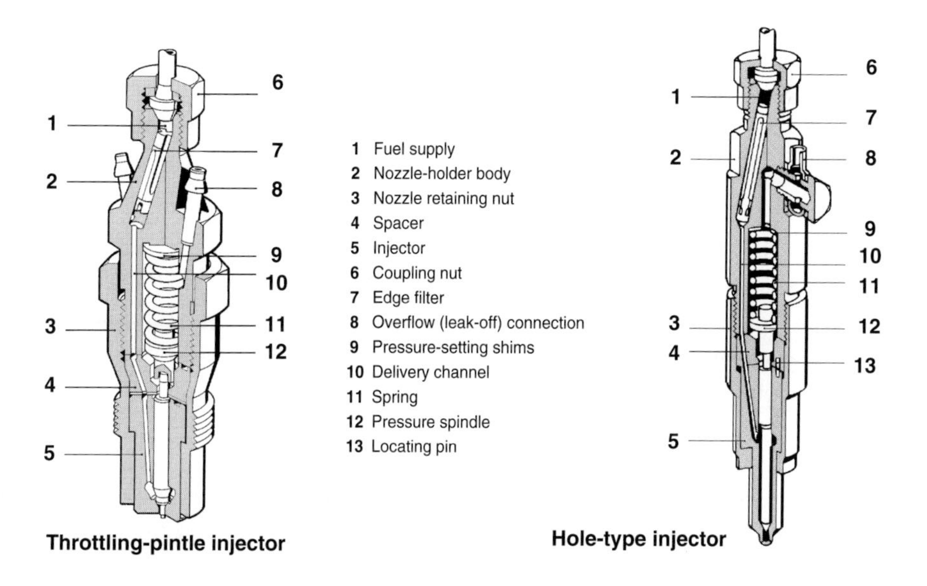

Figure 4.7 *Conventional nozzle-holder assemblies. (Source: Bosch 1991)*

B. Operation

With conventional fuel-injection systems, the fuel supplied under pressure from the injection pump is delivered to the pressure chamber by channels in the injector body (see Fig. 4.8). Because the diameter of the seating flange is greater than the needle, the fuel pressure lifts the needle from the seat when the pressure exceeds the pressure setting of the nozzle spring. The needle movement opens the orifices at the tip of the nozzle and the fuel is sprayed into the combustion chamber. As soon as the fuel pressure drops, the needle returns to its seat and blocks off the delivery orifices.

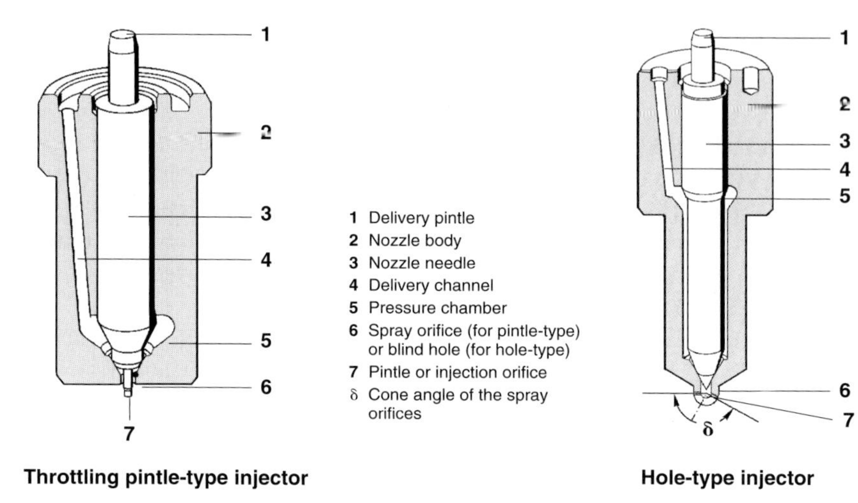

Figure 4.8 *Two types of injectors. (Source: Bosch 1985)*

The injection pressure setting can be pre-adjusted by the mechanical setting of the spring tension. If P_0 is the pressure setting and F is spring tension, then

$$F = P_0 \frac{\pi}{4} (D^2 - d^2) \qquad (1)$$

where

D = needle diameter

d = diameter of the seat (see Fig. 4.10)

Fuel pressure acts on the whole area of the needle at the start of injection, which means that it lifts very quickly; this obviously helps with atomization.

The injector closes at pressure P_F according to the following equation:

$$P_F \frac{\pi}{4} D^2 = F_1 \qquad (2)$$

The corresponding spring tension F_1 is almost equal to F. In fact, given the very-small movement of the needle (less than 1 mm), the stiffness of the spring is negligible. Thus

$$P_F = P_0 \frac{D^2 - d^2}{D^2} \qquad (3)$$

The instantaneous injector volume flow (Q_i) is expressed as follows according to Bernoulli's law,

$$Q_i = KS \left(\frac{P_i - P_c}{\rho} \right)^{1/2} \qquad (4)$$

where

K = a coefficient specific to the injector under study

S = orifice area

P_i = injector pressure (fuel)

P_c = combustion chamber pressure (air)

ρ = fuel density

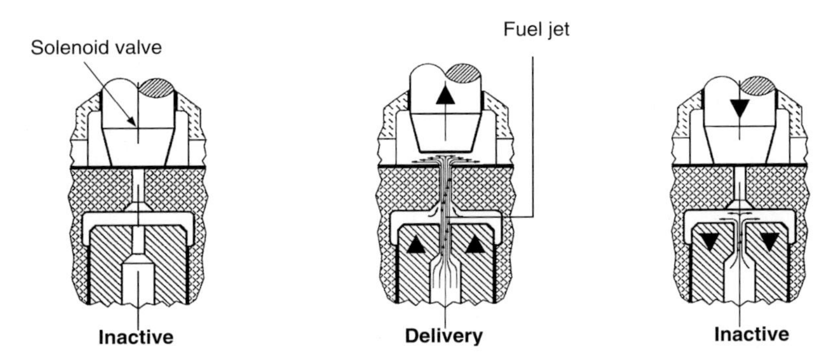

Figure 4.9 *Needle lifting mechanism.*

With **common rail** systems, needle raising is actuated electromagnetically, instead of mechanically. When the injector is inactive, the needle blocks the spray orifices; this needle position is maintained with the help of the fuel pressure acting on the head of the needle. When the solenoid valve opens, a pressure drop occurs in the fuel at the head of the needle, which causes it to lift and to begin fuel delivery to the combustion chamber. The fuel delivery ends when the solenoid valve closes and the injection needle returns to normal, blocking the injection orifices.

C. Various types of injectors

Diesel engines can be either direct or indirect injection, depending on whether the fuel is sprayed directly into the combustion chamber or into a small cavity referred to as a **pre-combustion chamber**, which is connected to the combustion chamber by a channel. Today, all industrial vehicles have direct-injection engines, whereas most diesel passenger cars have engines with divided combustion chambers. However, the development of direct-injection engines for light vehicles is currently an important topic; all the major European and Japanese manufacturers are offering, or will be offering, diesel passenger cars with direct-injection engines.

Note that each type of combustion chamber uses a certain type of injector: the injectors used in divided combustion chambers are usually pintle-type injectors whereas the direct-injection engines are equipped with hole-type injectors (see Fig. 4.10). The fundamental difference between these two injectors is the shape of the fuel jet and the injection pressure. The two types of injector are described schematically in Fig. 4.8 (Bosch 1985).

The **pintle-type injector** has a conical orifice; the pintle located at the end of the nozzle needle combined with the circular orifice provides a **ring-shaped area** that varies with the needle lift.

When the needle begins to lift, only a small flow area is available; a low-flowrate preliminary fuel spray is directed into the combustion chamber; when the needle lifts completely off its seat, the orifice opens fully, which results in the main full-flow spray.

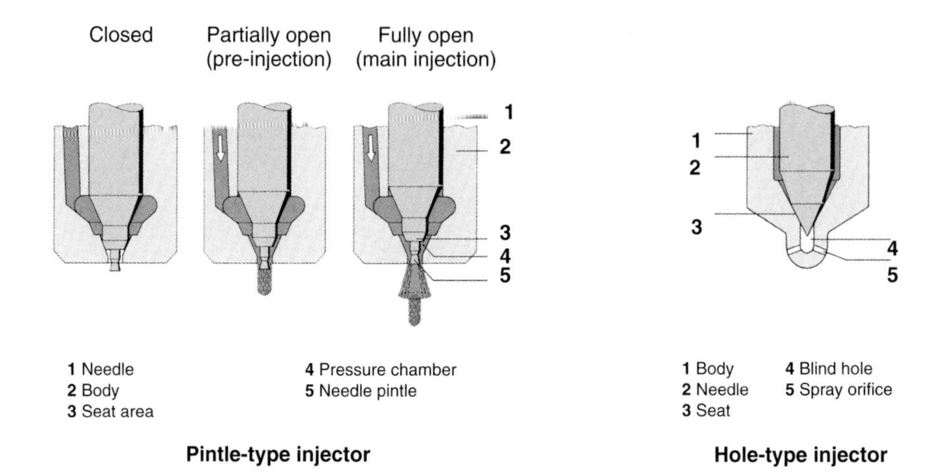

Figure 4.10 *Fuel-injection schematic description.*

Some pintles have a **flat section** to prevent irregular deposits from interfering with the flowrate of the spray (see 4.B). Also worth noting is the fact that injection pressure increases progressively, which results in quieter combustion. Finally, even if the spray pattern changes as the needle rises, it remains a deep cone with a fairly-tight base angle.

On the other hand, **hole-type injectors** have a **wide spray pattern**. They have several orifices that result in a more-open conical pattern. The number of orifices (3 to 7) depends on the shape of the combustion chamber and the air turbulence. The diameter (0.16 to 0.35 mm) and the length of the orifice depend on the desired atomization. To limit the production of unburned hydrocarbons, the volume of retained fuel between the edge of the needle seat and the nozzle orifice must be minimal; this is the reason why **seat-hole** nozzle injectors have been developed. This change has been implemented on the injectors of some industrial engines, but its introduction in the smaller systems used in passenger cars has been delayed. In fact, it is very difficult to make the small holes in the injector tip, which open directly onto the needle seat, at low cost and in volume production.

Another difference from pintle-type injectors is the high injection-pressure gradient, which results in relatively noisy combustion.

4.1.4 Establishing the injection rate for conventional systems

The "injection rate" describes the change in the flow of fuel injected into the combustion chamber as a function of the crankshaft and camshaft angles. Understanding the operation of the pump and the injectors leads to a determination of the fuel flow.

4.1.4.1 Simplified approach

A relationship should be established between the delivery flow of the pump (Q_r) and that of the injector (Q_i). The pipes between the pump and the injector act as a capacity-damper with volume V_t.

The change in pressure in the injection pipes (P_i) is determined by the following differential equation,

$$\frac{dP_i}{dt} = \frac{E}{V_t}(Q_r - Q_i) \tag{5}$$

where

E is the bulk modulus of diesel fuel (about 15 000 bar)

Q_i is related to P_i, which was previously described in equation (4)

A sample calculation of the change in P_i as a function of time is shown in Fig. 4.11. Fuel discharge from the pump begins at the injector closing pressure P_F; injection begins when the pressure rises to P_0. Pressure continues to rise in the delivery pipes because the injector orifices are too small to accommodate the normal flow from the pump. The delivery from the pump body stops when P_i reaches maximum, but flow continues from the injector until $P_i = P_F$. The value for Q_i in Eq. (4) is derived by applying the rule for the change in P_i as a function of time.

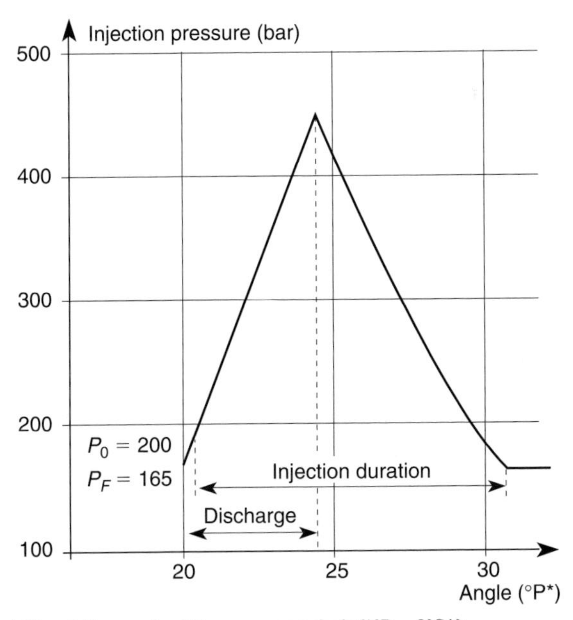

* °P: rotation angle of the pump camshaft (1°P = 2°CA).

Figure 4.11 *Sample calculation of injection-pressure change.*

Two important items of information can be derived from the results presented in Fig. 4.11. First, the pressure in the delivery pipes is about 500 bar, which is much greater than the injector pressure setting (100 to 300 bar). Second, the actual injection duration is much longer than the pump discharge duration. Yet, the most important parameter in the combustion process is the flow-rate of the diesel-fuel delivery to the combustion chamber.

However, this quick and simple method does not take into account other hydraulic phenomena (pressure oscillations, re-opening of the injector, and cavitation) that should be controlled.

4.1.4.2 Accounting for pressure waves

The previous discussion assumed that the pressure in the injection piping depended uniquely on time. The actual results shown in Fig. 4.12 demonstrate that the pressure, at any given time, is not the same at all points in the pipe.

The pressure rises first near the pump and then it proceeds toward the injector. This hydraulic phenomenon is similar to "water hammer" in pipes. It is based on pressure wave theory and it can be interpreted mathematically using the method of characteristics. This method deals with the equations of motion in a one-dimensional system—the variable being space—by considering two pressure waves as follows:

- one **incident wave** moving in the direction of the delivery at the speed of sound in the liquid (about 1400 m/s)
- one **reflection wave** moving from the injector to the pump, also at the speed of sound

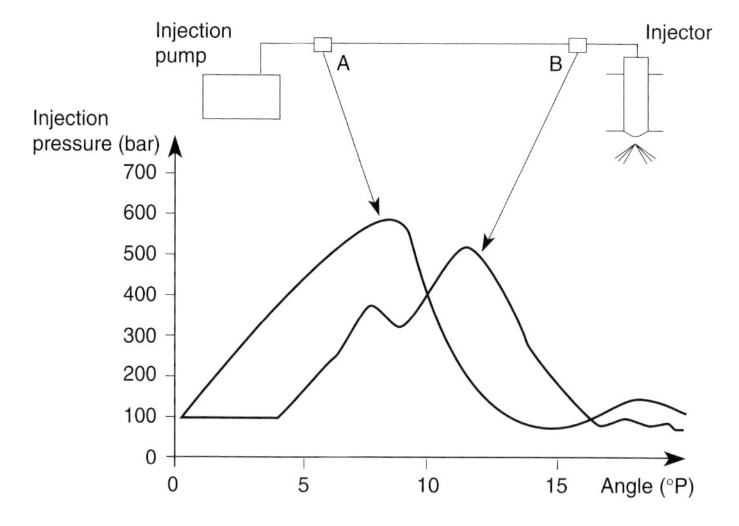

Figure 4.12 *Example of the difference in pressure at two points in an injection pipe.*

Along the length of these waves, the pressure (P) and flow (Q) are related by the following simple equations:

- for the incident wave

$$P = \alpha - \frac{C}{S} Q \qquad (6)$$

- for the reflection wave

$$P = \beta + \frac{C}{S} Q \qquad (7)$$

For these equations, C is the speed of sound, S is the cross-sectional area of the pipe, and α and β are constants defined by the initial and limit conditions.

A sample of the calculated change in injection pressure is shown in Fig. 4.13, where it is plotted as a function of time and as a function of its position between the pump and injector. Significant pressure fluctuations occur, especially at the end of an injection.

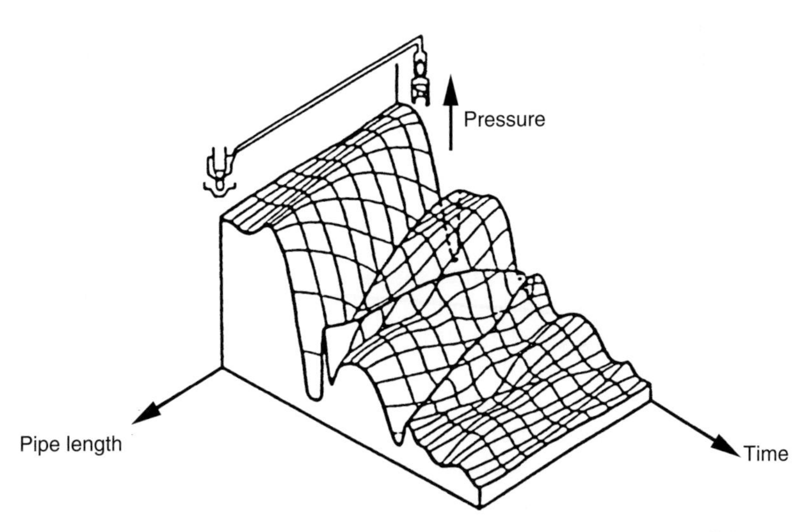

Figure 4.13 *An example of injection-pressure change as a function of time and location.*

4.1.4.3 Fuel flow curve

Based on P_i, the instantaneous flow of the injected fuel Q_i is easily determined, as is the total quantity of fuel supplied for a cycle:

$$Q = \int Q_i \, dt$$

Q is usually expressed in mm^3/stroke, where a stroke corresponds to one cycle of a pump plunger.

Fig. 4.14 shows an example of the flow curve Q_i with respect to time, which is expressed as the injection-pump camshaft angle. The calculated diameter of the fuel droplets at each point is also shown. Satisfactory atomization with fine

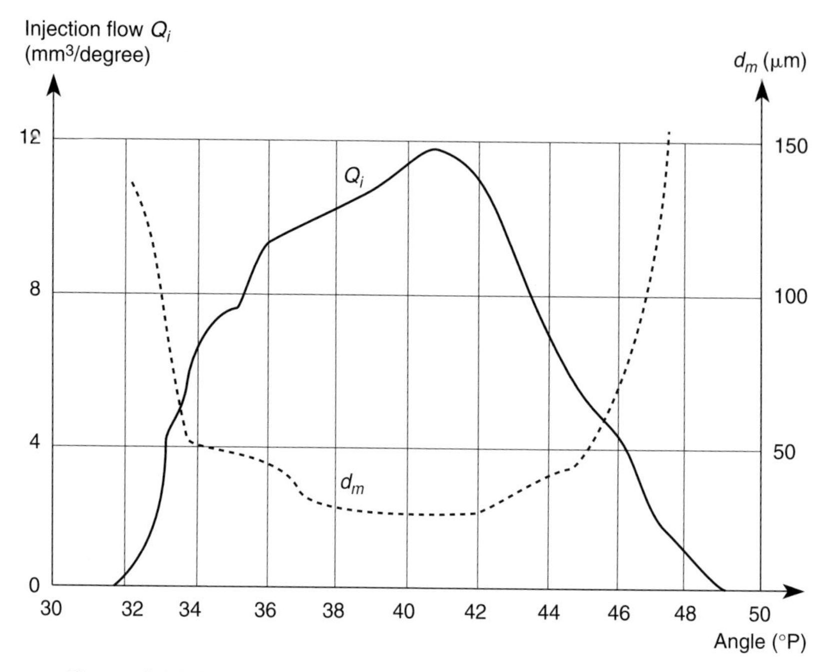

Figure 4.14 *Instantaneous diesel-fuel-injection flow rate and average droplet diameter during the course of injection.*

droplets occurs during the **median phase**, when the fuel pressure P_i is high. The opposite occurs at the beginning and at the end of injection, when the droplet size is much larger. This situation is essentially inconsequential when the injector opens, because the liquid fuel has a reasonably long time to vaporize and a sufficient quantity of air to burn. This is not the case however, as the **injector is nearly closed**; therefore, it is preferable to minimize the relative contribution of this poorly vaporized portion.

4.1.4.4 Parasitical phenomena

The major unwanted phenomenon is **after-injection**, which is well controlled with today's engines. The main injection-pressure wave is followed by another unwanted wave that oscillates between the pump and injector. This wave decays to a residual pressure that is less than the closing pressure of the injector. However, the amplitude of the first oscillation (see Fig. 4.15) can be high enough to cause the injector to re-open, which results in a late injection of low-pressure poorly-atomized fuel. This needle bounce must be avoided because it results in increased emissions of unburned hydrocarbons. One solution consists of using a **snubber valve,** which provides a restriction in the supply pipe when the injector closes.

The presence of "dead space" in the injector nozzle, beyond the needle seat, has the same effect as retarded injection. In effect, the fuel contained in the

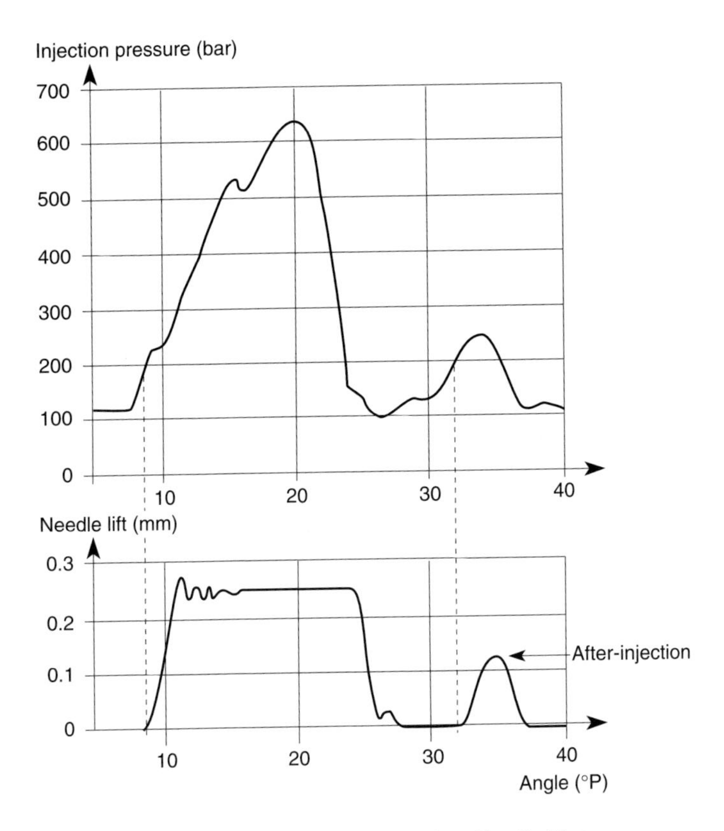

Figure 4.15 *Pressure wave in an injection pipe. Needle-lift trace showing after-injection.*

dead space, whether liquid or vapor, arrives late in the combustion process and it is not fully burned.

The quantity of fuel allowed to return through the valve must be carefully chosen to avoid gas bubbles in the fuel enclosed in the pipes. This problem, which is usually referred to as **cavitation**, is now well controlled; however, it may reappear with the advent of the new very-high-pressure systems being used on direct-injection engines. Cavitation appears on injection-pressure traces as oscillations resulting from the implosion of gas pockets and as a rapid drop in pressure when the injector nozzle closes. In fact, sound propagates more slowly in a heterogeneous liquid-gas mixture than in a liquid. Cavitation causes erosion and wear in fuel-injection systems and it must controlled.

4.2 Diesel engine combustion

Controlling the initiation and the progression of diesel-engine combustion is extremely complex for the following reasons: the air-fuel mixture is initially heterogeneous, which is inherent to the injection process; a diffusion flame exists; and several aerodynamic parameters, which are linked to technological details of the engine design, are involved. This section limits its description to some of the main phenomena, which are of interest due to their relationship to various fuel-quality criteria.

4.2.1 Ignition delay

In Chapter 3, which was devoted to gasoline, auto-ignition delay was defined as an intrinsic characteristic of a fuel that was related to temperature and pressure by the following equation,

$$\theta = AP^{-n} \exp\left(\frac{B}{T}\right)$$

where A, B, and n are numeric constants.

In the diesel engine, the auto-ignition delay (θ_d) has a far more **practical** significance and it represents the characteristics of the engine-fuel combination for an operating condition; θ_d represents the time lapse from the moment the first drops of fuel leave the injector to the instant when combustion begins. In fact, this delay includes **several phenomena** of very-different natures, which occur consecutively and simultaneously: heating of the fuel on contact with air, vaporization, formation of a homogeneous auto-ignitable mixture, and chemical preparation for auto-ignition. Moreover, the value of θ_d, which is expressed in milliseconds or in crankshaft degrees, depends on the chosen measurement technique.

The start of injection is easily determined by detecting the injector needle lift, however, attempting to determine the start of auto-ignition often runs into experimental problems. Two major measurement techniques are used: **pressure-diagram analysis** and **optical detection**. The first technique is frequently used because it is easy to implement—all that is required is a pressure-time diagram. The moment when combustion begins corresponds to a change in the slope of the measurement curve (see Fig. 4.16). This point is often difficult to locate, especially with indirect-injection engines, because their combustion evolves in a more progressive manner than that of direct-injection engines; therefore, the results derived using this technique are sometimes inaccurate.

The second technique for determining the auto-ignition delay is based on the detection of an optical signal (Claude et al. 1988). This technique provides accurate results but it requires sophisticated equipment: an **optical sensor** located in the combustion chamber is used to detect the light emitted when auto-ignition occurs. If the range of the detector is in the ultra-violet spectrum, the appearance of the first radicals are detected. If the range is in the visible spectrum, the signal corresponds to the first particles of soot.

With most current engines, the auto-ignition delay θ_d is on the order of a few °CA (1.5 to 10.0) for conventional diesel fuel. It varies with the operating parameters (speed, load, injection advance).

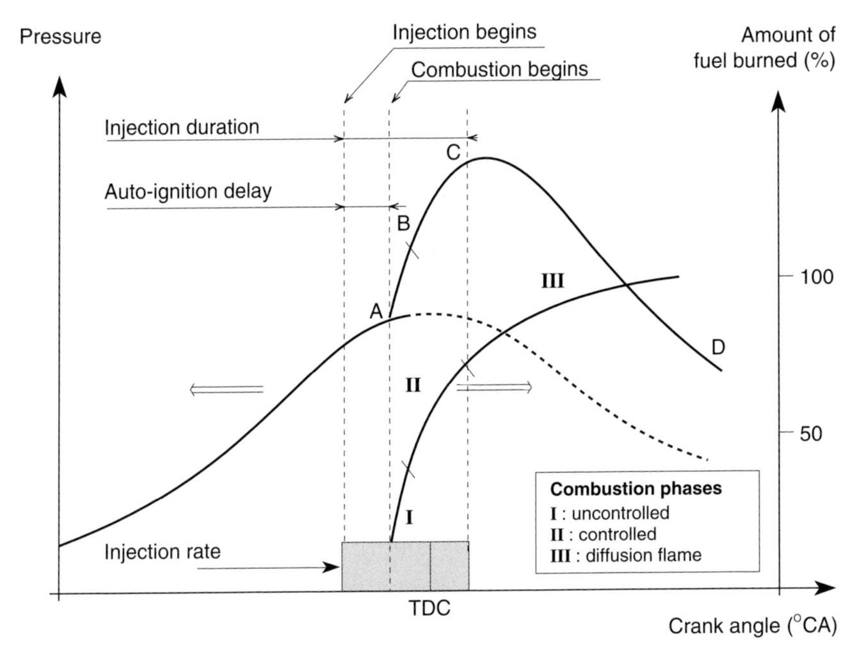

Figure 4.16 *Auto-ignition and combustion in the diesel engine. Analysis based on a pressure diagram. (The significance of points A, B, C, and D is described in 4.2.3)*

Fig. 4.17 shows the relative variation in delay, as a function of both engine speed and injection point, for a large number of engines of various designs. As the speed increases from 1000 to 4000 rpm, for example, the delay θ_d increases in °CA but decreases in real time. This change is enhanced by the pressure and the temperature, which increase with engine speed. There is also an optimal moment for injection that results in minimal delay. Premature fuel injection results in very-long delays because the thermodynamic conditions of pressure and temperature are not ideal for vaporization and auto-ignition.

To attempt an analysis of the various phenomena that precede the initiation of auto-ignition in a diesel engine, the total time θ_d can be separated into phases referred to respectively as the **physical** and the **chemical delay**. This distinction may appear to be artificial and arbitrary because the physical and chemical processes are unquestionably interlocked and mutually dependent. However, this approach is valid, at least from a theoretical standpoint.

4.2.1.1 Physical auto-ignition delay

This is the time between the moment when the jet of fuel begins to emerge from the injector and the instant that a auto-ignitable air-fuel mixture exists at a point in the combustion chamber. It is very difficult to analyze the physical changes in the fuel jet with precision. This can be represented schematically, as shown in Fig. 4.18, by a **center zone** of liquid surrounded by an **envelope**

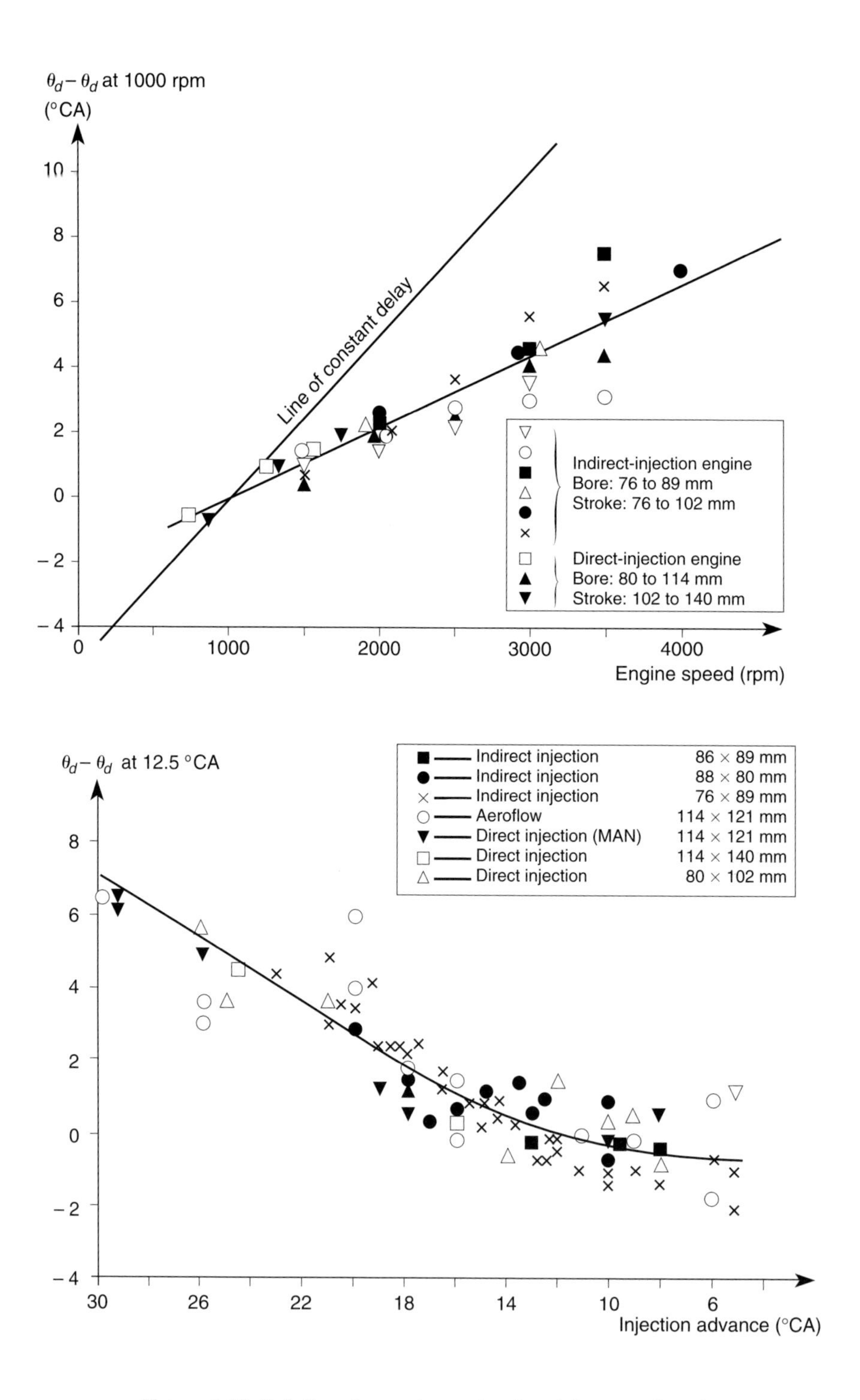

Figure 4.17 *Relative change in combustion delay as a function of engine speed and injection advance.*

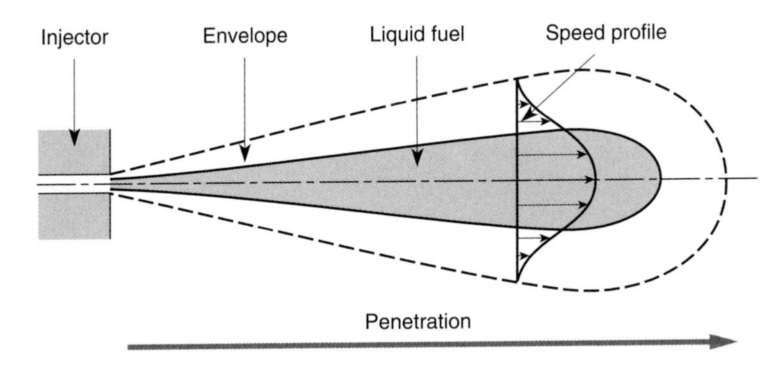

Figure 4.18 *Schematic representation of an injector fuel spray.*

consisting of a heterogeneous mixture of air and fuel that is partially vaporized. **Dispersed droplets**, which do not interact with others, can also exist at the extreme periphery of the jet.

It has often been conjectured in the past that the droplets detached from the jet could have an overriding involvement in the physical delay. The droplets change with respect to space and time, which depends on many parameters: diameter (itself a function of the fuel characteristics), initial speed, air pressure, air temperature, ... However, some relative orders of magnitude are available for a drop of fuel entering at a speed of 250 m/s into pure air at 540°C and 35 bar. It covers a distance of about 1.5 cm in 0.6 to 0.8 ms and then it stops. The fuel vaporizes progressively and the vapor diffuses into the air to form a mixture of decreasing equivalence ratio as function of distance (see Fig. 4.19).

However, this simplified approach does not correspond with reality; there is greater justification for treating the **entirety of the jet dynamics** by involv-

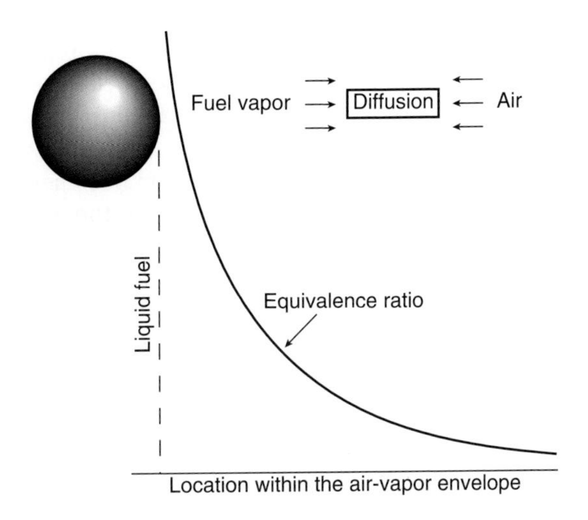

Figure 4.19 *Change in air-fuel mixture composition around a droplet undergoing evaporation. Schematic representation.*

ing the liquid-vapor equilibrium rule. Mixture zones appear as a function of time. Zones occur where the mixture composition depends mainly on the limitations imposed by the equilibrium rule, with a cooling tendency linked to the endothermic character of the state change. Zones also appear where the mixture is "superheated" and the fuel is completely vaporized.

A satisfactory model of the physical delay is still nonexistant, but the following qualitative conclusions can be drawn from the previous analysis:

- the physical delay is not negligible and it is certainly of the **same order of magnitude** as the chemical delay
- for conventional diesel fuels, minor changes to fuel volatility have little effect; in fact, the phenomena linked to **diffusion** and the aerodynamic characteristics of the environment seem to have a more determinant effect than the temperature of vaporization
- the physical delay affects the chemical delay in the way that it affects **the temperature and the air-fuel ratio** in the zones where auto-ignition begins

4.2.1.2 Chemical auto-ignition delay

The following conventional formula for the expression of auto-ignition delay can be applied in principle to the chemical processes leading to ignition in the diesel engine:

$$\theta = AP^{-n} \exp\left(\frac{B}{T}\right)$$

As with the prediction of knock in the spark-ignition engine, the coefficients A, n, and B could conceivably be determined from temperature and pressure data derived from various experimental conditions.

However, this approach runs into a major impediment: in the diesel engine, the air-fuel ratio that occurs in the zone where auto-ignition takes place is unknown. In fact, as the fuel evaporates, lean-mixture zones—where the temperature increases quickly—are the first to form and they are followed by lower-temperature stoichiometric or rich zones. These various regions all **compete** to initiate auto-ignition and the one that causes the decisive action cannot be predicted. It would seem, however, that the lean zones are the most likely to cause auto-ignition.

The difficulty of numerically modeling this system becomes apparent when the values of A, n, and B are narrowly dependent on the air-fuel ratio being considered.

However, because the chemical delay can be estimated, is has been observed to vary with the chemical structure of the fuel according to well-understood principles. Therefore, a paraffinic diesel fuel would, under fixed experimental conditions, have a shorter delay than a fuel that is rich in naphthenes or aromatics.

4.2.1.3 Numerical expression of the delay

The impediments mentioned previously have not deterred researchers from proposing formulas to correlate the auto-ignition delay θ_d in the diesel engine to easily measurable parameters: average pressure and temperature; engine speed; compression ratio; ... .

The auto-ignition models most often proposed take into account the **kinetic chemical phenomena in a homogeneous gas**.

The most elementary of the formulas is based on a single Arrhenius equation,

$$\theta_d = AP^{-n} \exp\left(\frac{B}{T}\right)$$

where the terms A, n, and B carry information relative to both the physical and chemical processes. These models have the advantage of being simple; however, they are not very precise.

Models are also based on the exhaustive Warnatz kinetic schema (Warnatz 1981, 1984). These models use **complex kinetics** consisting of a large number of reactants and products. Warnatz models are precise, but they require large amounts of memory and calculation time.

Finally, there are models based on a **simplified kinetic schema** with as few as 16 steps, and some models have only four steps (Peters 1992; Pinchon 1989); in the latter case, the steps include initiation, propagation, branching, and break away. These are intermediate models positioned between the two previous models; they provide a good compromise of precision, calculation time, and memory for three-dimensional codes of the Kiva type.

However, the auto-ignition delay is not governed exclusively by the rules of chemistry for homogeneous gas mixtures. The models mentioned previously do not take into account the physical-turbulence phenomena that occur during the auto-ignition delay. This is why **other types of models, which account not only for chemical kinetics but also for the turbulence of the environment**, are recently being developed (Linan 1993).

4.2.2 Auxiliary starting aids

Starting a diesel engine when the exterior air temperature is low usually requires the assistance of auxiliary starting aids.

With indirect-injection engines and with some small-displacement direct-injection engines, **pre-heating and sometimes post-heating glow plugs** are used (see Fig. 4.20). The plug is located in the center of the prechamber, as close as possible to the path of the fuel jet from the injector. The glow plugs currently in use consist of an electrical filament embedded in a magnesium-oxide ceramic powder inside a metallic glow tube. When the driver turns the key, the glow plug heats up quickly (temperatures can reach 1050°C in 12 seconds). The **hot spot** formed by the glow plug substitutes for the lack of energy in the chamber and makes the auto-ignition of the diesel fuel thermodynamically pos-

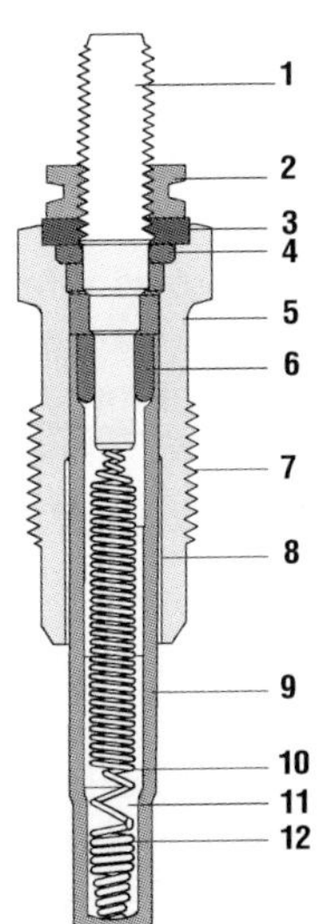

1 Terminal
2 Round nut
3 Insulating washer
4 Toroidal seal
5 Plug body
6 Seal
7 Thread
8 Annular gap
9 Glow tube
10 Control coil
11 Insulating powder
12 Heating coil

Figure 4.20
Beru-type self-regulating pencil glow plug.

sible. The pre-heat time varies from 4 to 10 seconds. The plug can also remain on after starting to improve combustion and reduce emissions (Chapter 5), particularly when the vehicle is operating at idle or at low loads.

The post-heat time varies from 10 seconds to 3 minutes. The current flow during starting is very high (20 to 22 A).

Large-displacement direct-injection engines are normally not equipped with glow plugs because they are not very useful. However, at cold temperatures it is sometimes necessary to **pre-heat the intake air**.

4.2.3 Progress of combustion

The speed of energy release in a diesel engine can be calculated from the pressure diagram. The hypothesis and the calculation method are not substantially different from those used for the gasoline engine (see 3.4.2.3). A schematic representation of the data and the results of such an analysis are shown in Fig. 4.16. Three phases can be distinguished between the beginning and the end of combustion.

A. *Phase I (part AB of the pressure curve)*

In phase I, the fuel that was injected during the auto-ignition delay burns **instantaneously** or very quickly. This results in a steep pressure rise on the order of 3 to 4 bar/°CA, which is the cause of the distinct diesel combustion noise. This short phase (about 5°CA) is determined by both the injection rate and the auto-ignition delay. Therefore, a relatively-long delay and a copious injector flow would lead to an accumulation of fuel in the combustion chamber. Auto-ignition of this large quantity of fuel would result in a significant rise in pressure. This must be avoided to prevent severe engine problems, both mechanical and thermal.

B. *Phase II (part BC of the pressure curve)*

Phase II represents the end of the injection period. The fuel enters a **burning environment** where it easily finds the oxygen required for combustion and burns quickly. At this point, the fuel flowrate of the injector determines the speed of energy release. However, combustion is restrained near the end of the injection period due to the difficulty of co-locating the air and the fuel.

C. *Phase III (part CD of the pressure curve)*

In phase III, the injection is complete. Unburned fuel is stirred by the gas movements inside the combustion chamber and combustion depends mainly on **diffusion phenomena**. It is worth noting that this terminal phase can be influenced in principle by much earlier events, for example, those related to auto-ignition. Thus, a relatively-long delay, which is usually not preferred, can cause a stronger swirl that enhances diffusion combustion.

Fig. 4.21 shows two typical examples that represent the speed of energy release in a diesel engine; one represents a direct-injection engine and the other represents an indirect-injection engine.

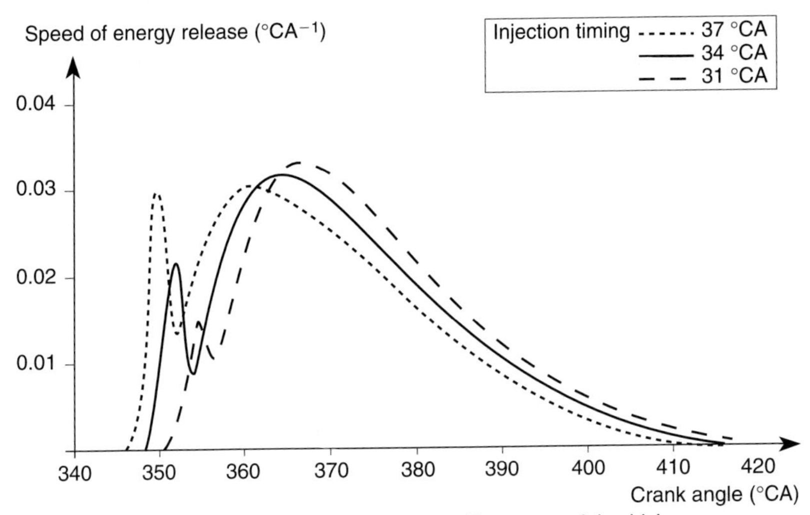

a. Direct-injection diesel engine used in commercial vehicles.
(MEP : 11 bar; speed : 2000 rpm)

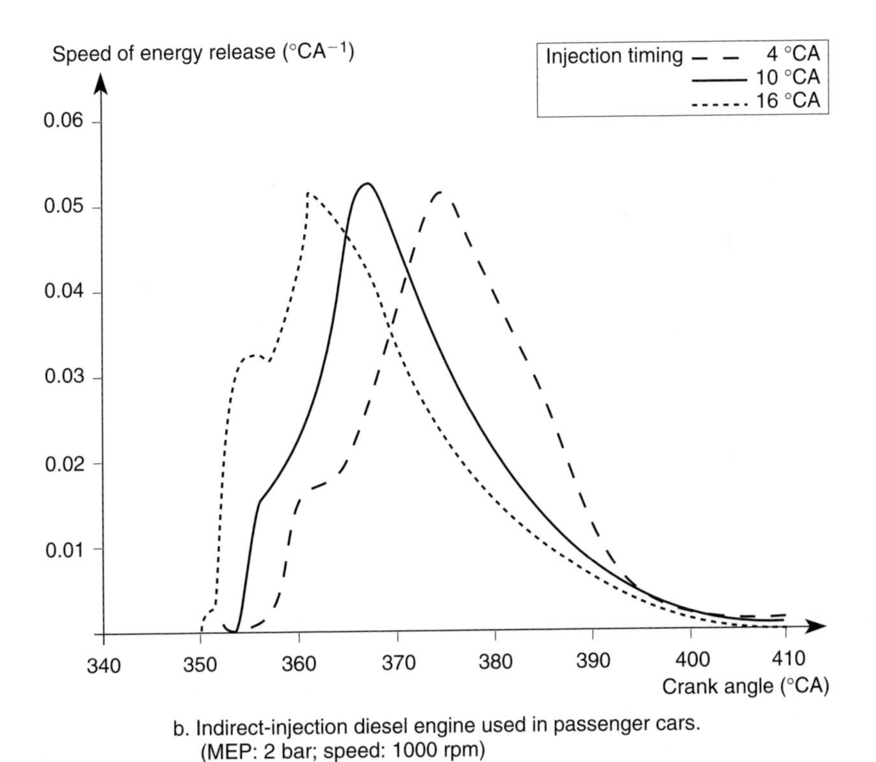

b. Indirect-injection diesel engine used in passenger cars.
(MEP: 2 bar; speed: 1000 rpm)

Figure 4.21 *Examples of the speed of energy release in diesel engines.*

In the first case (Fig. 4.21a), the fuel is sprayed directly into the combustion chamber. The first phase of rapid combustion is easily distinguished, which is followed by a slowing and then a renewed acceleration that represents the existence of a diffusion flame. In the second case (Fig. 4.21b), combustion occurs in a pre-combustion chamber and it evolves in a moderate and more progressive manner.

A characteristic of diesel combustion is the emission of **soot** and **smoke** from the exhaust (see 5.4.1). This phenomenon is caused by the presence of over-rich locally-distributed zones caused by insufficient air-fuel mixing. This occurs especially at high-power levels when the overall equivalence ratio is between 0.75 and 0.85, depending on the type of engine. If the fuel flowrate is increased beyond this point, a rapid increase in smoke occurs, power stagnates, and efficiency is significantly reduced. Therefore, this critical zone defines the "**full load**" of a diesel engine, which is obtained with a given mass of air that is not engaged in combustion, despite all the techniques that have been attempted to promote closer contact between the oxygen and the fuel.

4.3 Performance and technological achievements

Accounting for often-antagonistic constraints is required in the design of a diesel engine; therefore, obtaining proper operation throughout the full range of use requires compromises. For example, to promote ignition at **low loads** it is preferable to have a maximum amount of fuel concentrated at key points in the combustion chamber to quickly obtain zones of sufficient equivalence ratio. The opposite objective applies at **full load**, when it is preferable to have the fuel dispersed as widely as possible in the air to avoid the formation of smoke. The technological development of combustion chambers takes into account all the preceding criteria and other related requirements such as displacement, speed, the type of application expected, ...

Hence, this section covers, in a succinct way, the major types of diesel engines and their characteristics. The section also covers the supercharging process, which is being used more often with diesel technology. Finally, exhaust gas recycling (EGR) systems, which are used mainly to reduce oxides of nitrogen, are briefly described.

4.3.1 Direct-injection engines

With this engine design, fuel is supplied directly into a **single mass of air** in the combustion chamber by a hole-type injector (see Fig. 4.22). As a rule, to reduce exhaust emissions the fuel should not be sprayed on the combustion chamber walls. In this case, the injection parameters—location, type and number of jets, and degree of atomization—have a decisive effect.

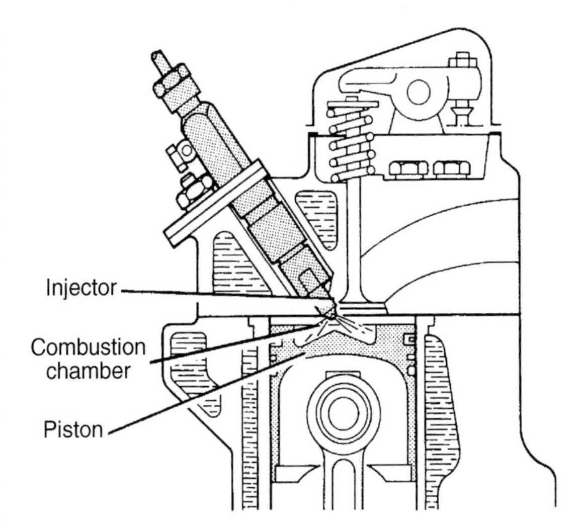

Figure 4.22 *Direct-injection diesel. (Source: Cédra 1990)*

Also, the formation of a turbulence zone in the piston, complimented by the swirling movement of the intake air, promotes mixing and provides complete combustion without smoke up to overall equivalence ratios on the order of 0.75 to 0.80. With high injection pressures (over 900 bar), the air is not usually subjected to a swirl movement, hence, these chambers are referred to as "**open**" chambers. If the injection pressures are lower, swirl is required to enhance mixture formation and to improve combustion. In this case the combustion chamber is deeper and **reentrant** to promote the development of swirl. Direct-injection engines have compression ratios on the order of 15 to 20:1 and speeds that usually do not exceed 4500 rpm.

The direct-injection technique provides a number advantages, especially efficiency. Direct injection is used extensively in medium- and high-powered engines (commercial vehicles, tractors, locomotives, and slow- or medium-speed stationary engines), but it is seldom used in passenger cars. In fact, the use of direct injection with low-displacement high-speed engines was delayed by technological problems related to the supply of fuel (atomization quality, control of the amount injected) and air to the combustion chamber. These problems are now on their way to being resolved.

The pattern of "isoconsumption" curves shown in Fig. 4.23 corresponds with those presented for the spark-ignition engine (see Chapter 3). However, there are two differences: the efficiency is **very high** for a heat engine (as a consequence, the specific fuel consumption is remarkably low) and the loss of efficiency at light loads is **not as outstanding** as that of a gasoline engine, because there is no significant contribution from the low-pressure loop.

The following **variant** of the direct-injection technique is worth mentioning. This combustion system, which was developed by the German firm *MAN*, uses a spherical combustion chamber in the top of the piston and a swirling intake charge. The fuel, which is injected tangentially into the swirl, hits the walls of the chamber where it **evaporates progressively**. This procedure provides better control over the auto-ignition delay and the energy release.

It also provides advantages in noise reduction and the eventual use of fuels that are not readily auto-ignitable.

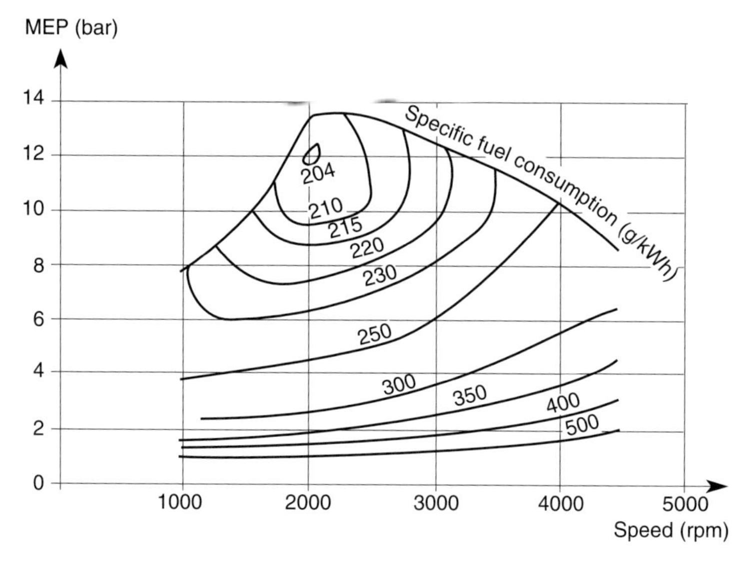

Figure 4.23 *Set of constant fuel-consumption curves for a direct-injection diesel engine (Displacement: 2.5 liters). (Source: Demel 1991)*

4.3.2 Indirect-injection engines

The indirect-injection technique is used on small high-speed engines and it is characterized by the use of a combustion chamber that is divided into two compartments (see Fig. 4.24). There are two different types of indirect-injection engines:

- **Pre-combustion chamber engines**: the pre-combustion chambers represent about 30% of the total dead space; in actual implementations of the technique, the chambers are centered as much as possible with respect to the main chamber and they are connected to it by a series of small orifices. This technique was currently used by *Mercedes*.
- **Induced-turbulence prechamber engines:** the induced-turbulence prechamber represents a dead space volume of more than 50% and it is located at the side of the main combustion chamber. The prechamber is connected to it by a **channel** that opens tangentially into the main chamber. This type of configuration, which is represented by the *Ricardo* (England) company's *Comet* chambers, is widely used in automobile diesel engines (England, France, Japan) but tends to be replaced by direct injection.

With both of these processes, injection and the start of combustion take place in the prechamber, which is **overpressured**. As injection progresses, the fuel is ejected into the main chamber, mixed with air, and burned progressively.

Indirect injection promotes the atomization of fuel by aerodynamic means and it is less demanding from a quality-of-injection standpoint. These cham-

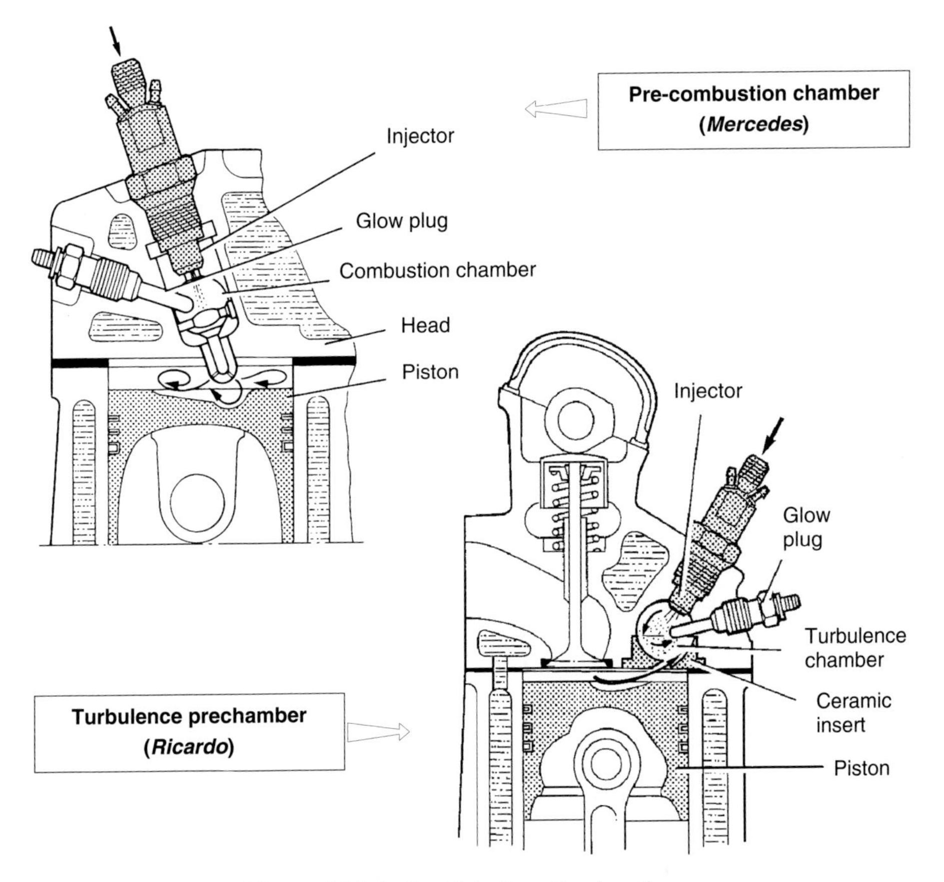

Figure 4.24 *Indirect-injection diesel engines.*

bers usually use pintle-type injectors, which have been adapted to suit each particular situation, especially with regard to the geometry of the fuel spray.

The concentration of the fuel in a small volume promotes ignition at low loads and at high speeds, as long as a high compression ratio (20 to 23) is used. These qualities, which are associated with satisfactory use of the intake air, explain why prechambers are often found in small diesel automotive engines.

However, the prechamber technique provides **lower performance** than direct-injection techniques. This results from the thermal losses inherent in the prechamber design (higher surface-to-volume ratio, higher turbulence) and the throttling losses through the chamber transfer channel.

The engine map shown in Fig. 4.25 provides the iso-consumption curves for a standard automobile diesel engine. The **maximum efficiency** is slightly less (about 5%) than a gasoline engine of the same power rating. The diesel engine's superiority resides in its **performance at part load**, which is much better than its gasoline-fueled rival.

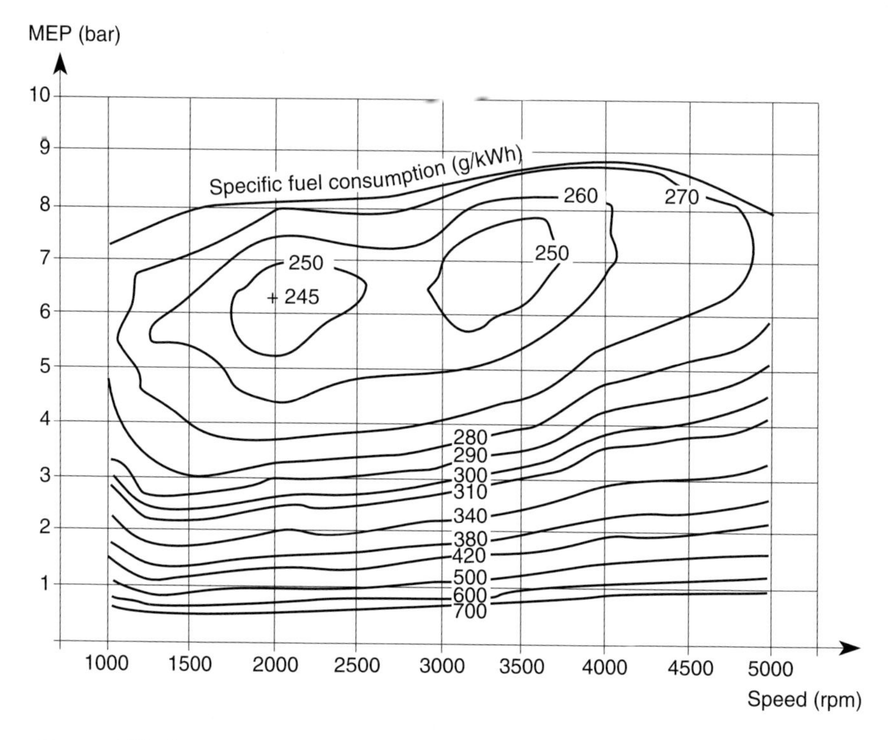

Figure 4.25 *Set of iso-consumption curves for an indirect-injection diesel engine (Displacement: 2.5 liters).*
(Source: Fortnagel 1993)

However, the automobile diesel engine also suffers from relatively high friction losses that are a direct result of the design technology and the characteristics of the metallic friction surfaces (crankshaft, bearings, pistons, cylinder walls, ...).

4.3.3 Supercharging diesel engines

There are two unfavorable factors that affect the diesel engine from the standpoint of specific power output: the need to constantly operate with excess air and the difficulty of obtaining high rotation speeds along with acceptable crank-angle timing of the combustion diagram. **Supercharging** is often used to overcome these problems.

Engine power increases with fuel flow and therefore with the density of the intake air. Supercharging, combined with increased displacement, enables a diesel engine to deliver torque and power outputs close to those obtained from gasoline engines (see Fig. 4.26).

Also, because air density is a function of temperature, combining supercharging with **charge-air cooling** (intercooling) provides even-better engine performance.

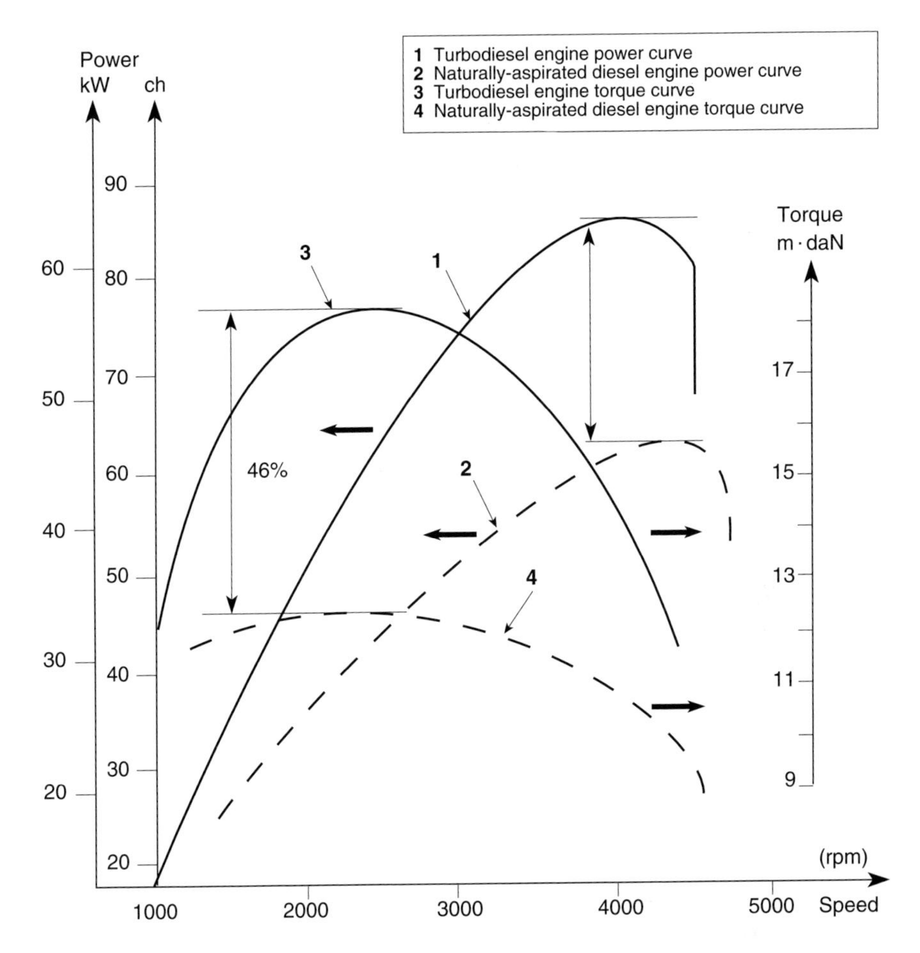

Figure 4.26 *Comparative performance curves for two diesel engines, one naturally aspirated and the other turbocharged and intercooled.*
(Application: passenger car).
(Source: Foy 1994)

Supercharging also allows the torque curve to be re-shaped and thus it provides, for the same power output, a **higher maximum torque at a lower engine speed**, which greatly improves driveability.

Supercharging can be provided by a volumetric device—either a compressor or a pump—which is driven mechanically or by means of a **turbine located in the exhaust stream**.

The turbocharger consists of a compressor and a turbine. The compressor is located in the engine intake system and the turbine, which drives the compressor, is located in the engine's exhaust stream (see Fig. 4.27). The wheels of the turbine and the compressor are mounted on the same shaft, which is supported by a system of bearings (see Plate 4.1). The assembly turns at very-high maximum speeds of 100 000 to 200 000 rpm. The turbine wheel is made

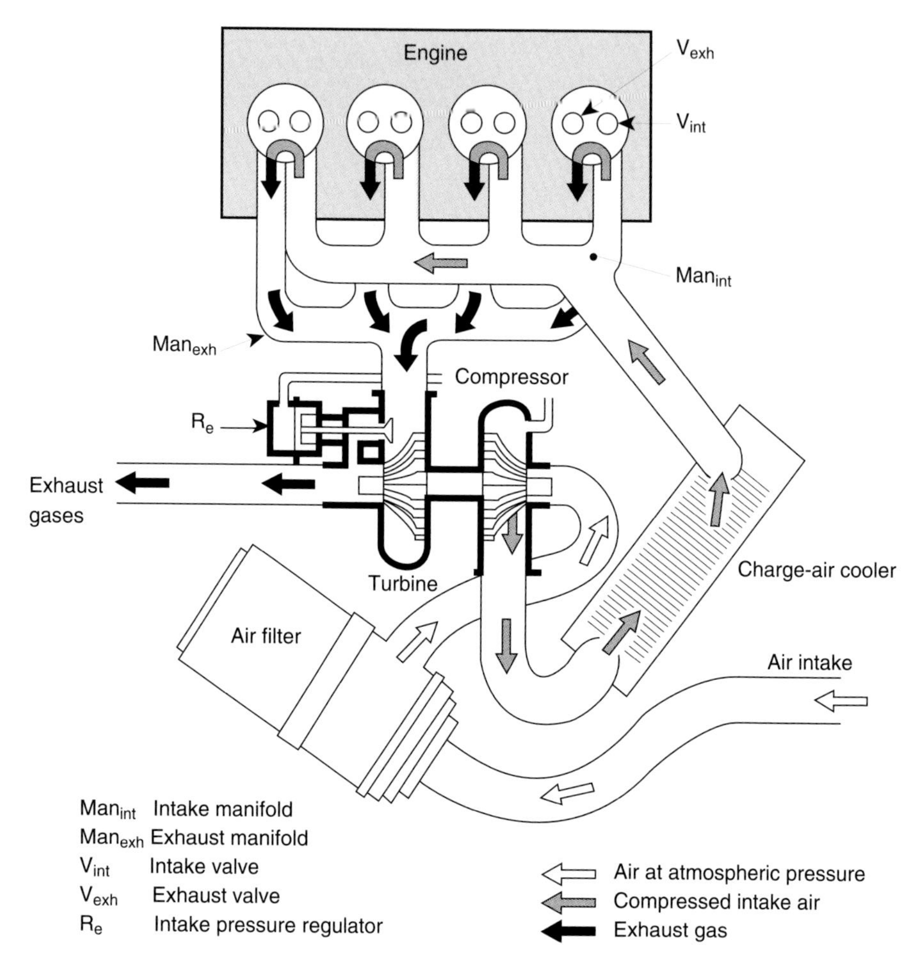

Figure 4.27 *Turbocharger installation on a passenger car. (Source: Renault)*

of special alloys (up to 75% nickel content), and the turbine case is made of nickel-rich iron, the composition of which depends on the operating temperatures.

The compressor wheel and case are made of aluminum alloy. The devices are joined at a bearing housing, which contains floating bearings lubricated by a large flow of high-pressure engine oil. To limit the supercharger boost pressure (2.0 bar for passenger cars and 3.0 bar for commercial vehicles), and to protect the turbine, an exhaust-gas bypass circuit is located upstream of the turbine.

This circuit is controlled by a valve referred to as a "**waste gate**", which opens when the boost pressure exceeds a predetermined value. When the valve opens, part of the exhaust gas bypasses the turbine and goes directly to the exhaust system.

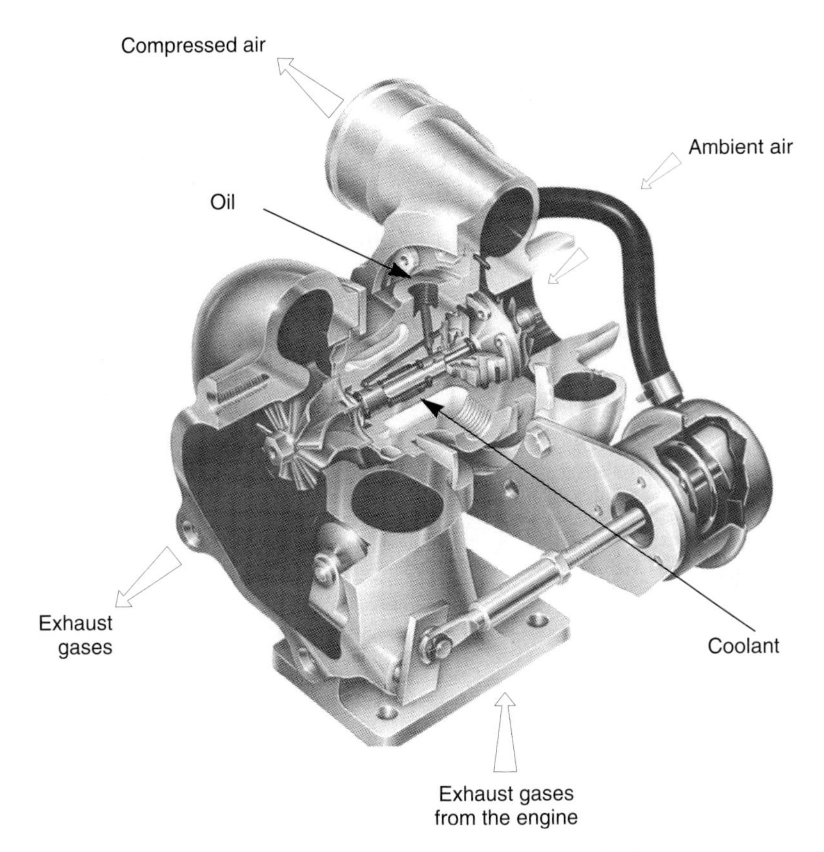

Plate 4.1 *Sectioned view of an automobile turbocharger.*
(Source: Garett)

At low loads, the energy absorbed by the compressor and the bearings is such that the turbocharger is not an active contributor to engine performance. The engine then behaves as a conventional naturally-aspirated diesel.

Variable geometry turbochargers also have a promising future. They are already proposed by several manufacturers. Their big advantage is their ability to continuously change the dynamic characteristics of the turbine and to make optimum use of the energy contained in the exhaust gases.

Supercharging can also be accomplished by using a **pressure-wave exchanger (or Comprex)**. This device consists of a cell-type compressor wheel driven from the crankshaft.

Each side of the compressor wheel has two ports, one each for the entry and the exit of gases. Exhaust gases are expanded on one side of the wheel, thus transferring their energy to compress the intake air on the other side. In operation, fresh air from the intake side fills a cell in the wheel while it is turning to face the exhaust gas ports. The pressure wave from the expanding exhaust gases forces the air in the cell back toward the air outlet port (toward the intake manifold), which opens as the wheel continues to turn. The process is repeated with fresh intake air for each cell in the wheel. This type of supercharger is currently being used on diesel-powered passenger cars produced by *Mazda*.

4.3.4 Exhaust gas recirculation

To satisfy the emission regulations for oxides of nitrogen, exhaust gas recirculation (EGR), which is already in use with indirect-injection diesel passenger cars, is now being applied to other diesel engines. The purpose of this technique is to recirculate a portion of the exhaust gas through the intake system to dilute the fuel-air mixture with **neutral gases**.

EGR is not a technique used exclusively on diesel engines. In fact, it is also used with lean-burn gasoline engines to reduce oxides of nitrogen emissions that cannot be treated with a catalytic converter. Chapter 5 (see 5.7.2) describes the implementation constraints, the details, and the effectiveness of EGR systems.

Characteristics and specifications of diesel fuel

Diesel fuels and diesel engines have been a satisfactory match for a long time. This situation was the result of the following two favorable situations:

- the diesel engine was only used in commercial vehicles (tractors and trucks), which were considered unsophisticated
- diesel fuel, which was obtained solely by direct distillation, provided a combination of characteristics that were very favorable for combustion by auto-ignition

Since the 1970s, however, several constraints have appeared: the development of small automobile diesel engines, the need to meet ever more stringent emission regulations, and the degradation of some diesel fuel properties due to changes in refining—especially the increased conversion of heavy fractions.

As a result, automobile and petroleum companies have been intensely preoccupied since 1990 with matching diesel fuels to diesel engines. The effects on energy consumption are not as significant as those of the spark-ignition engine, however, the implications are many and they relate to reliability, nuisance reduction (noise, emissions, ...), and driveability.

The quality characteristics and the criteria required for diesel fuel justify the thorough analysis that is provided in the following sections.

4.4 Density, volatility, and viscosity

The three characteristics of density, volatility, and viscosity, which are often interdependent, have a complex influence on fuel injection and on the preparation of an auto-ignitable mixture.

Establishing specifications consists of finding, for each physical property, the acceptable parameters that can be readily changed to optimize diesel-engine combustion.

4.4.1 Density

The density of diesel fuel is an important property, because the injection system—the pump and the injectors—are set to deliver a predetermined volume of fuel, whereas, in the combustion chamber the determining parameter is the air-fuel **mass** ratio.

4.4.1.1 Average values and variability

Diesel fuel distributed for road-use in Europe must have a density between 0.820 and 0.860 kg/dm^3 (standard EN 590). There are no official density specifications that apply in the United States. The average density of fuel sold in France is about 0.835 kg/dm³, which is the same as the rest of Europe and Japan (for conventional products only), whereas the density in the United States is higher (on the order of 0.843 kg/dm³) (Paramins 1994). Fuels sold in Scandinavian countries, which have characteristics that are closer to kerosene, have lower densities than those sold in other nations: densities are on the order of 0.823 kg/dm³ as a consequence of the local climate (see 4.5.2.1), which is very cold in winter. Otherwise, the variability of fuel density in areas of the world is between 0.811 and 0.857 kg/dm³ (Paramins 1994).

The establishment of a minimum value for fuel density is justified by the need to obtain **sufficient maximal power** from an engine that uses a fuel-injection pump with volume-based flow control. As well, assigning a maximum value to the density helps to avoid the formation of smoke at full load, which would result from the increase in average equivalence ratio in the combustion chamber.

As a final note, the main parameters affecting fuel density are the characteristics of the original crude oil, the broadness of the cut chosen for the diesel fraction, and the concentration of components from catalytic cracking.

4.4.1.2 Effects on injection and combustion

Changes in diesel fuel density can have effects on combustion. Some of these effects, like changes in the heating value and the equivalence ratio are known; others, like various changes in emissions are more difficult to determine. With regards to this subject, work was done in Europe within the EPEFE program to study the effect of diesel-fuel density on emissions and to determine the optimal diesel fuel based on environmental constraints (see 5.10.2).

With regards to the effect of fuel density on heating value and equivalence ratio, two extreme fuels with densities that differ by 7.25% (0.814 and 0.873 kg/dm³) are compared in Table 4.1. The increase in density results in a reduction in the gravimetric NHV (–3.6%), and in an increase in the volumetric NHV (+3.4%). At full load, the amount of energy introduced per cycle increases by 3.4% with the denser fuel, which unarguably provides a similar power gain at the same efficiency. In reality however, this cannot be used to

Table 4.1 *Characteristics and behaviors of two diesel fuels with very different densities.*

Characteristics	Diesel 1	Diesel 2	Relative variation (2/1 in %)
Density at 15°C (kg/dm³)	0.814	0.873	+7.25
Gravimetric NHV (kJ/kg)	43 486	41 922	–3.60
Volumetric NHV at 15°C (kJ/dm³)	35 398	36 598	+3.40
Stoichiometric air-fuel ratio	14.75	14.31	–3.00
Relative change in equivalence ratio at constant flow volume	–	–	+4.05

advantage because the increase in average equivalence ratio can result in the production of additional smoke.

4.4.2 Volatility

Fuel volatility is expressed by two characteristics: the **distillation curve** and the flash point. The **flash point** has no direct effect on combustion or engine performance, but it is an important safety consideration for storage and distribution operations. Flash point is described in paragraph 4.7, which covers diesel fuel properties that are related to operational situations.

The **distillation curve** of the fuel directly affects the evolution of combustion; the curve is established according to the test method described in Chapter 3 (ASTM D 86).

The European standard (EN 590) specifies three criteria related to the values of the distilled fractions at given temperatures. The initial and final distillation points are not specified because their determination is usually not very precise. Thus, the distilled fractions must be

- less than 65% at 250°C
- more than 85% at 350°C
- more than 95% at 370°C

American specifications specify two volatility classes for road-use diesel, depending on the intended application (ASTM D 975-94). The first class applies to the fuels used in passenger cars; these fuels must be 90% distilled at a temperature of 288°C. The second class applies to fuels used in commercial vehicles and specifies a 90% distilled point between 282 and 338°C.

The **heavy-fraction** portion cannot be too large, otherwise the atomization of the fuel into fine droplets is degraded on injection, which affects the optimal evolution of combustion. Besides, depending on the type of engine and the operating conditions, studies have shown that reductions in efficiency of 1 to 5% occur when the distillation interval increases from 185-370 to 185-440°C.

The heavy-fraction content in diesel fuel can also affect emissions (see Chapter 5). As well, although the final distillation point is not specified, it cannot in practice exceed a certain limit without causing a deterioration of other constraining characteristics such as **cold-temperature performance**. This why the final distillation point of diesel fuels distributed in Scandinavian countries is lower than the fuels available in warmer climates (see Table 4.2).

Conversely, lightening the fuel by adding lighter fractions like naphthas or kerosenes does not have any direct effect on engine performance. This technique is sometimes used to improve cold-temperature performance (see 4.5.4).

Engines demonstrate a certain degree of **tolerance** with respect to the distillation curve of diesel fuel, provided that the curve remains within the medium range of petroleum fractions. This finding suggests that an exaggerated degree of importance should not be attached to the apparent differences between Europe and the United States, which are provided as examples in Table 4.2.

Table 4.2 *Characteristics of European and American diesel fuels.*

Country	N*	Temperatures (°C) at which F% of the fuel has been distilled								
		F = 20%			F = 50%			F = 90%		
		min	avg	max	min	avg	max	min	avg	max
France	14	194	215	239	242	261	288	322	333	354
Scandinavia	11	191	210	226	220	239	264	258	292	323
European Union	89	191	220	256	220	261	277	258	327	358
US East Coast	20	209	221	229	242	253	264	305	312	322
US Midwest	45	209	233	256	236	266	283	297	315	327
US West Coast	6	203	212	233	229	246	278	305	315	335

* N: number of samples analyzed.

4.4.3 Viscosity

Viscosity must also meet precise specifications because it has a direct effect on engine operation. In effect, a fuel that is too viscous increases pumping losses in the injection pump and the injectors, which reduces injection pressure resulting in less atomization and ultimately affecting the combustion process.

The inverse—a reduction in fuel viscosity—results in an increase in leakage at the injection pump, which reduces the actual volume delivered at the injector; this also results in delayed needle lift.

In addition, insufficient viscosity can cause the injection pump to **seize**. This must be considered when diesel fuel is diluted with gasoline or kerosene during cold seasons: the addition of lighter fractions must be done in such a way that the viscosity meets the specifications. Also, in some countries with very-cold climates, an additive must be included with low-viscosity diesels to improve lubricity.

For a long time, official specifications indicated a single maximum kinematic viscosity (9.5 mm^2/s) at a temperature of 20°C. European standard EN 590 and the American specification ASTM D 975-94 now specify a range of possible viscosities at 40°C instead of 20°C, which is more representative of the functional temperature of a injection pump. In Europe, the diesel-fuel viscosity at 40°C must be between 2.0 and 4.5 mm^2/s; in the United States, the acceptable range varies from 1.3 to 2.4 mm^2/s for passenger-car diesels and from 1.9 to 4.1 mm^2/s for the other diesel vehicles.

4.5 Cold-temperature characteristics

The cold-temperature characteristics of diesel fuel have a greater effect on its **implementation** than on its behavior during combustion. However, this section describes their importance in providing satisfactory vehicle operation.

When diesel fuel moves through the fuel system of an engine it passes through a **fine-mesh filter** (micrometer porosity) before entering the injection pump. This pump is a very-precise mechanical device whose reliable operation can be affected by impurities and particles in suspension in the liquid. Unlike other petroleum fuels, such as gasolines and kerosene, diesel fuel can lose its **clarity** and **fluidity** at low temperatures (0°C). This is caused by the appearance of **crystals** in the fuel, which can cause various problems such as filter plugging, loss of fuel pump prime, ... These considerations justify the need to adopt very-strict specifications for cold-temperature behavior, even if some recently-introduced technological devices (heated fuel filters on some newer vehicles) have reduced the service risks.

4.5.1 Nature of the phenomena involved

The composition of diesel fuel depends on the original crude oil and the formulation methodology. In any case, it usually contains a broad range of **paraffinic hydrocarbons**—from C_{10} to C_{35}—which are mostly linear and do not readily stay in solution at temperatures below 0°C.

When diesel fuel is chilled, wax crystals begin to appear and affect the clarity of the liquid at a temperature threshold referred to as the **cloud point**. At lower temperatures, the crystals increase in size and form a structure that locks in the remaining liquid and prevents it from flowing; this point is referred to as the **pour point**.

Between these two widespread points, which can be from 10 to 25°C apart, depending on the situation, the fluidity of the fuel may not be sufficient to ensure that it will pass through the fuel **filters**. The partial or complete blockage of filters by paraffin wax crystals can slow or halt the fuel supply.

4.5.2 Methods of classifying diesel fuel

There are several standard procedures for measuring the cold-temperature behavior of diesel fuel. These procedures define the cloud point, the pour point, and the filter-plugging point.

4.5.2.1 Cloud point

The **cloud point (CP)** in European standard EN 590 is only specified for fuels distributed in countries with arctic climates (see Table 4.3). In the United States, the specifications for the cloud point vary depending on the region and change monthly during the winter (see Table 4.4 and Fig. 4.28). Worldwide, the cloud point usually varies from 0 to −15°C in temperate climates; it climbs to 14°C in some hot climates (Brazil) and sometimes drops to −40°C in Scandinavian countries (Paramins 1994).

The test methods that are used to determine the cloud point (ISO 3015, ASTM D 2500) rely on **visual** detection to establish the temperature at which

Table 4.3 *European specifications for diesel fuel (standard EN 590)*

Characteristics	Limiting values by class						Tests methods
	Temperate climate	Arctic climate					
		0	1	2	3	4	
Sulfur content (% mass) max	0.05						ISO 8754
CFPP (°C) max		−20	−26	−32	−38	−44	EN 116
Class A	+5						
Class B	0						
Class C	−5						
Class D	−10						
Class E	−15						
Class F	−20						
Cloud point (°C) max		−10	−16	−22	−28	−34	ISO 3015
Density at 15°C (kg/dm^3) min	0.820	0.800	0.800	0.800	0.800	0.800	ISO 3675
max	0.860	0.845	0.845	0.840	0.840	0.840	
Viscosity at 40°C (mm^2/s) min	2.0	1.5	1.5	1.5	1.4	1.2	ISO 3104
max	4.5	4.0	4.0	4.0	4.0	4.0	
Measured cetane number min	49	47	47	46	45	45	ISO 5165
Calculated cetane index min	46	46	46	46	43	43	ISO 4264
Distillation temperature (°C)							ISO 3405
E 250 max	65						
E 350 min	85						
E 370 min	95						
E 180 max		10	10	10	10	10	
E 340 min		95	95	95	95	95	
Flash point min	55						NF T 07-019

wax crystals, which are normally dissolved in solution with the other constituents, begin to separate and affect the clarity of the product.

A 40 cm^3 sample, which has been heated to a temperature 14°C above the suspected cloud point, is poured into a hermetically-sealed test tube equipped with a thermometer. The assembly is placed in a chilled bath that provides a temperature gradient of 0.5 to 1°C/min. For each change of 1°C, the tube is removed and examined until clouding or deposits appear in the liquid. Most laboratories are equipped with semi-automatic equipment that uses two fiber-optical devices to replace visual detection.

The first fiber is used to bounce a light signal off the reflective bottom of the test tube. The second fiber is used to detect the reflected signal. This reflected beam is diffracted the moment wax crystals appear. The repeatability of this test is about 2°C and the reproducibility is 4°C.

Table 4.4 *American specifications for diesel fuel (ASTM standard D 975-94).*

Characteristics	Limit values by class				Tests methods
	1-D low sulfur content	2-D low sulfur content	1-D	2-D	
Sulfur content (% mass) max	0.05	0.05	0.5	0.5	D 2622
Cloud point (°C) max	*	*	*	*	D 2500
Viscosity at 40°C (mm²/s) min	1.3	1.9	1.3	1.9	D 445
max	2.4	4.1	2.4	4.1	
Measured cetane number min	40	40	40	40	D 976
Calculated cetane index** min	40	40			D 976
Aromaticity** (% vol) max	35	35			D 1319
Distillation temperature (°C) min	288	282	288	282	D 976
(90% (vol) distilled) max		338		338	
Flash point (°C) min	38	52	38	52	D 93

* The cloud-point values are specific to each region of the United States and they vary monthly during the winter (see fig. 4.28).

** One characteristic or the other must meet the specifications.

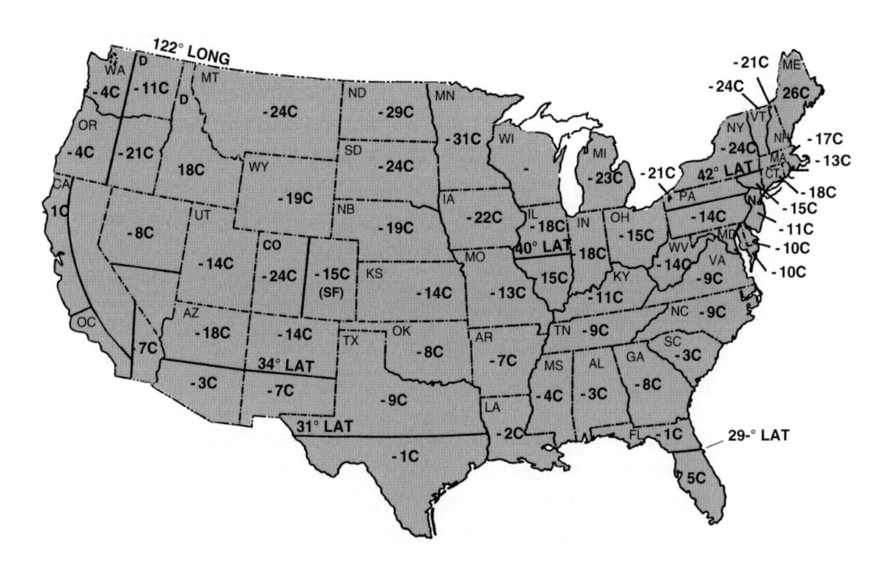

Figure 4.28 *Map showing the lowest temperatures obtainable with a 90% probability during the month of February in the United States (ASTM D 975). (C indicates °C; LONG and LAT indicate geographic longitude and latitude respectively)*

From a thermodynamic view, the cloud point is the visible manifestation of incipient **nucleation** and **crystallization**. In effect, diesel fuel can be considered as a solution in which paraffinic hydrocarbons are kept in a liquid state. The solubility varies with temperature according to the following formula,

$$RT \ln x = -\Delta H + T\Delta S$$

where

T = temperature

ΔH and ΔS = molar enthalpy and entropy of dissolution

R = universal gas constant

x = maximal molar fraction of the paraffin in solution

When the temperature drops and reaches the point of incipient crystallization, solid paraffins appear in the mixture. This phenomenon is accompanied by a small heat release that can be used effectively in a non-standardized technique called **differential calorimetric analysis** or DCA. This technique provides a precise means of detecting and evaluating the paraffin crystallization process in diesel fuels. However, the incipient-crystallization temperature is often lower than the cloud-point temperature.

4.5.2.2 Pour point

The **pour point** (PP) is the lowest temperature at which diesel fuel will flow. It varies according to country, from +4°C in India to −39°C in Sweden. However, it is most often found to be between −18 and −30°C (Paramins 1994).

The pour point (ISO standard 3016, ASTM test method D 97) is determined using very-rudimentary equipment: the sample is poured into a closed test tube equipped with a thermometer; it is then heated to 45°C and immersed in a chilled bath. When the sample reaches a temperature 9°C above the assumed pour point, the test begins. The test tube is inclined and the movement of the sample is observed. As long as the tube cannot be held **horizontal** for a period of 5 seconds without movement of the sample, the test is repeated at a temperature 3°C lower. The nominal pour point is obtained by adding 3°C to the last reading taken. The repeatability of this test is 3°C while the reproducibility is 6°C. As with the cloud-point test, there is a semi-automatic apparatus available for this test.

4.5.2.3 Cold filter plugging point

The cloud and pour points represent the two extremes of paraffin crystallization, but neither is directly related to problems that arise in actual service. To overcome this deficiency, another standardized test (EN 116) that is designed to assess the risk of filter plugging is now applied in Europe; it defines the **Cold Filter Plugging Point** (CFPP).

The filter-plugging-point temperature limit is part of European specification EN 590, which applies to diesel fuels distributed in temperate climates as well as arctic zones (see Table 4.3).

The CFPP range is relatively wide: in Europe during the winter, it is between –10 and –30°C. The lowest CFPPs in Europe are found in Germany and the Scandinavian countries (Octel 1994).

The test method for measuring the CFPP consists of determining the maximum temperature at which a specified volume (20 ml) of diesel fuel stops flowing through a specified filtration apparatus within a **given time limit**. The chilled sample is drawn through a filter from bottom to top under a vacuum of 20 mbar. The filter consists of a 15-mm diameter metallic disc made of bronze or stainless steel mesh with nominal 45-µm openings. The liquid is recovered in a 20-ml pipette.

The operation is repeated at decreasing 1°C intervals until the fuel does not pass through the filter or the pipette is not completely filled within 60 seconds. The CFPP is the temperature reading taken when the last filter pass began. Occasionally, after having been drawn through in less than 60 seconds, the fuel does not return normally to its original level and a portion of it remains in the pipette. This phenomenon, called the "critical filterability point", obviously indicates the test has ended. The CFPP is therefore the last temperature recorded.

The reproducibility and repeatability of this test method depends on the level of the CFPP: if it is close to 0°C, the repeatability approaches 1°C and reproducibility is 2.8°C, whereas a CFPP of about –35°C yields repeatability and reproducibility results of 2.1 and 6°C respectively.

In the 1960s, the CFPP results from test method EN 116 were very representative of the real behavior of diesel fuel in vehicles and were narrowly correlated with so called "**operability**" temperatures (thresholds beyond which the problem occurs). Since then, diesel-fuel qualities, like the diesel vehicles themselves, have undergone profound changes. As a result, the experimental conditions used for the CFPP differ considerably from vehicle conditions. The major distinctions are the **filter mesh diameter** (much more important on the laboratory filter), the **back pressure**, and the **chilling rate** (0.5°C/min!). Research is also underway to develop more representative procedures for the real behavior of diesel fuel, which will allow results to be well correlated with operability temperatures.

Two methods are currently proposed to meet this need: the Simulated Filter Plugging Point (SFPP) proposed by *Exxon Chemicals* and the Agelfi method favored by *Agip, Elf,* and *Fina.*

To reduce costs, the **SFPP method** uses a modified CFPP semi-automatic test apparatus; the modifications consist of replacing the metallic filter with a filter of smaller mesh diameter (25 µm), reducing the cooling rate, and changing the way the fuel is drawn through the filter (to represent the decantation tendency of the parafins in a vehicle fuel tank when parked outside during cold nights, the fuel is drawn from the top of the filter).

The **Agelfi method** follows the same principle, but it uses another type of apparatus as well as a new protocol. One precisely-weighed test sample, which has been chilled under controlled conditions, is pushed by a pressure of

660 mbar through a paper filter with a porosity of 25 µm. Successive tests are repeated at decreasing temperatures. When the quantity of paraffin solids is sufficient to stop or slow the flow (less than 90% of the sample passes through the filter in 5 min), the test is stopped. The Agelfi temperature is the last recorded temperature plus 1°C. In the medium or short term, either or both of these test methods could officially replace the CFPP.

In the United States, the CFPP results using test method EN 116 have never been representative of the **real behavior** of diesel fuel in vehicle service. Therefore, another test method was developed: the Low Temperature Flow Test (**LTFT**). This test is similar to the CFPP test, which it has in fact replaced. However, there is no ASTM test method describing the LTFT procedure. Following conventional protocol, a 200 cm^3 sample of diesel fuel is chilled progressively and at each 1°C interval it is drawn through a 17–µm filter by a vacuum of 200 mbar. The LTFT point is the lowest temperature at which 180 cm^3 of fuel passes through the filter in less than 60s. The repeatability and reproducibility are 1.4 and 5°C respectively. The LTFT point is usually reached a few degrees before the CFPP. Therefore, in the United States the average LTFT point is about −14°C, whereas the CFPP is lower at −18°C.

In Europe, diesel fuels are classified by their cold-temperature characteristics as listed in Table 4.3. The products are divided into **ten classes** (six for "temperate" climates and four for "arctic" zones). Each country uses one or more classes according to climatic conditions. Thus, France chooses classes B, E, and F respectively for summer, winter, and "severe cold". The first period extends from May 1 to October 31, the second from November 1 to April 30, and the third, as it stands, is not assigned in a formal way. The oil companies can thus take advantage of the opportunity provided to promote their brands.

In the United States, a specification also exists for cold-temperature properties (ASTM D 975-94) (see Table 4.4). Each fuel classification is defined by a cloud point whose value is specified for each region of the United States during the six winter months. An example of the map provided in the specification is reproduced in Fig. 4.28.

Each indicated value corresponds to the minimal regional temperature expected with a 90% probability for the month being considered.

The maximum cloud point is obtained by adding 6°C to the numbers on the map.

4.5.3 Cold-temperature performance of diesel vehicles

Now that the various cold-temperature characteristics of diesel fuel have been reviewed, the following section describes how these characteristics are related to problems that occur with vehicles in service.

4.5.3.1 Direct evaluation of the operability temperature limit

The operability temperature is defined for a fuel-vehicle combination; the **operability temperature limit** (OTL) is the lowest ambient operating temperature for the fuel-vehicle combination that does not present cold-temperature related problems. This characteristic, which is obtained from actual testing on a chassis dynamometer in a cold chamber or road testing at cold temperatures, depends not only on the chosen combination but also on the test method used to obtain the result. Various methods are in use today, because there is no standardized method. One of these methods, which is promoted by *Elf*, proceeds as follows:

- place the vehicle in the ambient temperature 24 hours before the test (fuel tanks full)
- start the vehicle
- operate the engine at idle speed for 30 s
- initiate the operating cycle
- accelerate to 60 km/h in fourth gear within 35 s and cruise for 3 min
- accelerate again to 100 km/h in fifth gear, and cruise for 30 min

The lowest temperature at which the vehicle can negotiate the test cycle without stalling is the OTL.

The OTL is a realistic way to evaluate the actual behavior of a given fuel-vehicle combination. Unfortunately, its implementation is lengthy and expensive, and the result cannot be applied to other vehicles using the same fuel. This explains why filterability temperature tests are still in use, even if they are less representative of real conditions.

4.5.3.2 Predicting the operability temperature limit by calculation

In the United States, a formula is used to predict the OTL by calculation, based on conventional cold-temperature characteristics. The result obtained by the calculations is called the Wax Precipitation Index (WPI); it is expressed in °C and it is obtained from the following equations:

$$\mathrm{WPI} = \mathrm{CP} - 1.3(\mathrm{CP} - \mathrm{PP} - 1.1)^{0.5} \qquad \text{if } \mathrm{CP} - \mathrm{PP} > 1.1^\circ\mathrm{C}$$

and: $$\mathrm{WPI} = \mathrm{CP} \qquad \text{if } \mathrm{CP} - \mathrm{PP} \leqslant 1.1^\circ\mathrm{C}$$

where

CP = cloud point (°C)

PP = pour point (°C)

The correlation between the WPI and operability of American vehicles is very good; in contrast, the application of the predictive formula to European vehicles has proven to be less satisfactory.

4.5.4 Improving the cold-temperature characteristics of diesel fuel

Like all refining operations, obtaining diesel fuel with good cold-temperature characteristics requires a techno-economic compromise. There are essentially two means to this end.

4.5.4.1 Selection and composition of the diesel fraction

The value obtained for the **final distillation point** must be as low as possible to minimize the heavy-fraction content in the fuel. In effect, these fractions contain the *n*-paraffins responsible for the formation of wax crystals at low temperatures.

A low **initial distillation point** enhances the retention in solution of heavy paraffins by the lighter fractions, which improves cold-temperature characteristics. In countries where winters are very severe, low initial points sometimes result in a pronounced overlap of diesel and kerosene cuts, which limits the refiner's production of kerosene. This type of "**winter diesel**" must also receive additives to increase its viscosity and thereby reduce injection-system wear caused by the presence of the light cuts.

The **chemical structure** of the crude oil also has an influence on the cold-temperature characteristics of the diesel fuel: aromatic and naphthenic crudes provide good cold-temperature performance, which is not the case with paraffinic crudes. All the same, cracked products with high aromatic content like LCO can improve the cold-temperature characteristics of diesel fuel, but their low cetane index allows only moderate use.

These technical capabilities, which are available to the refiner, are examined again in the section on diesel-fuel blending (see 4.9).

4.5.4.2 Incorporating additives

Incorporating additives into diesel fuel is a low-energy-cost, interesting, and effective way to supplement refining operations.

A. *Lowering the CFPP and the pour point*

Lowering the CFPP and the pour point can be accomplished using **flow improvers**. These improvers function by assisting the **dispersion of paraffin crystals** and preventing them from forming large structures that block filter pores. Conventional flow improvers act on the CFPP and the pour point, but they have little effect on the cloud point.

Flow improvers are usually **copolymers**, based on ethylene and vinyl acetate monomers, for example:

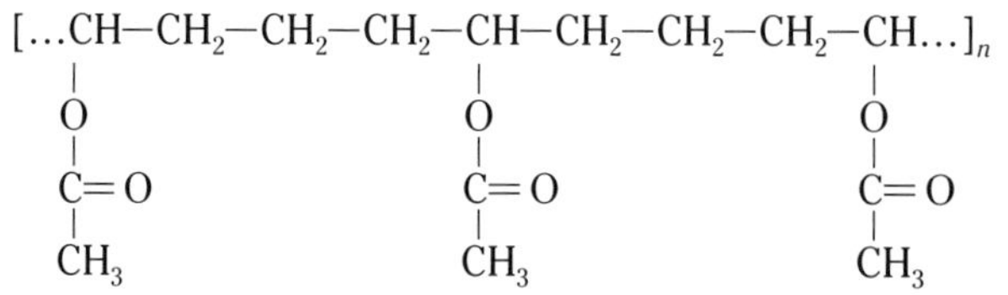

$$[\ldots \underset{\substack{|\\ O\\ |\\ C=O\\ |\\ CH_3}}{CH} - CH_2 - CH_2 - CH_2 - \underset{\substack{|\\ O\\ |\\ C=O\\ |\\ CH_3}}{CH} - CH_2 - CH_2 - CH_2 - \underset{\substack{|\\ O\\ |\\ C=O\\ |\\ CH_3}}{CH} \ldots]_n$$

In practice, the effectiveness of these additives depends very much on the origin and composition of the diesel fuels. Therefore, additives are less effective on the characteristics of North Sea crudes than they are on those of other crudes.

Fig. 4.29 shows a classic example of the effectiveness of flow improvers. The reductions of CFPP and pour-point temperatures can readily attain 6 to 12°C for additive concentrations between 200 and 600 ppm (by mass).

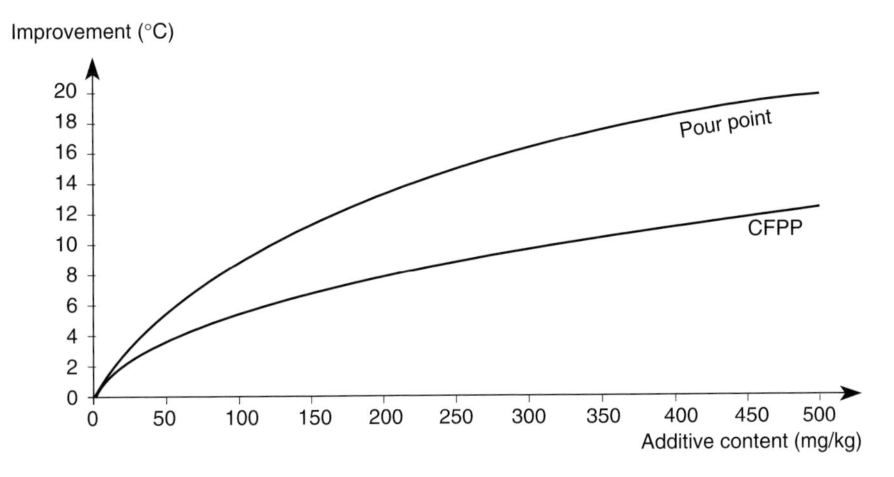

Figure 4.29 *Example of the effect of a flow improver on the CFPP and pour point temperatures. (Source: Elf Aquitaine)*

Additive treatment of diesel fuels is relatively low in cost: on the order of a few cents per liter of fuel. In practice, a diesel fuel containing a flow improver can be identified by the large difference (more than 10°C) between the cloud point and the CFPP.

B. Lowering the cloud point

There are cloud-point depressants (CPD) that act specifically on the cloud-point temperature. These additives are obtained from the radical polymerization of α-olefins and acrylic, vinyl, or unsaturated maleic compounds (Damin 1986).

These **macromolecules** block the formation of paraffin crystals in the solution. To form associations with the paraffins, they contain chemical groups similar to paraffins and they also contain solubility chemicals to retain the associations in solution. The improvement in the cloud-point temperature obtained with these additives is relatively small: on the order of 2 to 4°C for concentrations of 250 to 1000 ppm.

CPDs can also interfere with flow improvers and when they are used together they can reduce their respective effectiveness (Owen 1990, 1995).

C. Preventing settling

Parking a diesel vehicle at cold temperatures can cause wax crystals to form in the fuel. If the vehicle is parked for a long time the wax crystals **settle** to the bottom of the fuel tank and obstruct the intake to the engine fuel system. The smaller the wax crystals, the quicker this process occurs, which is the result when a high concentration of flow improvers is used.

The use of wax antisettling additives slows settling by dispersing the crystals and retaining them in solution. These additives consist of polymers, for example, that have **dispersing abilities** derived from groups of amines or amides; their usage concentrations are in the range of 100 to 500 ppm. A test to measure sedimentation tendencies is currently undergoing standardization in France.

4.5.5 Technologies to improve the cold-temperature operability of diesel engines

To improve the cold-temperature operability of diesel vehicles, many manufacturers are equipping their vehicles with fuel **heaters**. These devices, which are designed to retard or prevent the formation of paraffin wax crystals in diesel fuel, are currently used on commercial vehicles. Passenger-car applications are currently restricted to luxury cars sold in cold climates.

Coolant-based fuel heaters use the heat from the engine cooling system. They use very-little energy (which is rejected heat in any case), but they are technologically complex and ineffective at startup. **Electric fuel heaters** consist of a resistance heater included in the fuel filter. They are supplied by the battery, from which they draw a considerable amount of energy (150 W for a passenger car). Some heaters combine both processes (Adam 1990).

The use of additives to improve the CFPP and the pour point at the point of sale (service stations), as well as the dilution of diesel fuel with lighter kerosene-type products when refueling, is not recommended at this time.

In fact, given the problems that occur during very-cold winters, petroleum companies are now recommending "**winter**" diesel fuels adapted for the coldest climates.

4.6 Auto-ignition quality

Diesel fuel must have a chemical structure that facilitates auto-ignition. This characteristic is expressed by the cetane[1] number (or rating), which is obtained by comparing the behavior of a fuel in a standard engine with the behavior of **two reference hydrocarbons**.

This section describes the standardized method for measuring the **cetane rating** and the associated calculation formulas that allow the rating to be estimated based on the simple physical and chemical characteristics of a diesel fuel.

Following the presentation of some cetane ratings for pure hydrocarbons and commercial diesel fuels, the following section describes the importance of this characteristic as it relates to changes in refining and the demands of modern diesel engines. Finally, the use of additives to possibility improve cetane ratings is described.

1. A distinction is often made between "cetane number" (measured in an engine) and "cetane index" (a value calculated from physical and chemical properties). The term "cetane rating" (measured or calculated), which is analogous to the octane rating, is used herein.

4.6.1 Historical information

During the 1930s, G. D. Boerlage of the Delf Laboratory (The Netherlands) sought a procedure to determine the ignition quality of diesel fuels, based on the same techniques used to determine octane ratings. He selected two reference hydrocarbons with opposite behaviors:

- **ketene** (hexadec-1ene), which is characterized by a low auto-ignition delay due to its long straight carbon chain
- **α-methylnaphthalene**, which has two aromatic rings and is thus very resistant to oxidation

G. D. Boerlage proposed the determination of the **ketene rating** for a given fuel by finding the ketene content in the ketene-α-methylnaphthalene mixture that provided the same behavior.

The idea of a cetane rating was introduced later by the Americans who used the saturated hydrocarbon ***n*-cetane** ($C_{16}H_{34}$) instead of ketene.

The following relationship between the two indexes was proposed:

$$\text{Cetane rating} = 0.88 \cdot \text{ketene rating}$$

This method was used for several years. However, the main objective was to establish a definitive test method. One method proposed by Pope and Mardock consisted of finding the lowest compression ratio that would produce auto-ignition—using a conventional engine modified with a special piston, pump, and injector. This approach, which did not account for delay, proved to be too imprecise and it was abandoned in favor of the Boerlage and Broeze method, which uses a diesel CFR engine with a prechamber. The compression ratio was adjusted by moving a plunger inside the prechamber to obtain a constant ignition delay. Injection took place at 10°CA before TDC and ignition took place 1°CA after TDC.

A change was proposed by ASTM and it was accepted; this change consisted of extending the delay to 13°CA and having the ignition take place at TDC. The cetane rating was obtained by comparing compression ratios required to obtain a 13°CA delay with the fuel under test and the reference mixtures.

The important work of P. Dumanois (1885-1964) on the correlation of octane and cetane ratings is also worth mentioning. He proposed a calculation method for the cetane rating based on the octane rating of various diesel-fuel-gasoline mixtures. This method was called the dilution method and the diesel-fuel content of the gasoline varied from 0 to 20%.

4.6.2 Measuring the cetane rating using a CFR engine

The current CFR engine procedure was developed in 1941 and despite its imperfections, it continues to be the selection criteria that is in widespread use today for defining diesel-fuel auto-ignition quality.

4.6.2.1 Principle

The method described in the ISO 5164 and ASTM D 613 standards repeats the main proposals made by Boerlage and Broeze. The compression ratio of the CFR engine is varied to obtain the **standard ignition delay of 13°CA**. This value is then bracketed by the two other values obtained with the reference mixtures, whose theoretical cetane ratings differ by less than 5.5 numbers.

The primary reference fuels are theoretically defined as

- ***n*-cetane** (*n*-hexadecane) with a rating of 100

$$CH_3{-}(CH_2)_{14}{-}CH_3$$

- ***α*-methylnaphthalene** with an rating of 0

$$CH_3$$

Under these conditions, a fuel has a cetane rating of *X* if it behaves like a blend of *X*% (volume) *n*-cetane and (100-*X*)% *α*-methylnaphthalene.

In fact, *α*-methylnaphthalene is no longer used a reference, it having been replaced by **2,2,4,4,6,8,8-heptamethylnonane** (HMN), an isomer of *n*-cetane that has a cetane rating of 15 and the following semi-expanded formula:

$$\begin{array}{ccccccccccccc} & & CH_3 & & & & CH_3 & & & & & & CH_3 & & \\ & & | & & & & | & & & & & & | & & \\ CH_3 & - & C & - & CH_2 & - & C & - & CH_2 & - & CH & - & CH_2 & - & C & - & CH_3 \\ & & | & & & & | & & & & | & & & & | & & \\ & & CH_3 & & & & CH_3 & & & & CH_3 & & & & CH_3 & & \end{array}$$

Therefore, the reference blends are identified by the following formula for cetane rating (CR):

$$CR = X + 0.15\,(100 - X)$$

where *X* is the cetane level (% volume) of the cetane-HMN mixture.

Table 4.5 lists the purity criteria required for the two reference hydrocarbons. There are also secondary reference fuels designated by the letters T and U, which consist of hydrocarbon mixtures that are calibrated with respect to the primary references. These products, by way of their physical and chemi-

Table 4.5 *Specifications of primary reference fuels used to measure cetane ratings.*

Characteristics of *n*-Cetane	
Freezing point	≧ 16.2°C
Hydroxyl index	nil
Iodine index	nil
Color	clear
Sedimentation	nil
Distillation* 5%	286.6 ± 1°C
Maximum distillation range	6°C

Characteristics of heptamethylnonane	
Distillation 50%	246.9 ± 1°C
Maximum difference from 20 to 80%	4°C
Density at 20°C	0.78448 ± 0.00020
Refraction index n_D	1.43990 ± 0.00020
Bromine index	≦ 1.0

* Standardized distillation method for industrial aromatic hydrocarbons (ASTM D 850).

cal properties—distillation intervals of 150 to 220°C—are better matched to conventional diesel fuels and they are cheaper than the primary reference fuels.

4.6.2.2 Equipment

The diesel CFR engine is a single-cylinder engine with a **variable compression ratio** and geometric characteristics (bore and stroke) that are identical to the gasoline CFR engine. The head is equipped with a variable-volume high-swirl cylindrical prechamber. The diameter of the prechambe is 41 mm and the length varies from 9.5 to 69.8 mm. The movement of a plunger piston inside the prechamber allows the volume, and hence the compression ratio, to be changed at will.

The movement of the plunger is controlled by a micrometer screw mechanism.

The fuel system consists of a conventional injection pump and a pintle-type injector. The operating conditions are described in Table 4.6.

From a knowledge of the injected-fuel flow and the intake-air flow, the overall equivalence ratio is calculated and it is found to be about 0.60 during cetane rating measurement. This corresponds to the operation of a conventional engine at **about 2/3 of maximum load**. The other control parameters, notably speed, compression ratio, and aerodynamic conditions, are obviously very different from those found on a production engine.

The auto-ignition delay is measured by an electronic instrument called a "**delaymeter**". This device includes two electromagnetic sensors and two optical sensors that respectively measure

- the time of injector needle lift
- the start of combustion as detected on a pressure diagram
- two angular readings, one positioned at **12.5°CA** before TDC and the other at TDC

Table 4.6 *Cetane rating measurement: CFR engine operating conditions.*

Rotation speed (rpm)	900 ± 9
Injection timing (°CA before TDC)	13
Injection pressure (bar)	103.0 ± 3.4
Injection rate (cm^3/min)	13.0 ± 0.2
Needle lift (mm)	0.27 ± 0.025
Compression ratio	Variable from 8 to 36
Injector cooling water temperature (°C)	38 ± 3
Oil pressure (bar)	1.75 à 2.10
Oil temperature (°C)	57 ± 8
Coolant temperature (°C)	100 ± 2
Intake-air temperature (°C)	66.0 ± 0.5
Valve clearance (mm)	0.200 ± 0.025
Lubricating oil	SAE 30

4.6.2.3 Operating method

The engine is fueled with the test sample. The operator adjusts the flow rate of the injected fuel to 13.0 ± 0.2 cm³/min using a micrometer adjustment on the injection pump. With the same adjustment, the operator sets the injection advance to 13.0 ± 0.2°CA before TDC; this measurement is read directly from a digital delaymeter.

The operator then adjusts the position of the plunger in the engine prechamber to change the compression ratio, thereby changing the auto-ignition delay (see Fig. 4.30). When ignition occurs 13.0 ± 0.2°CA after injection (reading also taken from the delaymeter), the position of the plunger indicated by the micrometer is recorded.

The preceding operations are repeated for each of two reference blends whose cetane ratings differ by 5.5 numbers or less, and whose micrometer readings bracket the reading obtained for the fuel sample.

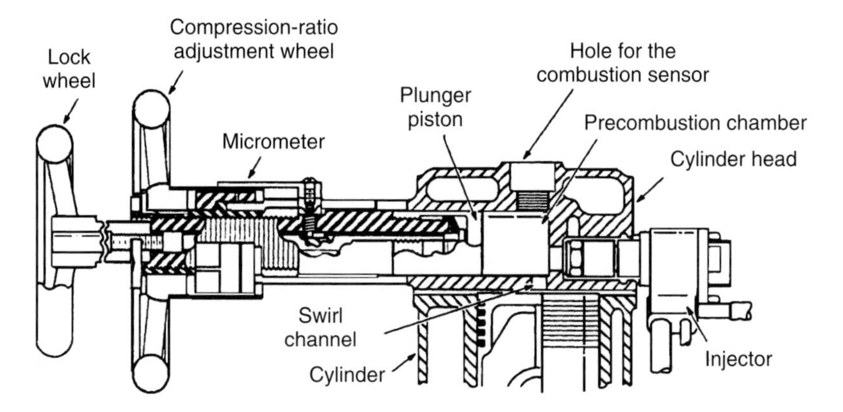

Figure 4.30 *CFR engine combustion chamber. (Source: ASTM D 613 1994)*

4.6.2.4 Reporting the results: degrees of precision

The cetane rating is obtained by linear interpolation.

The following table lists an example of the test results:

Product tested	Position of the micrometer screw
Fuel sample	1649
Reference fuel cetane 44.9	1575
Reference fuel cetane 49.9	1701

The cetane rating of the sample is

$$44.9 + 5\left(\frac{1649 - 1575}{1701 - 1575}\right) = 47.83$$

which rounds to 47.8.

Table 4.7 *Reliability of the cetane rating test method. (Source: ASTM)*

Cetane rating	Repeatability	Reproducibility
40	0.8	2.8
44	0.9	3.3
48	0.9	3.8
52	0.9	4.3
56	1.0	4.8

The test method states that all results must be rounded to nearest single decimal place.

Most conventional diesel fuels have **cetane ratings between 40 and 60**.

The repeatability and reproducibility of this test method for 95% probability are specified in ASTM D 613 and listed in Table 4.7.

These values indicate a **much-less satisfactory** degree of precision than those inherent in octane rating measurements.

This situation results from several causes. The lack of precision in the measuring apparatus is certainly involved, but it is not the only determining factor. The very principle of diesel combustion in a diffused environment, which depends on the physical parameters of evaporation, also has a significant effect. It is also interesting to note that the micrometer reading on the CFR engine fluctuates in a random way during a series of measurements taken for the same reference fuel. Table 4.8 lists some typical results to that effect, which were obtained on a carefully maintained and adjusted CFR engine.

Some efforts have been made to **improve the procedure** for determining cetane ratings or to find another test procedure that is more precise and discriminating. Choosing other test conditions would allow the sensitivity of the test method to be improved, but that would not appear, at this point, to change the relative classifications of the various fuels.

Table 4.8 *Measurement of the cetane rating of various reference fuels. Example of the variations in compression ratio for the same reference fuel.*

Cetane rating	Number of readings taken	Compression ratio values*				
		min	max	Average	Standard deviation σ	Relative standard deviation (%)
40	10	12.45	12.93	12.70	0.18	1.4
45	15	11.83	12.60	12.33	0.25	2.0
50	15	10.99	12.07	11.55	0.31	2.7
55	10	10.25	11.15	10.86	0.30	2.8

* The compression-ratio value t is calculated according to the following formula, $t = \frac{18}{u} + 1$, where u is the micrometer reading in inches.

The current version of the cetane rating is already a reasonably selective criteria. Hence, based on CFR engine tests, it can be shown that a 1°CA (out of 13°CA) ignition delay on the CFR engine corresponds with an approximate 2% change in cetane rating between two fuels. The same order of magnitude in sensitivity is obtained with octane ratings.

4.6.3 Correlation formulas

The measurement of cetane ratings using CFR engines is done less often than the measurement of octane ratings, due to the complexity and high cost of its implementation. Also, there are several lower-cost methods that can be used to estimate diesel-fuel cetane ratings, which are based on fuel characteristics and chemical structure.

4.6.3.1 Using a single temperature reference point (ASTM D 976)

The single-temperature reference point method is currently the best known low-cost method. The formula used with this method expresses the **calculated** cetane rating, in this case a cetane index, as function of density and the 50% distillation point on the ASTM D 86 distillation curve. The equation is

$$CCI = 454.74 - 1641.416\rho + 774.74\rho^2 - 0.554(T_{50}) + 97.803(\log T_{50})^2$$

where

CCI = calculated cetane index

ρ = density at 15°C in kg/dm^3 from the ASTM D 1298 test method

T_{50} = temperature in °C at the 50% distilled point according to ASTM D 86

This equation applies to both direct-distilled and catalytic-cracked diesel fuel, and to blends of these two different cuts. However, the equation can prove to be completely unsuitable when applied to light diesel fuels with a final distillation point of less than 260°C. Finally, **this equation does not predict the behavior of diesel fuels treated with cetane improving additives**.

This method's precision depends to a large measure on the precision of the parameters that characterize the chosen fuel. Examples of the correlation between measured and calculated cetane ratings are provided only as a guide. On average, if the measured cetane number is between 30 and 60, the calculated cetane index differs by more than 2 values in 75% of the cases. This is shown in Fig. 4.31, which compares fuel samples collected from various regions of the world in 1994.

4.6.3.2 Using three temperature reference points (ASTM D 4737 or ISO 4264)

The three-temperature method is progressively replacing the previous single-temperature method. Both are based on the same correlation; however, the equation used in standard ASTM D 4737 considers **three** points on the distil-

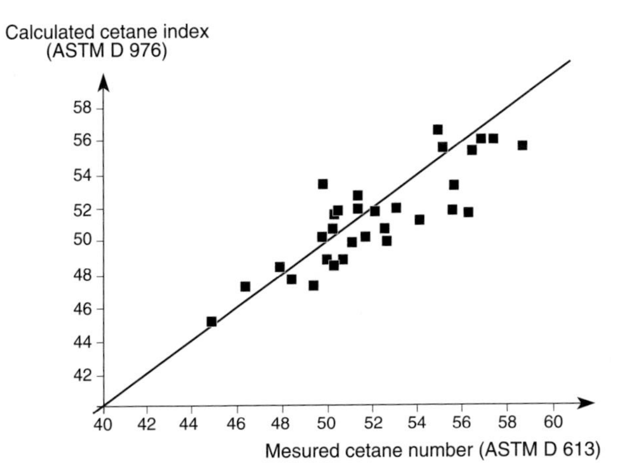

Figure 4.31 *Correlation between measured and calculated cetane ratings. (Source: Exxon Paramins 1994)*

lation curve instead of the single point used ASTM D 86. The equation is

$$CCI = 45.2 + 0.0892T_{10N} + (0.131 + 0.901B)T_{50N} + (0.0523 - 0.420B)T_{90N} + 0.00049({T^2}_{10N} - {T^2}_{90N}) + 107B + 60B^2$$

where

CCI = calculated cetane index

ρ = density at 15°C in kg/dm³ from the ASTM D 1298 test method

B = exp (–3.5 (ρ –0.85)) –1

T_{10} = temperature in °C at 10% distillation

T_{10N} = T_{10} –215

T_{50} = temperature in °C at 50% distillation

T_{50N} = T_{50} –260

T_{90} = temperature in °C at 90% distillation

T_{90N} = T_{90} –310

Like the previous method, this method applies to direct-distilled and catalytic-cracked diesel fuels, and to blends of these two different cuts. It can also be used to determine the calculated cetane indexes of products with 90% distillation points that are lower than 382°C.

According to the standard, if the measured cetane number of a product is between 32.5 and 56.5, the calculated cetane index will differ by less than 2 values in 65% of the cases. In fact, as shown in Fig. 4.32, the correlation seems to be better than predicted.

4.6.3.3 Diesel Index

The Diesel Index (DI) has been used for a long time to express the auto-ignition quality of diesel fuels. The DI is defined by the following equation:

$$DI = \frac{\text{Aniline point (°F)} \cdot \text{Density (°API)}}{100}$$

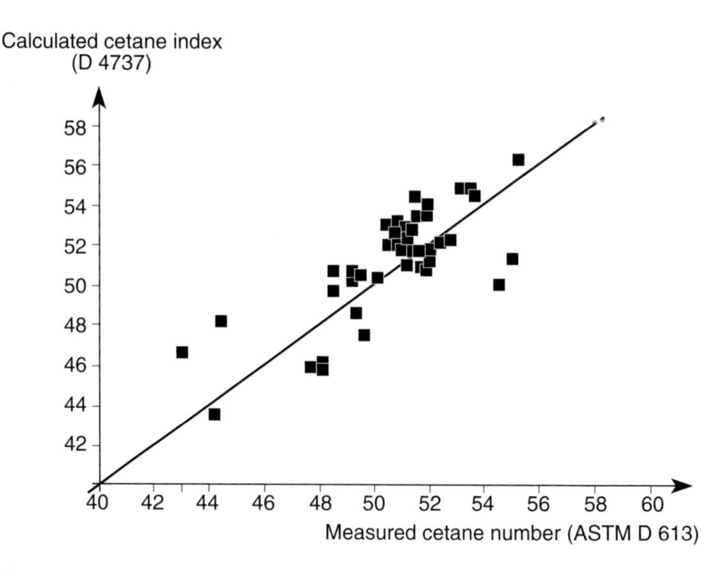

Figure 4.32 *Correlation between measured and calculated cetane ratings.*

The **aniline point** (AP) is the temperature at which a mixture of equal volumes of aniline and fuel becomes **homogeneous**. The solubility of petroleum products, in an aromatic compound such as aniline, depends on the chemical structure of the products. In effect, low-solubility paraffins have a high aniline point, which is not the case with aromatics.

Therefore, the DI is directly related to the paraffin content of a fuel by two intermediary parameters (aniline point and density) and so it represents the auto-ignition tendency. For example, a paraffinic diesel, which has a low density and a low solubility in aniline, will have a high DI.

Some empirical formulas have been proposed to relate cetane ratings to the DI or even directly to the aniline point. Two of these formulas are reproduced here for reference:

- cetane rating = 0.72DI + 10
- cetane rating = AP −15.5 (AP is in°C)

Other techniques for predicting cetane rating use **chemical analysis**. Gas chromatography, nuclear magnetic resonance, or mass spectrometry (see 4.10.2) can also be used as predictors.

4.6.4 Cetane ratings of pure hydrocarbons and diesel fuels

4.6.4.1 Pure hydrocarbons

Table 4.9 lists the measured cetane numbers of some pure hydrocarbons. The relationship of properties and structures relates to the study of octane ratings, which was covered previously. Therefore, the following statements logically apply:

- the cetane rating of paraffins increases with the **length of the major chain**, but it decreases with the number and complexity of the **branches**
- olefins have **lower** cetane ratings than the corresponding paraffins
- ring structures (naphthenes, cyclenes) tend to **reduce** cetane rating
- aromatics exhibit **mediocre** cetane behavior, which is mitigated somewhat as the lateral chains increase in length

4.6.4.2 Commercial diesel fuels

The cetane index of commercial diesel fuel varies from 40 to 60, depending on the country where it is distributed (see Fig. 4.33). One of the reasons for this disparity is the lack of harmonization of legislations between countries (see Tables 4.3 and 4.4). Thus, European standard EN 590 requires a minimum cetane rating of 49, except in Scandinavian countries where it is 46. In the United States, the minimum value required is 40 (ASTM D 975), except in California where it is 48. The cetane ratings of commercial products are often higher than the minimum specified requirements. Hence, the average values are on the order of 51 in Europe, 47 in the United States, and 54 in South-East Asia (Paramins 1994).

Another factor that indirectly affects the cetane ratings of commercial diesel fuel is, undoubtedly, the climate: in the tropics, paraffin content is not limited by cold-temperature problems and cetane ratings are particularly high. The problems that refiners have in formulating a diesel fuel with both a high cetane rating and good cold-temperature behavior are covered in subsequent sections (see 4.10).

Table 4.9 *Cetane numbers for some pure hydrocarbons.*

Hydrocarbon	Cetane number
Paraffins	
2-Methylpentane	33
3-Methylpentane	30
n-Heptane	56
2,2,4-Trimethylpentane	12
n-Decane	76
n-Dodecane	80
3-Ethyldecane	48
4,5-Diethyloctane	20
2,3,4,5,6-Pentamethylheptane	9
n-Tridecane	88
2,5-Dimethylundecane	58

Table 4.9 *Cetane numbers for some pure hydrocarbons. (continued)*

Hydrocarbon	Cetane number
4-Propyldecane	39
5-Butylnonane	53
n-Tetradecane	93
2,7-Dimethyl-4,5-diethyloctane	39
n-Pentadecane	95
n-Hexadecane	100
Heptamethylnonane	15
5-Butyldodecane	45
7,8-Dimethyltetradecane	40
n-Heptadecane	105
7-Butyltridecane	70
n-Octadecane	110
9-Methylheptadecane	66
8-Propylpentadecane	48
7,8-Diethyltetradecane	67
5,6-Dibutyldecane	30
n-Nonadecane	110
n-Eicosadecane	110
9,10-Dimethyloctadecane	59
7-Hexylpentadecane	83
2,9-Dimethyl-5,6-diisoamyldecane	48
9,10-Dipropyloctadecane	47
10,13-Dimethyldocosane	56
9-Heptylheptadecane	87
Olefins	
Diisobutylene	10
Tetradec-1-ene	79
Hexadec-1-ene	88
4-Butyldodec-4-ene	45
Tetraisobutylene	4
2,6,7-Trimethyltrideca-2,6-diene	24
7-Butyltridecene	36
9-Methylheptadec-9-ene	66
7,10-Dimethylhexadec-8-ene	43
8-Propylpentadec-8-ene	45
3,12-Diethyltetradeca-3,11-diene	26
7-Hexylpentadec-7-ene	47
10,13-Dimethyldoeicos-11-ene	56

Table 4.9 *Cetane numbers for some pure hydrocarbons. (continued)*

Hydrocarbon	Cetane number
Naphthenes	
Cyclohexane	13
Methylcyclohexane	20
Decaline	48
Dicyclohexyle	53
3-Cyclohexylhexane	36
n-Propyldecaline	35
n-Propyltetraline	8
n-Butyldecaline	31
sec-Butyldecaline	34
t-Butyldecaline	24
n-Butyltetraline	18
sec-Butyltetraline	7
t-Butyltetraline	17
2-Methyl-3-cyclohexylnonane	70
n-Octyldecaline	31
4-Methyl-4-decalylheptane	21
n-Octyltetraline	18
1-Methyl-3-docecylcyclohexane	70
2-Cyclohexyltetradecane	57
3-Methyl-3-decalylnonane	18
2-Methyl-2-Decalyldecane	37
2-Methyl-2-cyclohexylpentadecane	45
1,2,4-Trimethyl-5-hexadecylcyclohexane	42
5-Cyclohexyleicosane	66
di-*n*-Octyltetraline	26
Aromatics	
n-Pentylbenzene	8
1-Methylnaphtalene (or α-Methylnaphtalene)	0
n-Hexylbenzene	26
Diphenyle	21
n-Heptylbenzene	35
Diphenylmethane	11
n-Octylbenzene	31
2-Phenyloctane	33
1,2-Diphenylethane	1
1-*n*-Butylnaphtalene	6
2-*t*-Butylnaphtalene	3

Table 4.9 *Cetane numbers for some pure hydrocarbons. (end) (Source: Rose et al. 1977)*

Hydrocarbon	Cetane number
n-Nonylbenzene	50
n-Octylxylene	20
2-Phenylundecane	51
2-Phenylundec-2-ene	23
2-Methyl-2-(2-naphthyl)hexane	10
n-Dodecylbenzene	68
4-Phenyldodecane	42
2-*n*-Octylnaphthalene	18
4-Methyl-4-(2-naphthyl)heptane	9
7-Phenyltridecane	41
n-Tetradecylbenzene	72
2-Phenyltetradecane	49
3,6-Dimethyl-3-(2-naphthyl)octane	18
5-Methyl-5-(2-naphthyl)nonane	12
2-Methyl-2-(2-naphthyl)decane	18
3-Ethyl-3-(2-naphthyl)nonane	13
2-Methyl-4-isobutyl-4-phenylundecane	38
2-Methyl-2-phenylpentadecane	39
5-Butyl-5-phenyltetradecane	58
1,2,4-Trimethyl-5-hexadecylbenzene	42
5-Phenyleicosane	39

4.6.5 The effect of cetane ratings on engine performance

The cetane rating does not play the same vital role as the octane rating in the optimization of engine and fuel interaction; in particular, it has no **direct effect** on engine efficiency. Nevertheless, petroleum companies and motorists grant it considerable importance. This stems from the effect, direct or indirect, that it has on emissions, noise, and vehicle driveability.

4.6.5.1 Cetane rating and combustion parameters

As indicated previously, the cetane rating is a measure of a fuel's aptitude for auto-ignition.

For road diesel engines, with either direct or indirect injection, it is estimated that a cetane rating of about 50 is the **minimum acceptable value** needed to provide satisfactory results during the various phases of combustion (delay, progressive energy release, ...).

The auto-ignition delay increases almost linearly as the cetane rating gradually drops from 55 to 35. The amplitude of the recorded variations depends on the mode of combustion: it is usually more pronounced on a direct-injec-

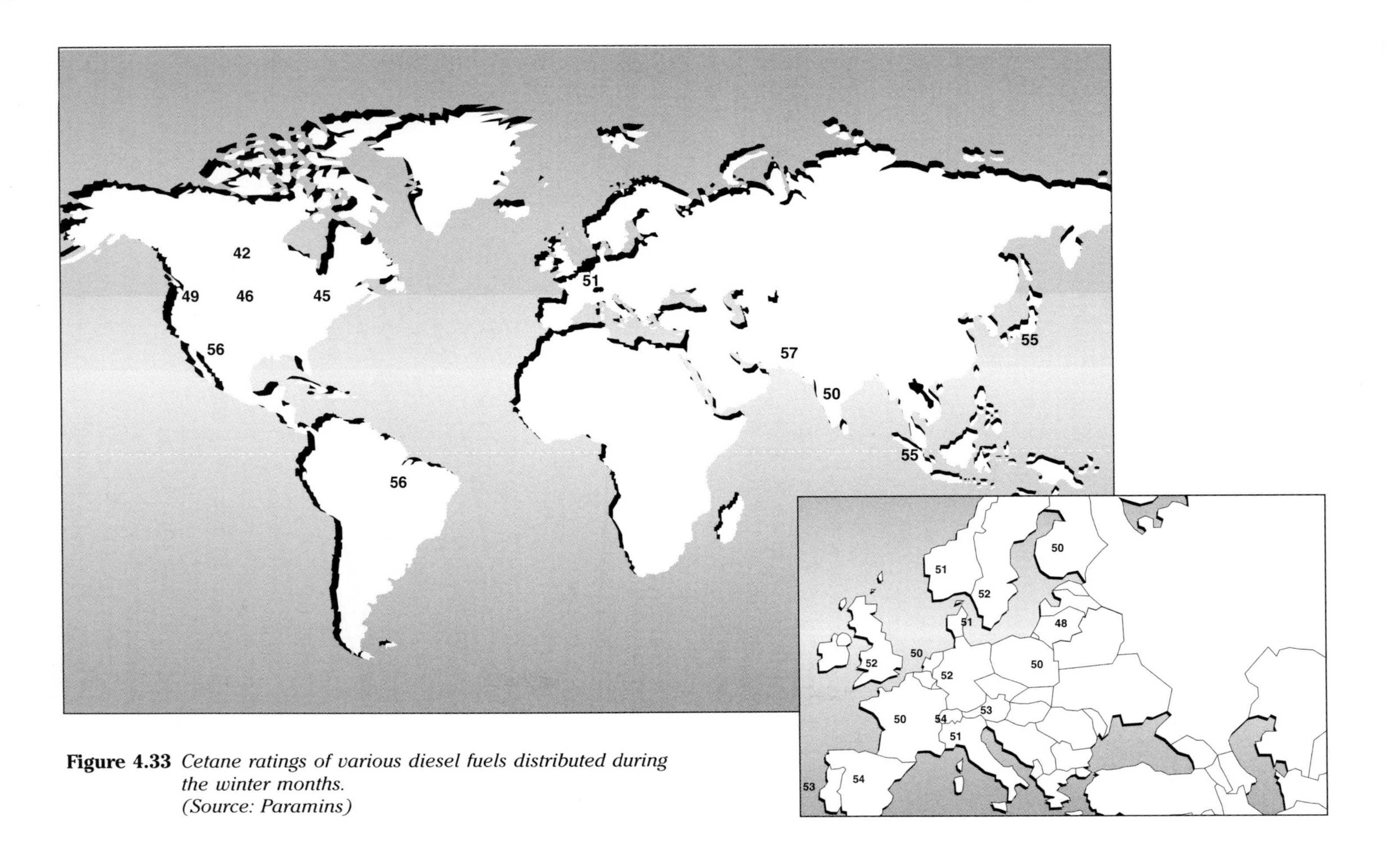

Figure 4.33 *Cetane ratings of various diesel fuels distributed during the winter months.*
(Source: Paramins)

tion engine than on a indirect-injection engine. This amplitude of variation also expands as the load increases (see Fig. 4.34).

Also, because the drop in cetane rating causes an increase in the auto-ignition delay, the amount of fuel present in the combustion chamber when ignition occurs is also greater. Combustion then takes place very-quickly causing a **high pressure gradient** (see Fig. 4.35). This results in increased noise (see 4.6.5.2), but the immediate effects on engine performance are minimal. However, this change in the combustion pressure diagram is a real problem because it can affect the mechanical integrity and durability of the engine. This is especially true with supercharged engines, which are already subjected to high maximum pressures.

4.6.5.2 Cetane rating and driveability

A change in engine noise is directly related to the pressure gradient in the combustion chamber; a low cetane rating therefore causes an increase in noise (see Fig. 4.36). The effect occurs under all operating conditions, but it is especially apparent at idle, both cold and hot, or at full load.

Direct-injection diesels demonstrate the clearest relationship between noise and cetane rating. When the cetane rating changes by 10, the noise level changes between 1.5 dB(A) and 6 dB(A), depending on the observation and measurement conditions (Russel et al. 1985; Waters et al. 1989).

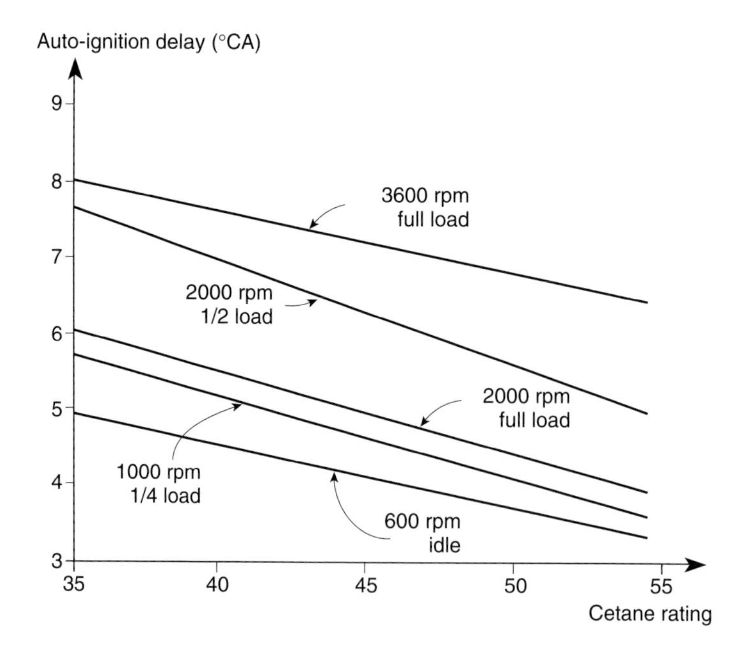

Figure 4.34 *Effect of cetane rating on auto-ignition delay. (Source: Olree et al. 1984)*

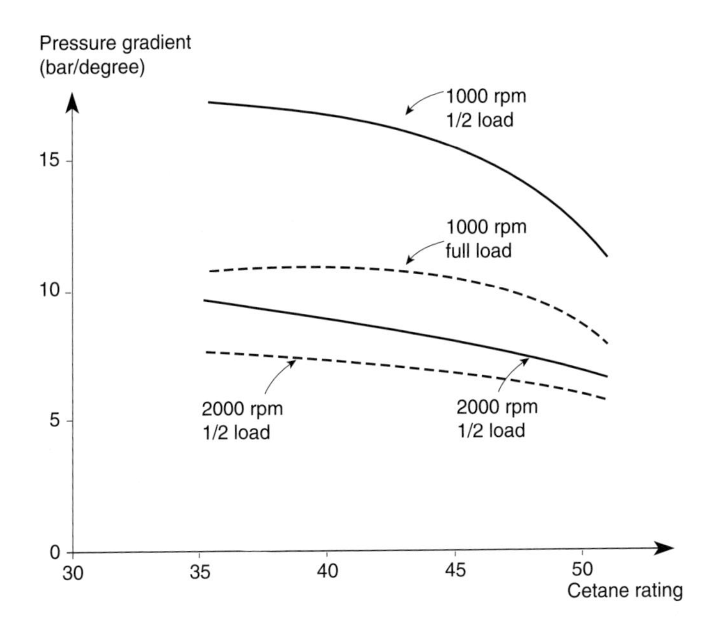

Figure 4.35 *Effect of cetane rating on the combustion pressure gradient (turbocharged and aftercooled heavy-truck engine).*
(Source: Bidault et al. 1987)

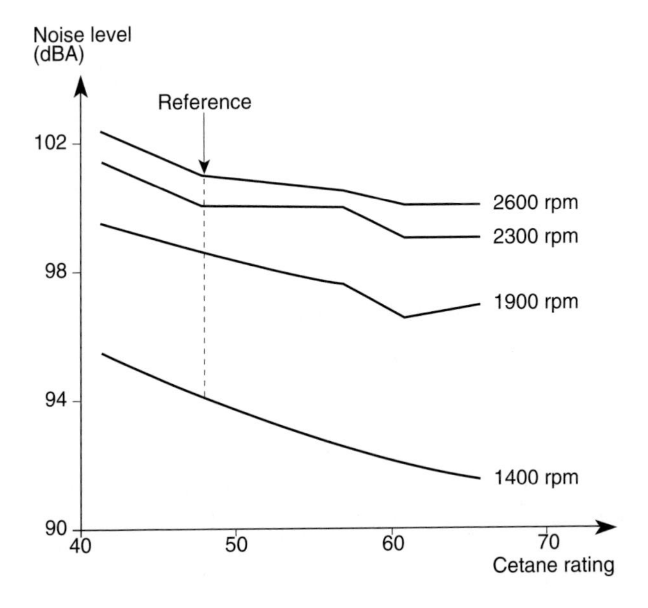

Figure 4.36 *Example of the relationship between the engine noise level at various speeds and the cetane rating of the fuel (direct-injection diesel engine running at full load; noise readings taken at 1 m from the front of the engine).*
(Source: Waters et al. 1989)

With **indirect-injection** engines, the effects are usually not as great (1 to 2 dB(A) for a cetane rating change of 10).

With light-duty diesel vehicles (passenger cars or utility vehicles), the operator also perceives—from a subjective but indisputable point of view—better cold-temperature driveability, and faster smoother warm-up periods, when the cetane rating increases.

Also, the important subject of the relationship between cetane rating and the emissions from diesel-powered vehicles is covered in Chapter 5.

4.6.6 Cetane rating and refining constraints

The relationship between cetane ratings and refining constraints is covered in greater detail in the following section devoted to the blending of diesel fuel. However, the main avenues available to maintain or improve cetane ratings are discussed here.

4.6.6.1 Current trends

Major changes in worldwide refining are combining to drive cetane ratings **down**. In Europe, increasing demands for diesel fuel are forcing refiners to blend conversion products (catalytic cracked) into atmospherically-distilled diesel fuel. In the United States, the demand for gasoline is changing the orientation of refinery schemas: they are arranged in such a way that heavy cuts are the only diesel cuts available. Therefore, in either case, the problem of **using Light Cycle Oil** (LCO) in the diesel pool arises. However, the amount of LCO that can be introduced is very limited (5 to 10% maximum) because it has a low cetane rating (about 20) and it contains high levels of aromatics, sulfur, and nitrogen. Other refinery processes (hydrocracking) provide diesel cuts with high cetane ratings, but their high cost makes them uneconomic.

When refinery schemas yield only low-cetane diesel fuel, cetane improvers are sometimes required to bring fuels into specification.

4.6.6.2 Improving cetane ratings with additives

The additives used to improve cetane ratings, which are referred to as **cetane improvers**, are very-active **oxidizers** that generate free radicals when they decompose. These improvers become involved in the early stages of the oxidation that precedes ignition, thus contributing to a reduction in the auto-ignition delay and increasing the cetane rating. Two families of cetane improvers are currently in use: nitrates and peroxides.

A. *Alkyl nitrates*

Alkyl nitrates, especially **2-ethylhexyl nitrate** are practically the only cetane improvers in use, due mainly to their superior combination of cost and performance. An example of their effectiveness is shown in Fig. 4.37; note that the increases in cetane rating depend on the composition and the characteristics of the recipient fuels; **their effect is directly proportional to the initial fuel's cetane level** (Unzelman 1984).

With respect to concentration, the improvements that are usually sought, when a base stock has a cetane rating between 45 and 48, are about 3 to 5 values. Under these conditions, concentrations of 300 to 1000 ppm (mass) are sufficient. Treatment is cost-effective at the rate of a few cents per liter.

There is now a new generation of nitrate-based cetane improvers appearing on the market; these are **alkyl or poly-alcohol dinitrates** (triethyleneglycol dinitrate) compounds marketed by *Imperial Chemical Industries* (ICI) under the name AVOCET. Their high cost limits their application to special uses (as additives for ethanol used in diesel-cycle engines).

However, their future appearance in the cetane improver market for conventional diesel fuels is possible.

B. Peroxides

The cetane improvement potential of peroxides has been known for some time, however, their commercialization has been slow to develop. The first reason is related to their insufficient oxidation stability when mixed with diesel fuel; the second reason is the generally mediocre cost/performance ratio they pro-

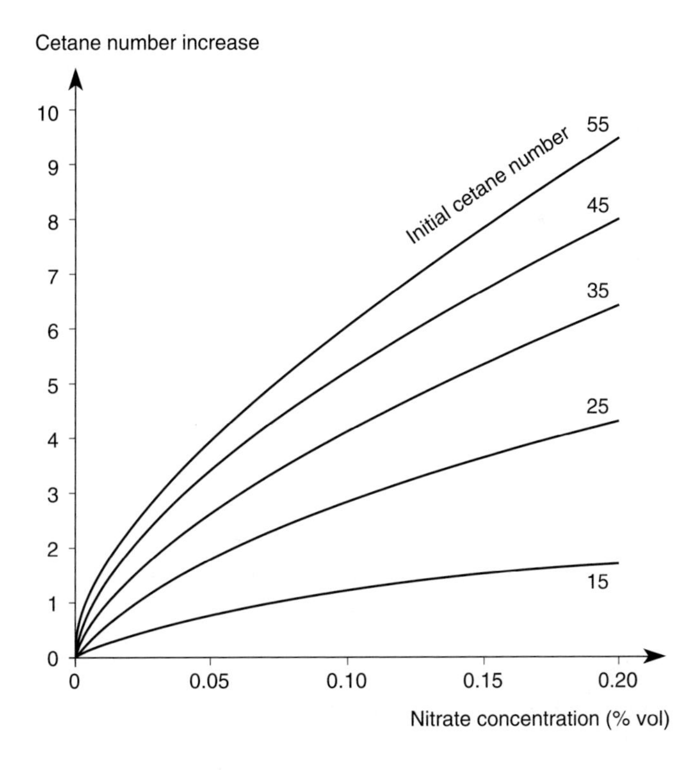

Figure 4.37 *Effectiveness of 2-ethylhexyl nitrate on cetane number. (Source: Unzelman 1984)*

vide. Nevertheless, a recent study of di-t-butyl peroxides provided encouraging results (see Table 4.10). Hence, the problems of storage instability seem to be resolved and their performance, like their price, is approaching those of nitrates. In addition, this study shows that both types of additives are compatible, which opens up some interesting commercial possibilities for peroxides (Nandi et al. 1994).

The improvement in cetane rating by the use of additives delivers improved engine behavior, as would be expected from improved combustion (less noise and improved driveability, especially at cold temperatures).

The addition of cetane improvers to diesel fuel results in a **measured** cetane improvement, whereas the **calculated** cetane index does not change because the physical and chemical characteristics used to obtain it are effectively unchanged (see 4.6.3).

Automobile manufacturers are certainly in favor of cetane improvers for diesel fuel, but they hope that they will be used in products that already have an initially **acceptable cetane rating**, given the relationship between chemical composition and emissions (see Chapter 5).

This concern explains why European standard EN 590 lists a minimum value for **both** a measured cetane number and a calculated cetane index (49 and 46 respectively for temperate regions, 46 and 45 for Arctic regions).

Table 4.10 *Example comparing the effectiveness of two types of cetane improvers (changes in cetane number).*
(Source: Nandi 1994)

DTBP (% vol)	EHN (% vol)			
	0.00	**0.33**	**0.66**	**1.00**
0.00	42.0	50.1	54.8	57.1
0.33	48.1	53.2	56.0	
0.66	52.3	55.2		
1.00	53.4			

Each cetane number corresponds to a concentration of EHN (row) and DTBP (column).
EHN = 2-ethylhexyl nitrate
DTBP = di-t-butyl peroxide

4.7 Properties related to the storage and distribution of diesel fuel

The most important property with respect to the use of diesel fuel is its behavior at **low temperatures**. This point was covered previously because it affects the winter operation of diesel engines.

Other product quality criteria, which are related to handling safety and stability over time, must also be met as described in the following sections.

4.7.1 Flash point

The distribution of the various types of fuels requires adherence to stringent safety rules in the refinery, in the storage areas, and at the service stations. To classify the various products from the safety standpoint, the concept of a **flash point** is used.

The flash point of a liquid petroleum product is the minimum temperature to which the fuel must be heated to produce vapors that ignite spontaneously in the presence of a flame. For diesel fuel, the flash point is determined in a closed vessel by following a well-defined protocol. The result is always given in association with the name of the protocol, because of its relevance is important.

Using the **Luchaire** method (standard NF T 60-103), the product is placed in a standard metal crucible and heated progressively without interruption in such a way that the temperature rises at 2 to 3°C/min. There are two holes in the crucible cover, one for a thermometer and a second hole equipped with a stack, which has a pilot light located at the top. A small but unmistakable deflagration signals the occurrence of combustion. However, the repeatability of this method is less than excellent, with differences of 2 to 3°C for the same product occurring regularly.

Test methods ASTM D 93 and ISO 2719 follow almost the same protocol, but they apply only to diesel fuels with flash points greater than 50°C and they use the **Pensky-Martens** apparatus. This apparatus consists of a covered crucible with an orifice that is opened only when a test flame is applied.

The second test method is used to determine the flash points quoted in the standards. In regulation EN 590, European diesel fuels must have flash points higher than 55°C. In the United States, the minimum value is 38°C for passenger car diesel (Class 1-D) and 52°C for commercial vehicles (Class 2-D) (ASTM D 975). California regulations are even more severe with a minimum threshold of 55°C.

The flash point is narrowly dependent on the initial distillation point. The following empirical equation is often used to express the relationship,

$$FP = IP - 100$$

where the flash point (FP) and the initial point (IP) are expressed in °C.

The flash points of gasolines are usually not measured, but they are understood to be much less than 0°C.

Consequently, it is obvious that adding gasoline to diesel fuel, which is sometimes done as a means of improving cold-temperature characteristics, results in a **unacceptable** reduction of the flash point and therefore should be prohibited.

The following explanation describes a potential **in-use safety** problem with fuels in the vehicle environment. With gasoline-fueled vehicles, the atmosphere above the liquid in the fuel tank consists of a fuel-air mixture that is **too rich** to ignite when subjected to a severe impact accompanied by sparks or hot spots. With diesel-fueled vehicles, given the low volatility of diesel fuel, the atmosphere is, on the contrary, **too lean**. However, the use of diesel fuel mixed with gasoline can result in an **inflammable** atmosphere above the liquid in the tank.

Furthermore, jet fuel has a flash point that is very close to that diesel fuel (see 7.3.3.1). It can therefore be used safely as a diesel-fuel diluant and it can even be used under special circumstances as diesel fuel—in military trucks and other motorized vehicles for example.

4.7.2 Fuel stability

Between the time that it leaves the refinery and it is burned in an engine, diesel fuel undergoes a number of storage and transportation operations. Even on board a vehicle, the fuel is heated (injector return circuit, parking in the sun) and it is in contact with oxygen and air-borne humidity (fuel tank vents). These conditions can lead to fuel degradation that results in fairly serious consequences. The general term **stability** is used to describe a fuel's ability to retain its original composition and characteristics over time. This section broadly describes the possible mechanisms that can alter diesel fuel during storage, the procedures used to detect them, and the means used to control them.

4.7.2.1 Reaction mechanisms

The longterm changes that occur in diesel fuel are the result of the chemical reactions that occur within the product and lead to the formation of gums and sediments.

Gums result from the polymerization and oxidation reactions that were previously described with respect to gasoline (see 3.7.1); these reactions develop slowly over weeks and even months. The precursors to gum formation are mainly the **mono- and di-olefins** that are present in trace amounts in the fuel; they affect stability at the beginning of storage. Complex aromatic structures can also generate gums, but only under longer-term and higher-temperature conditions.

Sediments are solids that are the result of very-complex chemical reactions (Pedley et al. 1988). The precursors are mainly heterocyclic compounds (Batts et al. 1991) containing nitrogen, sulfur, or oxygen. The following products and their derivatives are the ones most likely to be involved:

- pyrrole   indole 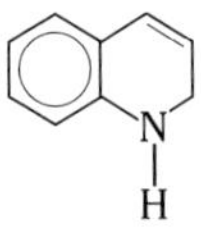   carbazole

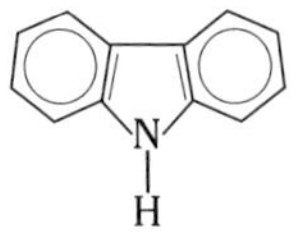

In particular, 2,5-Dimethylpyrrole

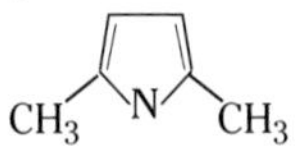

would be very active.

- thiophenol

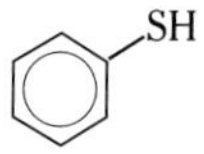

SH

- organic acids and thiophenes ...

One probable mechanism is an **esterification,** wherein the acid and alcohol functions are accompanied by aromatic rings, nitrogenated heterocycles, and thiophenes.

Sediments would have the following structure,

$$A-\underset{\underset{O}{\|}}{C}-O-R-B-\underset{\underset{O}{\|}}{C}-O-R-D$$

where

- A represents an aromatic compound
- B and D represent nitrogenated and sulfurized heterocycles
- R represents alkyl groups

Also, aromatic compounds with olefinic bonds can act on nitrogenated heterocycles of the indole type. For example:

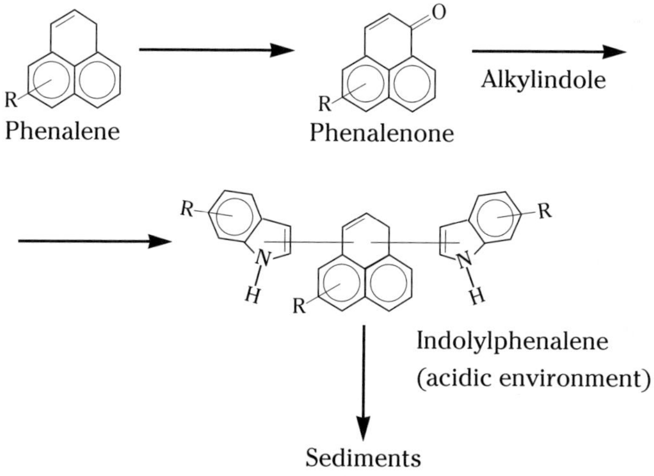

R Phenalene → R Phenalenone —Alkylindole→

→ R, N H, R, N H, R

Indolylphenalene (acidic environment)

↓

Sediments

Any of these reactions can be **catalyzed** by trace metals, especially **copper**, which becomes active at concentrations of about 0.01 ppm. Therefore, it is important to be very attentive to the presence of certain **metallic impurities** in diesel fuel.

4.7.2.2 Relative stability of various fuel fractions

The relative instability of diesel fuel is definitively related to the presence of **cracked products**. It is known that LCO—which is rich in aromatic compounds—and the fuels from **visbreaking** and **coking** that include unsaturated compounds, are particularly unstable. For example, Table 4.11 shows that the

Table 4.11 *Example of the quantities of gum and sediment precursors found in various diesel fuel cuts.*
(Source: Schrepfer et al. 1983)

Precursors	Operation				
	Direct distillation		Thermal cracking	Catalytic cracking	
	*	**		*	**
Alkenes (%)	0.5-5	0.5-5	8-30	3-13	3-30
Phenols (ppm)	1-50	1-50	10-5000	10-1000	100-7000
Thiophenols (ppm)	1-10	1-10	10-500	10-500	10-2000
Acid index (mg KOH/g)	0-1.3	0-5	0-0.2	0-0.1	0-0.5
Pyrroles (ppm)	10	10	100	200	200
Mercaptans (ppm)	200	400	400	100	200

* Normal severity.
** Increased severity.

various precursors of gums and sediments are far more numerous in these baseproducts than in straight-run distillate.

Also, increasing the severity of catalytic cracking or distillation—by increasing the final distillation point of the cut—increases the risk of gum and sediment formation in diesel fuels.

4.7.2.3 Classification methods

A. Diesel fuel color

A change in color toward a **darker hue** often indicates a chemical deterioration of the product. There are two methods of determining color:

- The **Saybolt method** (ASTM D 156 64 and NF M 07-003):
  This technique consists of gradually reducing the height of a sample column of fuel until its color has evidently become clearer than a measurement standard. The Saybolt color is expressed as a number related to the height of the column between +30 (almost colorless) to −16 (lightly colored). Strictly speaking, the Saybolt scale is most often used for clear medium fractions (jet fuel, hydrotreated distillates).
- The **conventional colorimeter method**:
  This technique is carried out using a colorimeter (standards NF T 60-104 and ASTM D 1500) to compare a fuel sample with colored-glass standards. The scale ranges from 0.5 to 8. In general, the result should not be greater than 5 (orange-brown hue); the results are usually a 1 or a 2 (straw yellow).

Automatic colorimeters are currently being developed to cover the full range of colors used in these two methods.

B. Oxidation stability

The two techniques most often used on a worldwide basis (ASTM D 2274 and NF M 07-047) both apply **accelerated** aging for 16 h at 95°C using oxygen bubbling. At the end of the test, the gums and sediments are trapped in a filter membrane (mesh hole diameter 0.8 µm) or in a filtering crucible with a 20- to 40-µm porosity—depending on whether the standard being followed is American or French. The filtrate is dried and weighed. The maximum permissible levels are 25 g/m^3 in Europe and 15 g/m^3 in the United States.

The correlation of these methods to actual storage behavior at ambient temperature is not very well established. However, despite these limitations, this test is in widespread use as a general indicator of fuel stability.

There is another technique (ASTM D 4625) that appears to correlate much better with field results. However, the test conditions (maintain at 43°C in the presence of air for 3 months) are very constraining.

C. Sediment and ash content

Determining the sediment content (standard DIN 51 419) consists of a series of solvent extractions. The unextracted **residue** is dried and weighed. The European specifications for diesel fuel set 24 mg/kg as the maximum quantity of sediment permitted.

The diesel fuels distributed in the United States are not subject to regulations requiring strict sediment content. Nevertheless, the combination of contaminants, both sediment and water, must not exceed 0.05% by volume when subjected to ASTM D 1796 (see 4.7.3.2a).

The ash content (EN 26 245 and ASTM D 482) is also obtained by burning a sample at 800°C in a platinum crucible and weighing the residue. The ash content must be less than 0.01% for both European and American diesel fuel.

D. Direct filterability measurement

The direct filterability technique, which was still unstandardized in 1998, consists of flowing a sample through an cellulose ester filter (0.5 µm nominal porosity). The total filtration time is recorded. If the filter becomes blocked, the volume of fuel filtered in 30 min is recorded. The sediments trapped by the filter can also be weighed as an adjunct to the test.

E. Acidity index

The total acidity index is the quantity of base, expressed in miligrams of potassium hydroxide, required to neutralize the acids in 1 g of product. In some cases, the **strong acidity index**, which is obtained after the boiling-water extraction of the strong acids (for example, HCl), can be distinguished from the **total acidity index**.

In any case, according to all specifications, the strong- or total-acid index of diesel fuels must by exactly nil. This measurement is therefore used as a control. It can also be applied to other refinery basestocks to ensure their suitability before adding them to the pool of finished product.

F. *Carbon residue*

The carbon residue procedure is not actually part of the array of ways to characterize the stability of diesel fuels at ambient temperatures. In fact, it provides information about a fuel's tendency to **coke** after pyrolysis. Hence, it is a test of a fuel's thermal stability, which provides information about its behavior when injected, particularly in relation to its tendency to form **deposits on injectors**.

The "micro" method (ISO 10 370—ASTM D 4530) uses the fraction between 90 and 100% distilled, or about 10% of the basestock, to determine the **carbon residue**. A sample is heated to 500°C in an inert atmosphere for a predetermined time. The residue is weighed after cooling and the results are expressed as the % mass with respect to the sample, which itself represents 10% of the total petroleum fraction.

European specifications limit the carbon residue to 0.3% maximum (for a fuel without cetane improvers). This value is rarely attained, or even approached, because all analyzed samples deliver results of less than 0.1%.

The "micro" method described here replaces the method used to determine the **Conradson carbon** residue, but they are equivalent.

There is another method that is in widespread use in the United States, which leads to the determination of the **Ramsbottom residue** (ISO 4262 and ASTM D 524). It follows a similar method to the one described here, but the sample is heated in air instead of nitrogen. The requirements for American diesel fuels limit the residue to 0.15% (mass) for passenger car fuels or 0.35% (mass) for commercial vehicle fuels.

The results obtained with the Ramsbottom method are not the same as the Conradson carbon results. There is no simple relationship between the two characteristics. However, there is a nomograph that provides a conversion in the ISO 4262 standard.

Information provided in Chapter 7 indicates that carbon residue, although low in value and not very discriminating for medium cuts, becomes much higher in value and provides a lot of useful information about the behavior of heavy cuts.

4.7.2.4 Results of diesel fuel instability

Two results stem from diesel fuel instability: filter blocking (other than the filter blocking caused by wax crystals at low temperature) and injector fouling.

A. *Filter blocking*

Filter blocking is a relatively rare occurrence; it requires very-unstable diesel fuel crammed with sediment. The resulting pressure drop across the filter (*General Motors* 1986) is shown in Fig. 4.38, which causes major operating disorders that can lead to a complete breakdown after just a few hours of operation.

Current diesel specifications, in principle, protect users from this type of problem.

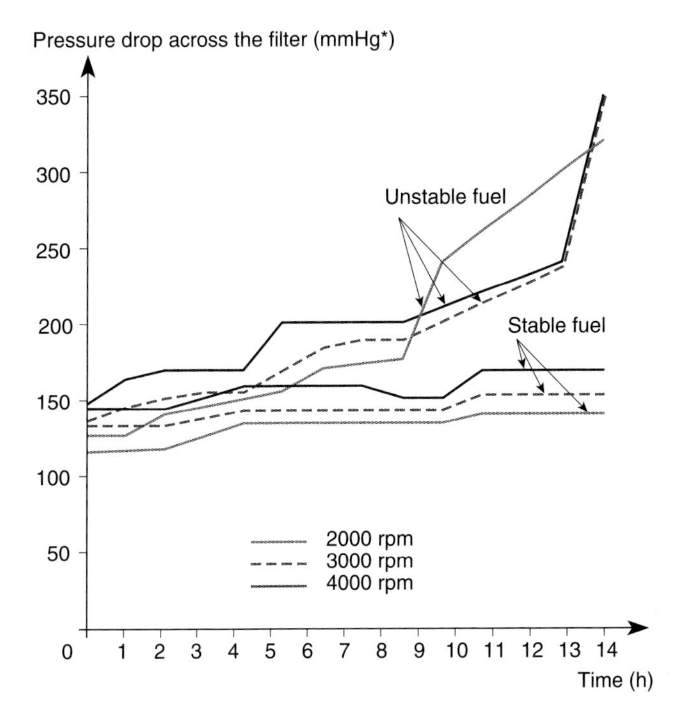

Figure 4.38 *Example of filter blocking by an unstable diesel fuel.*
** 100 mmHg = 13.3 kPa = 133 mbar*

B. *Injector fouling*

It can take only a few hours of operation for deposits originating from diesel fuel to form on certain areas of the injectors. This phenomenon is neither pronounced nor an impediment for high-pressure hole-type injectors. This is in contrast to the pintle-type injectors used in prechamber engines—especially those with exhaust gas recirculation—where deposits can accumulate in a thin layer (see Fig. 4.39) on the **needle shaft** and on the **walls of the injector body**, mainly in the conical-shaped zone used to control the injection flow.

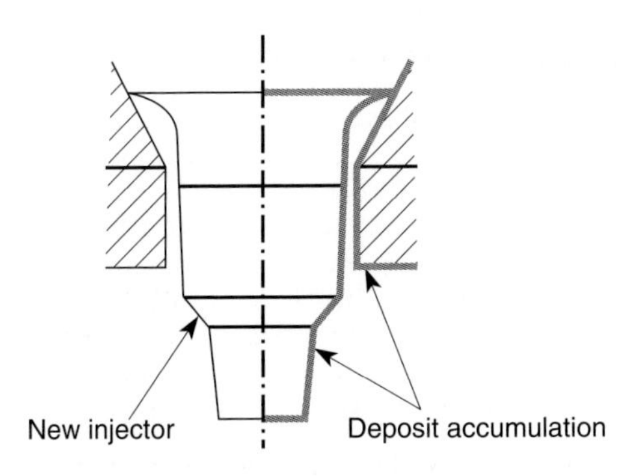

Figure 4.39 *Fouling of a pintle-type injector.*

At this point, the annular clearance between the pintle and the body is relatively small (15 to 18 µm) and the deposits reduce the effective area available for fluid flow. This causes a **change in the injection rate** (Montagne et al. 1987) that most often leads to a degradation in engine behavior: increased noise and speed fluctuations, higher emissions, and sometimes a loss of efficiency.

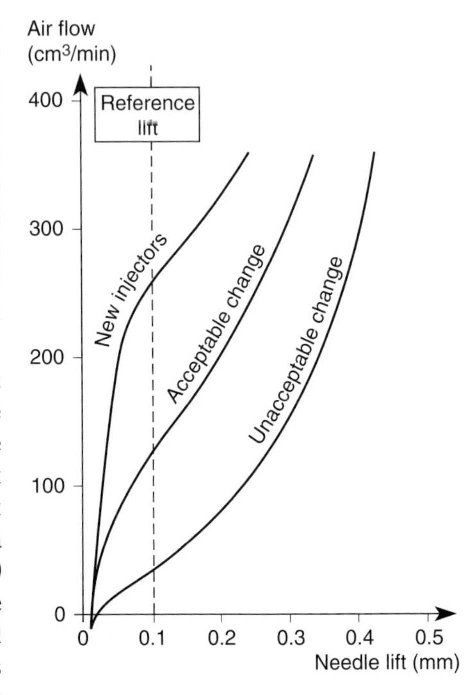

Figure 4.40 *Diesel-fuel injector flow rate at a pressure drop of 0.6 bar.*

There is a non-standardized but well-defined European **procedure** (CEC PF 26) for characterizing the fouled state of injectors. The test lasts for 6 hours and it is carried out using a diesel automobile on a dynamometer. After a run-up of 20 min (2000 rpm, 30% load), the engine is run continuously at 3000 rpm and heavy load (75%). Next, the injectors are removed and the air flow through the injectors is measured as a function of needle lift, according to a standard procedure (ISO 4010) at atmospheric pressure.

Fig. 4.40 shows the results obtained for new injectors, lightly-fouled injectors, and heavily-fouled injectors.

Even with perfectly-stable diesel fuels, there is always a drift in the flow rate after a few hours, which is acceptable because the engine manufacturer has taken it into account in tuning the engine.

Consequently, a major change as indicated in Fig. 4.41 would be unacceptable because it would cause the operational degradation mentioned previously.

The extent of injector fouling is most often provided in terms of the residual flow (RF), which is expressed with respect to the flow for a new injector at a given needle lift. In Europe, the automobile industry requires a minimum RF of 15% for a needle lift of 0.1 mm.

Obtaining this level of RF is greatly facilitated by the use of **technological arrangements**. As a result, the general use of **flattened nozzles** has considerably reduced fouling problems. However, this precaution alone is insufficient and it must always be accompanied by control and treatment of fuels.

4.7.2.5 Methods of improving diesel fuel stability

The available methods for improving diesel-fuel stability include the simultaneous or separate application of refinery procedures and the use of additives.

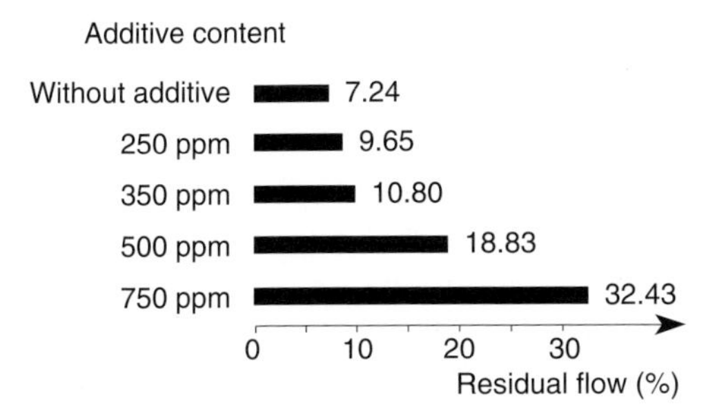

Figure 4.41 *Controlling diesel-engine injector fouling. Residual flow (RF) for various treat rates.*

A. Refining procedures

Fairly-severe **hydrotreatment** is one of the best techniques for improving the stability of diesel fuel. The stability behavior of some cuts, before and after hydrotreatment, is listed in Table 4.12. Incidentally, hydrotreatment is absolutely essential for cuts derived from thermal cracking processes (visbreaking, coking) because their instability. The most obvious indication of the improvements obtained is the reduction in the content of **nitrogen constituents** that always results from hydrotreatment (Martin et al. 1990). While this is not the only factor to take into consideration, the **nitrogen content** of a diesel fuel can be considered a general indicator of its stability. In this regard, a maximum level of 100 ppm should be considered as a valid reference.

Deep hydrodesulfurization, which is essential in this context, usually has a positive effect as well on the stability of diesel fuel because it is accompanied by denitrification. However, when the desulfurization occurs at high temperature (over 370°C), instability reappears as evidenced by a change in color (see 4.10.3).

Finally, there is another very-effective process that is not often used today: **caustic washing**, which can neutralize a large number of gum and sediment precursors like thiophenes, mercaptans, phenols, and acids. This procedure is reserved for special situations, due to its high cost and difficult implementation.

B. Additives

Several types of additives can be used to improve the stability of diesel fuels (Waynick 1994). One of the most effective products is N,N-dimethylcyclohexylamine, but other **tertiary amines** or other chemical structures can have the same beneficial effect. This effect consists of **inhibiting** the chemical reactions between the nitrogenated heterocycles and the oxygenated compounds (organic acids, phenalenone ...). Some formulations also contain **metal deactivators**.

All of these additives reduce the mass of sediment formed, but usually do not change the immediate color or the change in color over time.

Finally, although **detergent additives**—either dispersants or surfactants—do not act directly on the intrinsic stability of fuels, they reduce or eliminate

Table 4.12 *Oxidation stability of hydrotreated diesel fuels. (Stability test: ASTM D 4625)*

Characteristics	Diesel fuels											
	LCO				LCO/Straight run blend				Straight run/Visbreaking blend			
	Ref.*	25 bar	35 bar	45 bar	Ref.*	25 bar	35 bar	45 bar	Ref.*	25 bar	35 bar	45 bar
Density (kg/dm^3)	0.941	0.909	0.905	0.900	0.895	0.873	0.870	0.868	0.847	0.834	0.833	0.832
HDS rate (%)		89.7	91.4	93.6		91.6	93.2	95.6		92.0	994.5	96.3
HDN rate (%)		58.0	70.0	82.1		48.4	67.7	82.0		46.9	66.0	78.4
Stability test:												
Color before the test	2.5	2	1.5	1.5	1.5	1.5	1.5	1.5	2.5	0.5	0.5	0.5
Color after the test	8	3	2	2	8	2	2	2	3	0.5	0.5	0.5
Sediments (mg/100 ml)												
free	0.9	0.1	0.2	0.2	9.0	0.3	0.2	0.2	0.1	0.1	0.2	0.1
adhering (to container walls)	28.0	0.2	0.3	0.4	7.4	2.3	0.9	0.9	1.7	1.7	1.0	0.4
total	28.9	0.3	0.5	0.6	16.4	2.6	1.1	1.1	1.8	1.8	1.2	0.5

* Non-hydrotreated reference.
The pressures shown are those provided during hydrotreatment.

engine problems. This is shown in Fig. 4.41, where the results of injector fouling tests with various amounts of detergent additives are provided. The previously defined residual flow (RF) increases in step with the additive content.

In Europe, detergent additives are widely used to control injector fouling in diesel passenger cars.

4.7.3 Preventing bacteriological contamination

Bacteriological contamination is a problem mainly with medium and heavy fractions: jet fuel, diesel fuel, and domestic fuel oil; it can have serious consequences and therefore merits careful attention.

4.7.3.1 Explaining the phenomenon and the consequences

The natural venting of fuel storage tanks and reservoirs introduces some **water**, which usually decants at the bottom of the containers. This water, as well as the ambient air, contain **microorganisms**—bacteria, yeasts, or fungi—that find these containers to be a hospitable environment for their proliferation; this environment includes both oxygen and organic matter. Thus, the bacteria grow by breaking down certain organic molecules using the oxygen atoms from the air dissolved in the water or from oxygenated molecules, depending on whether the process is **aerobic** or **anaerobic**. These bacteria have very-high metabolic rates and some of them can double their population in 20 min!

Furthermore, these microorganisms usually grow under moderate temperatures, generally between 20 and 40°C. This why bacteriological contamination problems occur mainly during the **summer**.

The resulting organic matter can migrate into the fuel causing serious operational problems: blocked filters, corrosion of the fuel system by acids and other aggressive organic substances, deteriorated injection, and possibly seized pistons.

These types of serious problems are reasonably frequent occurrences with marine diesels in port areas, but they occur less often with road vehicles.

4.7.3.2 Preventive or curative measures

These measures consist of minimizing a diesel fuel's contact with water, taking some precautions during storage, and using biocide additives in high-risk situations.

A. Controlling the presence of water in diesel fuel

The solubility of water in diesel fuel is very **low**. Solubility depends on the temperature and the nature of the product, but it does not exceed 100 ppm under normal storage conditions. However, diesel fuel can also contain free water in suspension.

The test method for water content most often used in Europe is the **Karl Fischer** method (standards NF T 60-154 and ASTM D 1744). The method can be used to precisely measure the total quantity of water in the fuel, both free and dissolved. It consists of an electrometric titration method using a specific reagent based on iodine, pyridine, methanol, and sulfur dioxide. European diesel-fuel specifications allow a maximum water content of 200 mg/kg. This limit is high and it is **rarely attained** or even approached in practice.

American requirements specify a more general method that provides both the water and sediment content of the fuel (ASTM D 1796). A fuel sample is mixed with equal volumes of toluene in a graduate; the water and sediments are then separated using a centrifuge. The maximum rate permitted in the United States is 0.05% by volume.

B. Cleanliness of storage installations

Storage installations must be as **anhydrous** as possible, which implies the use of desiccation, decantation, even filtration and centrifuging. Some diesel fuel characteristics other than water content can encourage bacteriological contamination. Thus, the presence of sediments, especially **iron oxides** from metal reservoir walls, provides microorganisms with a favorable development environment. As well, **surfactant additives**, which are very useful for controlling fouling, can have unfavorable effects on the spread of bacteria.

In effect, they diminish surface tension between the water and the fuel, thereby keeping water and other impurities in suspension.

C. Use of additives

The additives that can be used to control bacteria are either **biocides** or **bacteriostats**; the former kill microorganisms and the later prevent their reproduction. Depending on the situation, either the diesel fuel itself or the water, which is almost always present in the bottom of fuel storage tanks, can be treated.

The following biocides are soluble in diesel fuel:

- derivatives of **quinoline**

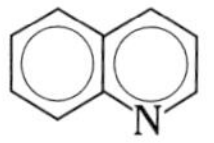

N

- **cyclic amines**

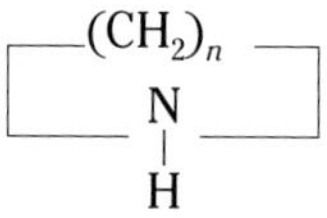

$(CH_2)_n$

N

H

- **N-alkylpropanediamine**

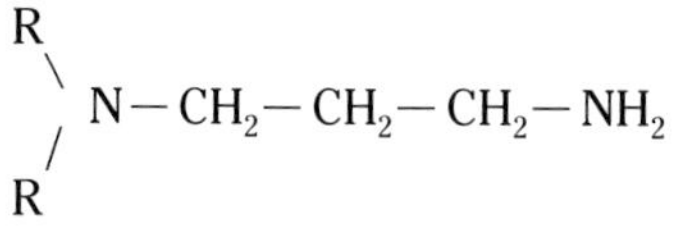

$R_2N-CH_2-CH_2-CH_2-NH_2$

- **imidazolines**

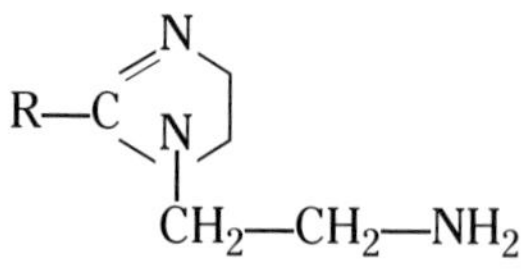

R—C N N

$CH_2-CH_2-NH_2$

The only additives that are water soluble are biocides that are based on formaldehyde ($H-\underset{\displaystyle O}{\overset{\|}{C}}-H$).

Additives based on **boron**, which are soluble in both the aqueous and organic phases, are also used occasionally. Among these are bacteriostats based on boron amides and esters.

The **treatment procedure** is very important. It is generally thought that the most appropriate additives are those that act on the aqueous phase itself. This requires less product and it is therefore less costly. However, the users with the greatest exposure to bacterial contamination (navies, commercial fishing, ...) also choose to treat the diesel fuel.

The reader is reminded that the use of a biocide intended exclusively for the aqueous phase must not be used directly in diesel fuel.

Such a product would be insoluble in diesel fuel and it would remain in suspension, causing serious damage to the engine's fuel supply system (filters, pumps, ...).

The treatment rate for biocides and bacteriostats varies widely depending on the chemical structure of the product: treatment rates typically range from 100 to 400 ppm.

4.7.4 Controlling other characteristics

Controlling other characteristics consists of minimizing some minor, but sometimes troublesome, problems that can occur during distribution.

4.7.4.1 Suppressing emulsification

Diesel fuel can sometimes appear hazy due to the presence of finely-dispersed water droplets. This usually occurs when the products are churned through the distribution chain. The use of detergent additives accentuates the problem. Therefore, the use of small amounts (about 10 ppm) of **demulsifiers** like alkyloxypolyglycol or arylsulfonates is recommended. Their effectiveness can be tested using a simple process: a small quantity of water is added to a fuel sample in a test tube and the mixture is shaken vigorously. The sample is placed at rest and the appearance and volume of the fuel/water interface is examined.

4.7.4.2 Reducing foaming

Diesel fuel has a tendency to **foam**, especially when refilling passenger car fuel tanks. This can be an aggravating situation if the fuel splashes on the hands and clothes of users, or spills on the ground. Foaming can be reduced significantly with the use of **silicone-based** additives (polysiloxanes) that remain in suspension in the fuel. The treatment rates are very low: about 10 to 20 ppm of additive, which is 1 to 2 ppm of silicon.

4.7.4.3 Controlling odor

It is well-known that even desulfurized diesel fuel has a specific lingering odor, which is usually considered to be disagreeable, especially in contact with the skin or clothing. Petroleum companies have developed **odor masks** to make diesel fuel more ... sociable.

Two approaches to the problem are possible:

- using a product that conveys its own odor ... as discreetly as possible, within the range of conventional perfumes
- covering up or neutralizing the various odors of some of the typical components of diesel fuel within the overall distillation range. This approach appears to be the preferred one, although it is more difficult to implement

The odor masks most often used are derived from natural odorous substances, or they contain **ketones** or **synthetic esters**.

Incidentally, special diesel fuels that have been severely hydrotreated, or have been derived from the Fischer-Tropsch process, are practically odor-free.

4.7.5 Preventing unintended use

Unfortunately, there are no devices available in Europe that would prevent a distracted user from inadvertently adding diesel fuel to a gasoline-powered vehicle, or conversely, gasoline to a diesel vehicle.

The contamination of diesel fuel with a small quantity of gasoline can be momentarily accommodated, if need be, by a diesel engine; on the contrary, however, the addition of diesel fuel to gasoline, in sufficient quantity to reduce the octane rating and cause knock, could have very-serious consequences.

Also, it is usually illegal to mix diesel fuel with a similar product that carries a lower level of taxation, such as domestic fuel oil.

To detect and suppress attempts at tax evasion, European authorities have set up **marking** and **tracing** additives. These additives consist of a scarlet red **dye** (*o*-toluene-azo-2-naphthol), which is added in concentrations of 1 g/hl, and two **tracers**: diphenylamine and furfural in concentrations of 5 g/hl and 1 g/hl respectively.

The two tracers can be easily detected using simple tests. In effect, aniline acetate turns bright red in the presence of furfural and a sulphochromic reagent turns bright blue in the presence of diphenylamine.

Potential tax evaders—although there are certainly none among the readers of this book—should be advised that these tests are very effective and reliable.

4.8 Sulfur content

The sulfur content of diesel fuel is subject to precise regulation everywhere. The objective of these regulations is to limit tailpipe emissions of SO_2 and SO_3. In addition, reducing sulfur content brings other indirect benefits (reduced particle emissions, improved operation of post-treatment catalysts) that are described in Chapter 5.

Hence, this section only describes the effects that the presence of sulfur in diesel fuel has on engine behavior.

4.8.1 Sulfur-content limits and measurement methods

The maximum diesel-fuel sulfur content permitted in Europe, since October 1, 1996, is 0.05% by mass (500 ppm). This limit will be lowered at 0.035% (350 ppm) in 2000 and at 0.005% (50 ppm) in 2005. The United States (since 1994) and Japan (1997) also apply the 500 ppm limit. Other countries are also moving, with varying schedules, toward this 0.05% maximum limit.

The sulfur content can be measured by X-ray fluorescence spectrometry (ISO 8754 and ASTM D 2622 test methods). A fuel sample, when placed in an X-ray beam, emits characteristic radiation. By evaluating an absorption line from the appropriate spectrum, the sulfur content of the sample can be determined. This technique works well for low sulfur-content fuels (up to 10 ppm using ASTM D 2622).

Another longer but more precise method is known as the **Wickbold technique** (EN 24 260). It consists of the complete combustion of the sample, followed by the total oxygenation of the SO_2 and SO_3 into sulfuric acid by absorption into a solution of oxygenated water. The sulfate ions are then measured using conventional techniques (volumetric, conductimetry). Remarkable levels of precision, down to very-low sulfur levels on the order of a few ppm, can be achieved using the Wickbold technique. It can be used as a reference method, given that other more practical, but sometimes less reliable, techniques are available.

There are also other methods using ultraviolet fluorescence (NF M07-059) that can be used to measure sulfur content.

4.8.2 The effect of sulfur on engine wear

Until recently, reducing the sulfur content in diesel fuel was considered beneficial in reducing corrosive engine wear. However, with the introduction of very low-sulfur diesel fuels other wear factors have begun to appear in the fuel-injection system, instead of the engine.

4.8.2.1 Engine corrosion

The sulfur oxides SO_2 and SO_3, which occur during combustion, and the water vapor in the burned gases can combine to form the sulfurous and sulfuric acids H_2SO_3 and H_2SO_4.

The most probable reaction schema is the following one:

$$\text{Organic sulfide compounds} + O_2 \longrightarrow SO_2 \quad (1)$$

$$2\,SO_2 + O_2 \longrightarrow 2SO_3 \quad (2)$$

$$SO_3 + H_2O \longrightarrow H_2SO_4 \quad (3)$$

Step (1) is almost completed during combustion of the fuel, while reactions (2) and (3) are thermodynamically enhanced at **low temperatures**.

In particular, sulfuric acid can form on the cylinder walls as soon as they reach the temperature for the liquefaction of the acid-water mixture to occur (**dew point**). The result is condensation on the metal, which is followed eventually by a **corrosive attack**.

Similar processes have been observed in the past on diesel engines using fuels with very-high sulfur contents (up to 0.8 or 1%). At present, significant corrosion has not been observed for sulfur content levels lower than 0.2%. Additional protection is also provided by lubricant formulations that have sufficient alkalinity (Total Basic Number [TBN]).

Finally, with large marine diesel engines that use heavy fuels, which are often high in sulfur (up to 4%), the risk of corrosive wear can be very great and it must be controlled by technological means (see 7.10.6).

4.8.2.2 Wear in the fuel-injection system

This phenomenon occurs mainly with very low-sulfur diesel fuels (less than 100 ppm). It has occurred in countries (Scandinavia, Canada) where very low-sulfur diesel fuels have characteristics similar to jet fuels, particularly the viscosity and the distillation interval.

The incidents include **seizures** of injection pumps caused by an adhesive-abrasive wear that occurs after short periods of use (15 000 km) (Lacey et al. 1992).

It has not been decisively determined if the wear results from low fuel viscosity, a reduction in lubricity due to the disappearance of the sulfur itself, or even the absence of the aromatic and naphthenic rings that normally include the sulfur atoms. **Ring structures** are known to have certain polar characteristics that allow them to be easily adsorbed onto metal surfaces.

The test methods used to classify the wear behavior of diesel fuels make use of **tribology devices**; the best known one is the Ball on Cylinder Lubricity Evaluator (BOCLE) (Jenkins et al. 1993). The BOCLE apparatus is used in a standard procedure (ASTM D 5001) that is applied mainly to jet fuels. Work has recently been completed to adapt this test method for use with diesel fuels.

Another type of tribology device, known as the High Frequency Reciprocation Rig (HFRR) can be used to study the wear phenomenon attributable to diesel fuel quality (Spikes et al. 1994). Fig. 4.42 shows a schematic representation of the device. A ball immersed in diesel fuel is subjected to reciprocating motion by an electronically controlled vibrator. The amplitude (0.02 to 2 mm), the frequency (20 to 200 Hz), the temperature, and the length of the test can be directly controlled. The device can be used to differentiate between adhesive wear and contact corrosion.

The test results are expressed as the "**wear scar diameter**", which is the diameter of the circular wear pattern. If the pattern is not a perfect circle, the wear scar diameter (WSD) is expressed by the following equation:

$$\text{WSD} = \frac{\text{Large diameter} + \text{Small diameter}}{2}$$

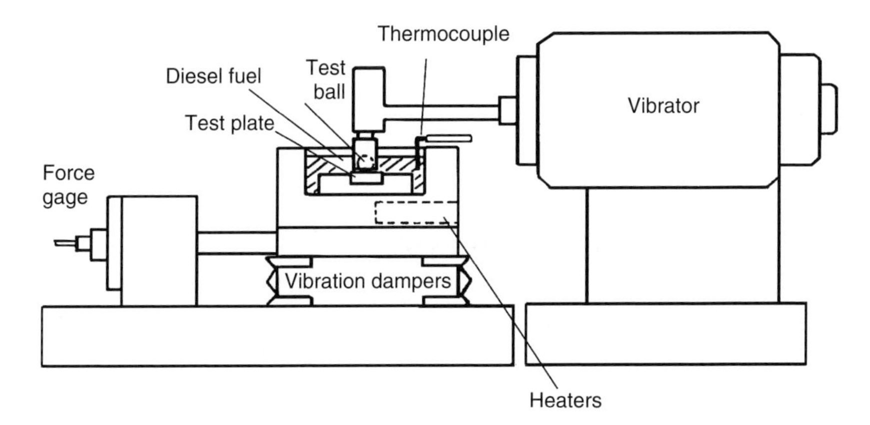

Figure 4.42 *Schematic representation of an HFRR tribology device.*

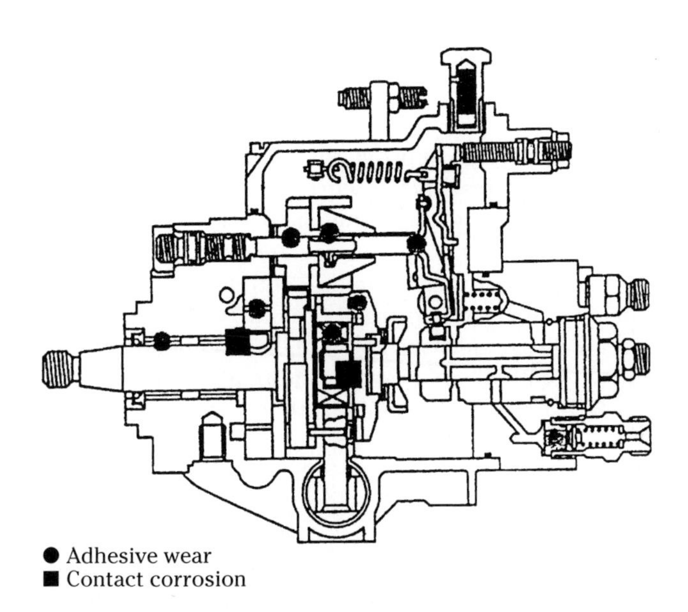

Figure 4.43 *Wear-sensitive areas in a Bosch rotary pump.*

After 1 or 2 hours of testing, the WSD can attain several hundred micrometers.

There are also long-term tests (several thousand hours) that are carried out directly on fuel-injection pumps. The various areas subject to wear within an injection pump are shown in Fig. 4.43.

The wear phenomenon can be controlled in two ways:

- **Technological means** can be applied by choosing suitable metals and coatings. These techniques are effective but they cannot be applied to equipment already in service.

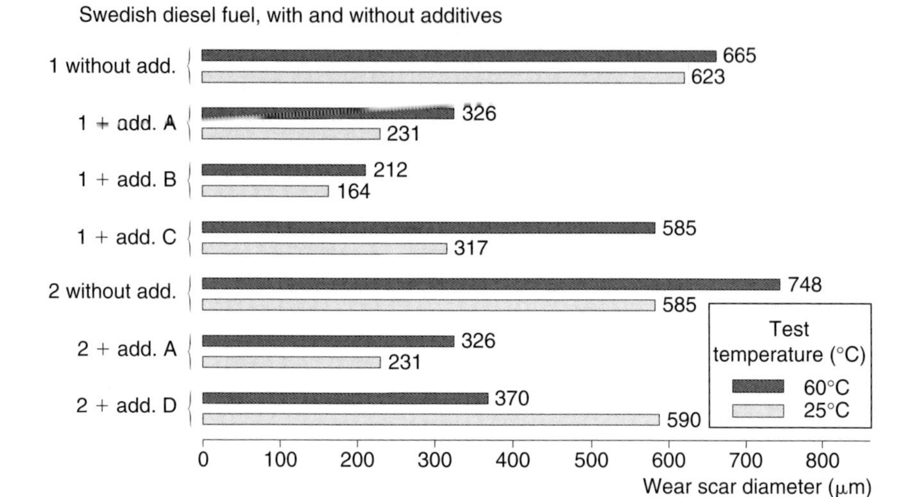

Figure 4.44 *Wear tests of Swedish diesel fuels. (HFRR tribology device)*

- **Additives** can be used to improve the **lubricity** of diesel fuel (Caprotti et al. 1992). Important research is currently under way to determine the most effective products and the most appropriate concentrations. Products currently in widespread use are ester type structures derived from fatty acids and carboxylic acids. In other respects, the required concentration for jet fuels are no greater than 15 to 20 ppm, whereas the concentration for diesel fuels are much higher (50 to 300 ppm). The results shown in Fig. 4.44 indicate that the addition of appropriate additives to wear-prone diesel fuels can significantly improve their performance.

Consequently, the deep desulfurization of diesel fuels, which is becoming common practice, leads one to question their lubricity and the need to include a sufficient level of quality to address that point. It is likely that future specifications will take this criteria into account. Diesel fuel distributors in Europe have already agreed not to exceed a WSD of 460 μm.

The formulation of diesel fuel

This section provides descriptions and rationales for the various diesel-fuel quality criteria and describes both the general and the specific problems associated with its production in the refinery.

Complementary information on the same theme is provided in Chapter 2 and in Chapter 7.

4.9 Diesel fuel's role in the refinery schema

For many years, diesel fuel was produced by the **simple atmospheric distillation** of crude oil. The cut between 180 and 360°C provided a product with characteristics that usually met all specifications. The only additional treatment was light desulfurization, if the crude had a high sulfur content.

Today, the situation has changed considerably with the introduction of **heavy-product conversion operations** that all yield medium cuts. These cuts can be included in the diesel pool, but they must undergo prior hydrodesulfurization and hydrotreating. At the same time, recent observations in several countries, particularly in Western Europe, indicate that market demands for quality and quantity are changing rapidly.

As a result, the formulation of diesel fuel has become as complex and multifaceted as that of gasoline.

4.9.1 Diesel fuel's place among the medium fractions

Fig. 4.45 shows both the observed and anticipated changes in the diesel-fuel portion of the medium fractions available in Western Europe between 1980 and 2000. The diesel- and jet-fuel portions continue to grow while the production of domestic fuel oil (DFO) is gradually diminishing.

Consequently, the diesel-fuel portion cannot be extended into the jet-fuel portion, which is another product that is in high demand and subject to very-severe quality constraints (see Chapter 7). On the other hand, extending the diesel-fuel portion toward the domestic-fuel-oil portion is possible, but this requires a considerable improvement in quality. Table 4.13 provides an example of diesel fuel and DFO characteristics. For the refiner, it is obviously far-more difficult to achieve diesel-fuel quality levels than DFO quality levels.

It is apparent that the market is demanding increasing quantities of products with high-quality requirements (diesel fuel) and diminishing quantities of less-sophisticated products (DFO).

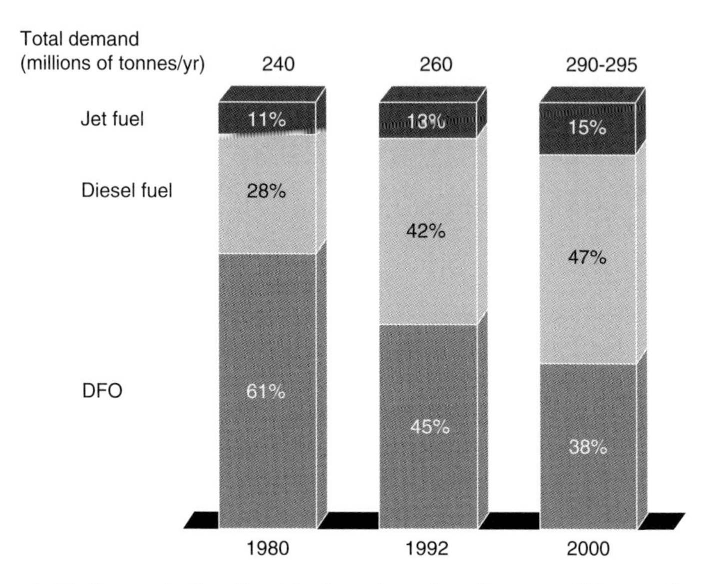

Figure 4.45 *Change in the diesel fuel portion of medium cuts (Western Europe).*

Table 4.13 *Comparative characteristics of diesel fuel and DFO (winter extreme).*

Characteristics	Products	
	Diesel fuel	**DFO**
Density at 15°C (kg/dm³)	0.824	0.860
Kinematic viscosity at 20°C (mm²/s)	3.47	6.20
Sulfur mass content (%)	0.05	0.200
Cloud point (°C)	–8	+3
CFPP (°C)	–23	–3
Pour point (°C)	–27	–21
Distillation IP (°C)	167	185
FP (°C)	364	374
Cetane rating	50.5	48.0
NHV_m (kJ/kg)	42 825	42 220
NHV_v (kJ/dm³)	35 285	36 310
Mass composition : Carbon (%)	86.60	87.1
Hydrogen (%)	13.35	12.7

4.9.2 Characteristics of available basestocks

Table 4.14 shows some of the physical and chemical characteristics of major refinery products that can be used to make up the diesel pool. The mass yield for each basestock, which is the quantity obtained with respect to the feed used in the refinery process, has also been included.

Table 4.14 *Examples of basestocks used in the formulation of diesel fuel.*

Characteristics	Basestock									
	Paraffinic crude			Naphthenic crude		Vacuum distillate		Vacuum residue		Deasphalted atmospheric residue
	Atmospheric distillation			Atmospheric distillation		FCC	Hydro-cracking	Vis-breaking	Coking	Hydrocracking
Yield (% mass)	30.3	32.8	36.7	29.2	47.2	10-15	30-40	5-15	35	20
Density at 15°C (kg/dm^3)	0.835	0.825	0.843	0.827	0.856	0.930	0.814	0.845	0.900	0.807
Distillation (°C)										
IP	170	180	170	180	170	170	220	170	170	260
FP	370	375	400	350	370	370	370	370	370	380
Cloud point* (°C)	−5	−2	+1	−10	−20	−5	−17	−4	−8	−13
Pour point* (°C)	−12	−9	−6	−18	−33	−14	−20	−18	−20	−18
Cetane rating	50	51	54	54	43	24	64	40	28	70
Sulfur content (% mass)	0.12	0.04	0.83	0.80	0.09	2.8	0.001	2.33	2.10	0.0005

(Source: Elf, Total)

* The results shown here are based on laboratory data. During formulation in the refinery, the cold-temperature characteristics will be markedly worse (by about 6°C), if the same crude yield is expected.

The properties of primary interest to the refiner are

- the low-temperature behavior
- the cetane rating
- the sulfur content

Note that the products obtained from direct distillation have characteristics that are highly dependent on the nature of the crude oil and the span of the cut (initial point and final point). In contrast, the composition and properties of the diesel fuel from conversion processes are very dependent on the feedstock and the operating conditions of the equipment. A refinery schema of the various processes that contribute to diesel cuts is shown in Fig. 4.46.

4.10 Refining constraints and flexibility

Diesel fuel production requires the simultaneous creation of the quantities and qualities required by the market using blends of different basestocks. This section describes the respective problems that result from meeting cold-temperature specifications, cetane ratings, and sulfur content.

4.10.1 Cold-temperature characteristics

Even thought the cloud point is not a specification requirement, it provides refiners with an **important key** to meeting the required cold-temperature characteristics. In fact, obtaining a required CFPP by the use of additives only is neither desirable nor always possible. In practice, to obtain a diesel fuel for extreme cold-temperature use (–20°C and colder), a cloud point less than or equal to –10°C is essential.

To meet this objective, the refiner can act on the **source of the crude** or on the **distillation range**, or on both together.

Naphthenic and aromatic crudes are known to yield distillates with excellent cold-temperature characteristics. The inverse is true of the paraffinic cuts. There is also a clear correlation between the cloud point and the final distillation temperature. Consequently, a drop in the cloud point to less than 0°C implies, in practice, a final distillation point of less than 360-370°C. This situation results in serious constraints: on average, a reduction of 1°C in the cloud point results in a 0.5% (mass) **loss in crude-oil yield**.

At this point, the compromise between quality and quantity is particularly difficult to establish.

Besides, a low cloud point signals the presence of abundant naphthenic and aromatic constituents, which usually means **low cetane numbers**, thus further complicating the search for an optimal formulation.

The adjustment of the cold-temperature characteristics of the final product, as indicated previously (see 4.5.4.2), is done by adding flow improvers. The resulting improvements depend on the type and concentration of the additive used, as well as the composition of the treated product. For example, it is

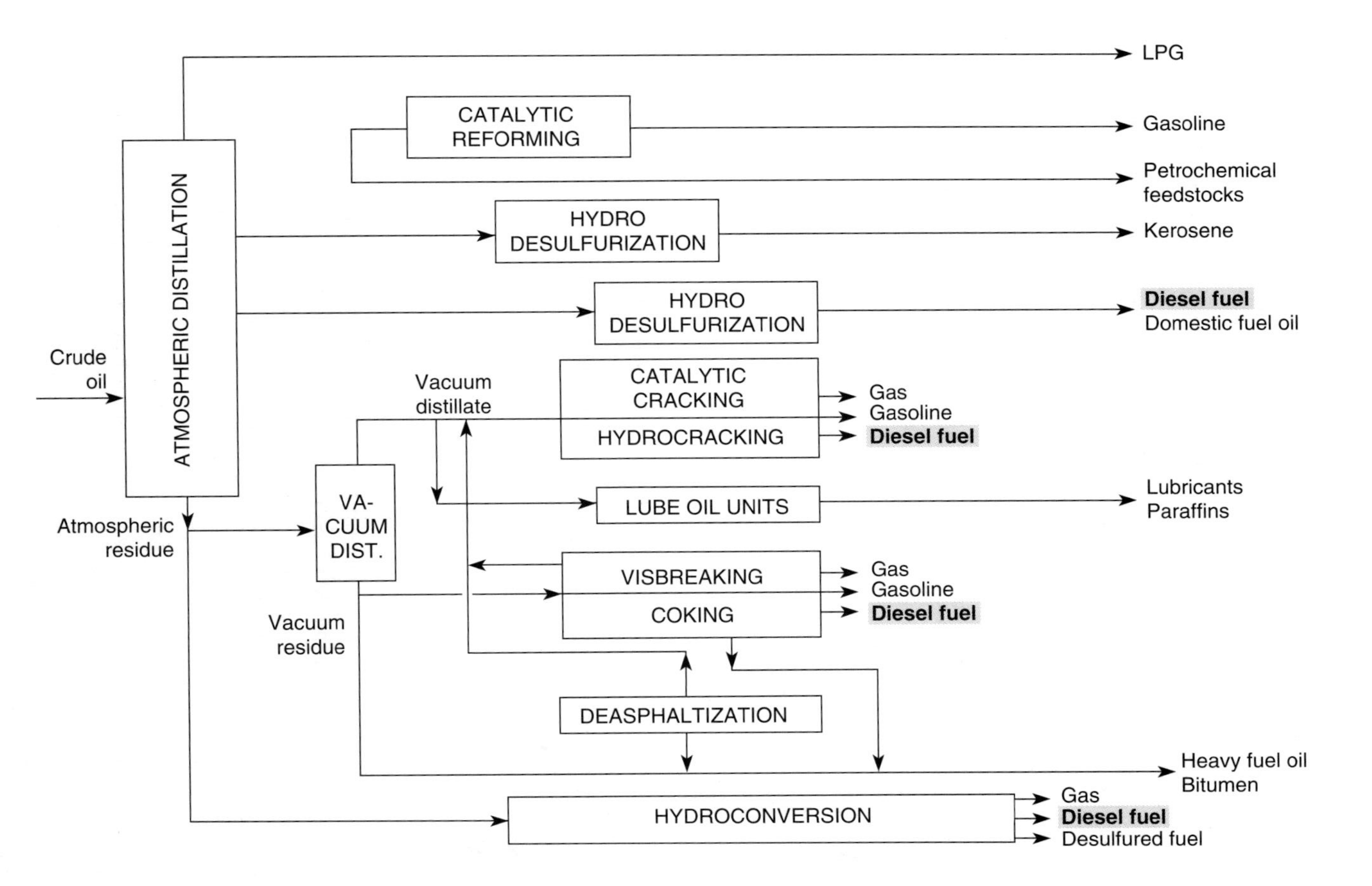

Figure 4.46 *Synoptic view of the refinery processes involved in the formulation of diesel fuel.*

known that certain paraffinic crude oils, like those from the North Sea, yield diesel cuts that are not very sensitive to flow improvers. Refiners integrate and use all these available data in the preparation of the final product.

4.10.2 Cetane rating

Table 4.14 indicates that the basestocks with the best cetane ratings are mainly the direct-distillation cuts from paraffinic crudes, or ultimately, naphthenic crudes, as well as the products from hydrocracking. Conversely, all the products from conversion units, particularly catalytic cracking, have low cetane ratings of between 20 and 40, depending on the supply conditions and process operations.

This presents the following question: by analogy to gasoline octane ratings, do blends of several diesel fuel cuts behave linearly with respect to cetane number? Recent research work (Quignard 1992) has shown that **linearity** does occur in almost all cases that are within the measurement precision of a CFR engine. Although, as previously indicated, this measurement does demonstrate considerable dispersion.

With a view to guiding the fine tuning of new processes and the optimization of diesel-fuel formulation, various techniques to predict the cetane index based on chemical analysis have been developed (Glavinceski et al. 1984; Pande et al. 1990). Gas chromatography, nuclear magnetic resonance (NMR), and even mass spectrometry can be used.

With **gas chromatography**, the columns available do not allow a total separation of the diesel cut; there appears to be a mass of constituents from C_{12}-C_{13} that cannot be resolved. Despite this problem, it is possible to obtain enough information from a chromatogram to estimate the final cetane number and the cetane profile of the distillation curve. An example for a direct-distillation diesel fuel is shown in Fig. 4.47.

For **spectroscopic methods** of cetane-number prediction (NMR, mass spectrometry), correlations have been established for a large number samples.

Carbon-13 or proton NMR can be used. In terms of ease of use, analysis time (15 min), and prediction accuracy (average of 1.4 standard deviations from the measured number), **proton NMR** seems to provide the most advantages.

With **mass spectrometry**, the degree of precision is not as high as NMR (average of 2.8 standard deviations), however, a comparatively smaller sample size (2 cm^3) is required. A very-good correlation between the measured and calculated values is shown in Fig. 4.48, for a range of cetane ratings between 20 and 60.

This shows that refiners have a number of tools at their disposal, both theoretical and experimental, to produce diesel fuels with the required cetane ratings. Adjustments can be made by resorting to **cetane improvers**; but this avenue always has limitations, because the cetane index calculated according to ASTM D 976 excludes additives, and the cetane rating must attain a minimal threshold value (49 in Western Europe).

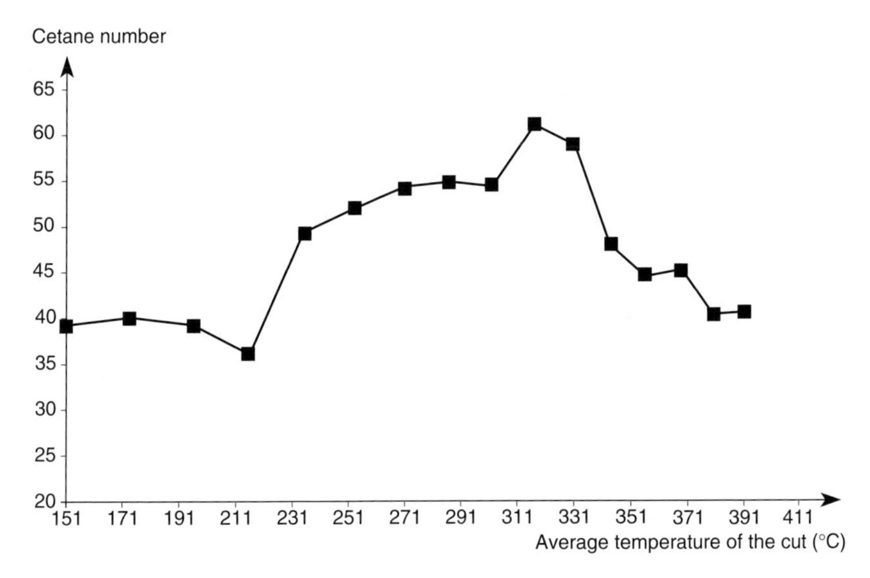

Figure 4.47 *Cetane-rating profile of a straight-run diesel fuel.*

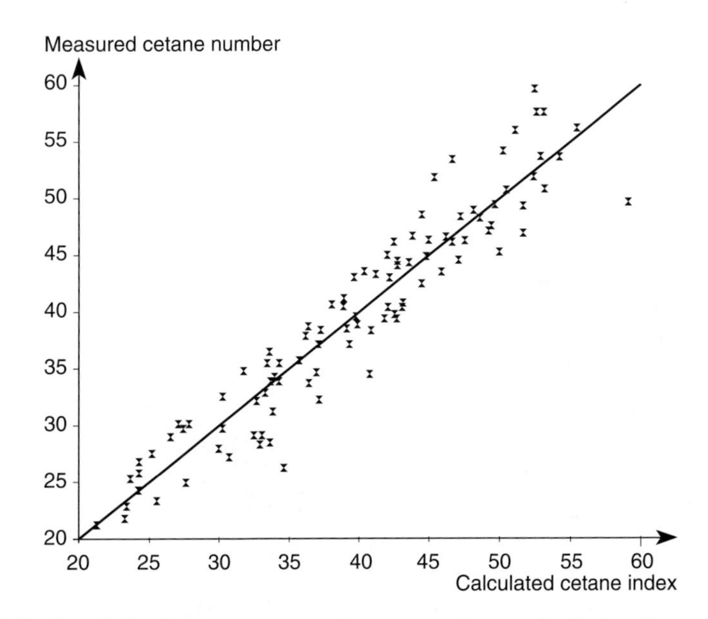

Figure 4.48 *Correlation between measured cetane numbers and calculated cetane indexes.*

In practice, the diesel fuels sold in Western Europe meet the cetane rating specifications by a comfortable margin. As a result, the average European (Scandinavia excepted) cetane rating for diesel fuels in 1998 was 51 when the specifications required a minimal rating of 49. It is estimated that 20 to 30% of the diesel fuels distributed by service stations contained cetane improvers, which used treat rates of 100 to 300 ppm.

4.10.3 Sulfur content

The sulfur content of straight-run fractions from atmospheric distillation depends heavily on the content of the crude oil, as indicated in Fig. 4.49. In fact, it is generally estimated that medium distillates have about half the sulfur content of the crude oil from which they originated (Girard et al. 1993).

Until 1994, the maximum sulfur content allowed in diesel fuel specifications was about 0.2 to 0.3%. Attaining this level required the use of processes that operated under relatively low-severity conditions (see 2.2.2.3).

The reduction in overall sulfur content to 0.05% requires changes in the operating conditions of existing units as well as the installation of new hydrodesulfurization units (Heinrich et al. 1993). The following measures can be applied, separately or in combination:

- **increasing the reaction temperature**
- **prolonging the residence time**; in effect, obtaining a higher desulfurization rate implies a lower gas-hourly space velocity (GHSV) through the catalyst (see Fig. 4.50); therefore, the size of the equipment must be increased to maintain the same production capacity
- **increasing the hydrogen partial pressure**
- **using more-active catalysts**

Judgment must be used in the application of these procedures, to limit costs and to control the combination of diesel fuel properties.

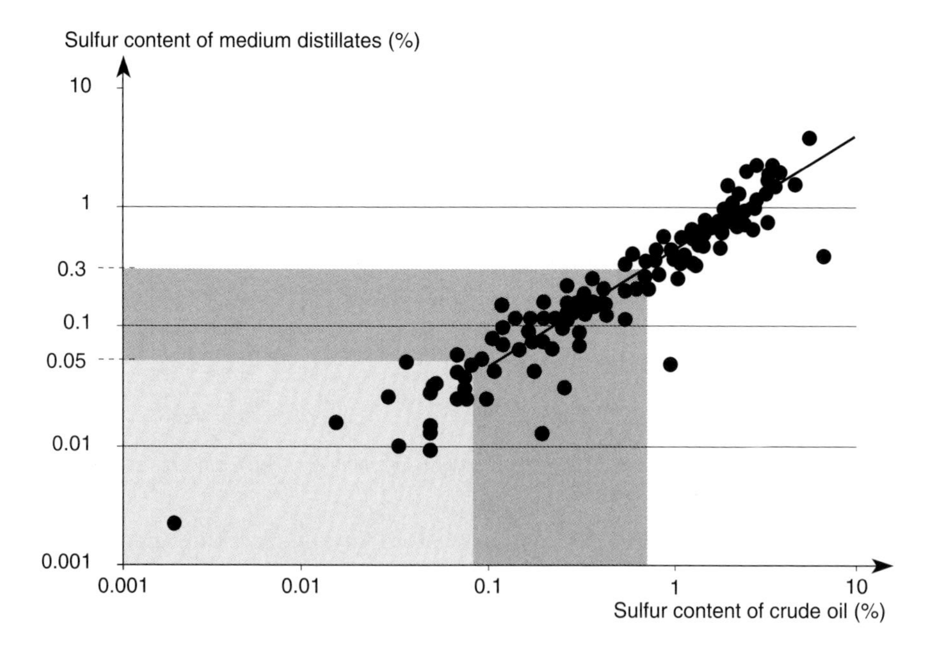

Figure 4.49 *Correlation between the sulfur content of straight-run diesel fuel and the sulfur content of crude oil.*

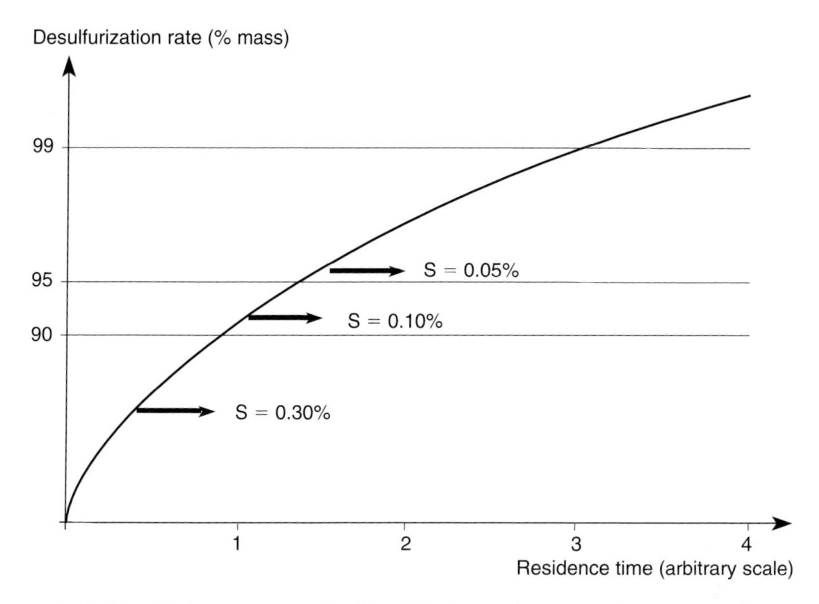

Figure 4.50 *Desulfurization rate of a diesel fuel as a function of reactor residence time.*

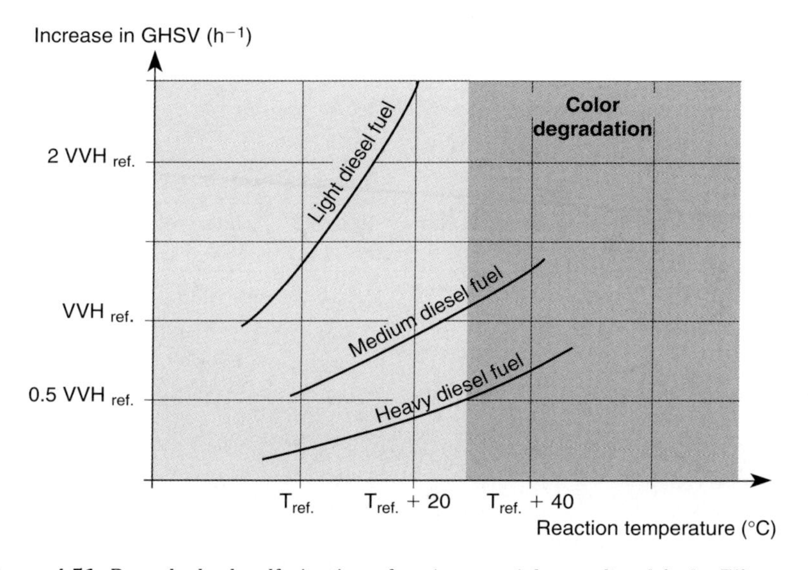

Figure 4.51 *Deep hydrodesulfurization of various straight-run diesel fuels. Effect of reaction temperature on the color and on the GHSV required to obtain a diesel fuel with a sulfur content of 500 ppm.*

Thus, Fig. 4.51 shows that increasing the temperature can cause a deterioration in the **color of diesel fuel**, which renders it unacceptable and out of specification. However, this change can be countered by increasing the hydrogen partial pressure (see Fig. 4.52).

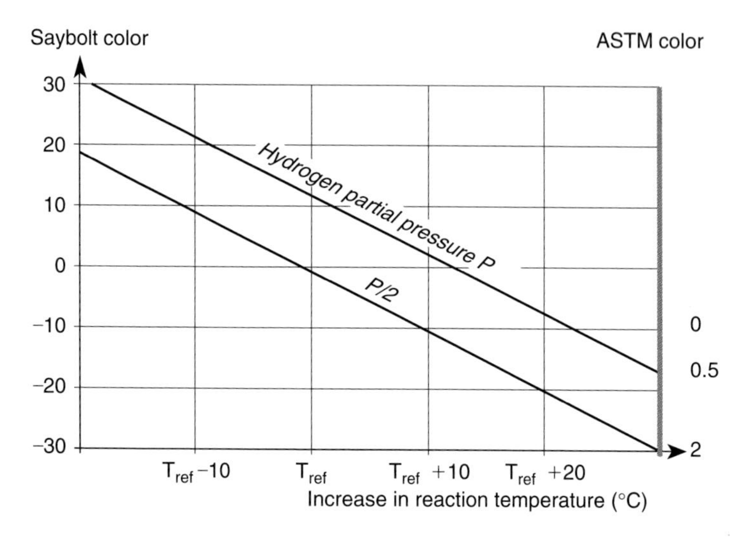

Figure 4.52 *Deep hydrotreatment of diesel fuels. Effect of temperature and hydrogen partial pressure on the color of the output.*

Table 4.15 *Characteristics of various diesel cuts before and after deep desulfurization.*

	Characteristics	Diesel fuel cuts			
		Straight Run (SR)	80% SR 20% LCO	50% SR 50% LCO	LCO
Feedstock	Sulfur (% mass)	1.31	1.16	1.86	2.24
	Nitrogen (ppm)	70	220	540	940
	Aromatics (% mass)	26.7	39.2	54.3	78.4
	Cetane rating	53	47	39	20
Operational variables	Hydrogen partial pressure (bar)	20	20	20	35
	GHSV (h^{-1})	Ref.	0.85 · Ref.	0.67 · Ref.	0.5 · Ref.
	Hydrogen consumption (m^3/m^3)	37	45	70	150
	Deep hydrodesulfurization (sulfur content = 0.05% [mass])				
Products	Nitrogen (ppm)	45	150	400	50
	Aromatics (% mass)	25.7	39.0	52.5	71
	Cetane rating	56	51	42	32

Table 4.16 *Detailed characteristics of a straight-run diesel fuel before and after deep hydrodesulfurization.*

Characteristics	Feedstock (SR)	Products*	
		Scenario 1	Scenario 2
Density at 15°C (kg/dm^3)	0.846	0.834	0.833
Viscosity (mm^2/s)			
20°C	6.14	5.58	5.52
50°C	2.99	2.81	2.79
Distillation (°C)			
IP	217	221	221
50%	294	285	285
90%	341	329	329
FP	358	350	349
Sulfur content (% mass)	1.31	0.070	0.015
Nitrogen content (ppm)	65	60	54
Cetane rating	54.8	56.4	57.6
Chemical composition (% mass)			
Paraffins	40.7	43.2	44.0
Naphthenes	32.6	31.1	30.9
Monoaromatics	11.0	18.2	17.6
Di-aromatics	7.4	5.9	6.5
Tri-aromatics	1.0	0.5	0.4
Benzothiophenes	5.4	0.5	0.4
Dibenzothiophenes	1.9	0.6	0.2

* The treatment severity (temperature) is higher for scenario 2.

Also, for the same type of feedstock (for example, straight-run diesel fuel), hydrodesulfurization becomes more difficult as the molecular weight of the fraction increases (see Fig. 4.51).

Catalytic-cracked diesel fuels, or light cycle oils (LCO), which are becoming more available on the market, are equally good candidates for the diesel fuel pool. However, desulfurizing them is very difficult. This is why they are more often pre-mixed with straight-run (SR) products. Table 4.15 lists the various characteristics of some SR-LCO blends and the treatment conditions required in each case to obtain a final sulfur content of 0.05%.

An example of the changes in diesel-fuel characteristics and composition after deep hydrodesulfurization is shown in Table 4.16. As predicted, there is a detectable reduction in the **benzothiophene** and **dibenzothiophene** content. The breaking of ring structures results in an increase in the monoaromatic content. As to di- and tri-aromatics, their concentration tends to diminish slightly.

These changes are accompanied by a **slight increase in cetane rating**. Admittedly, the reduction in diesel-fuel sulfur content from 0.3 to 0.05% results in an increase of **1 to 2 numbers** in the cetane rating of the whole pool.

4.11 New trends in the refining of high-quality diesel fuels

The medium- and long-term trend in diesel-fuel quality is to simultaneously reduce the **sulfur** and the **aromatic** content. This change is already occurring in the Unites States where the maximum aromatic content is set at 10% in the state of California. The origin of this regulation and its link to environmental concerns are explained subsequently (see 5.13).

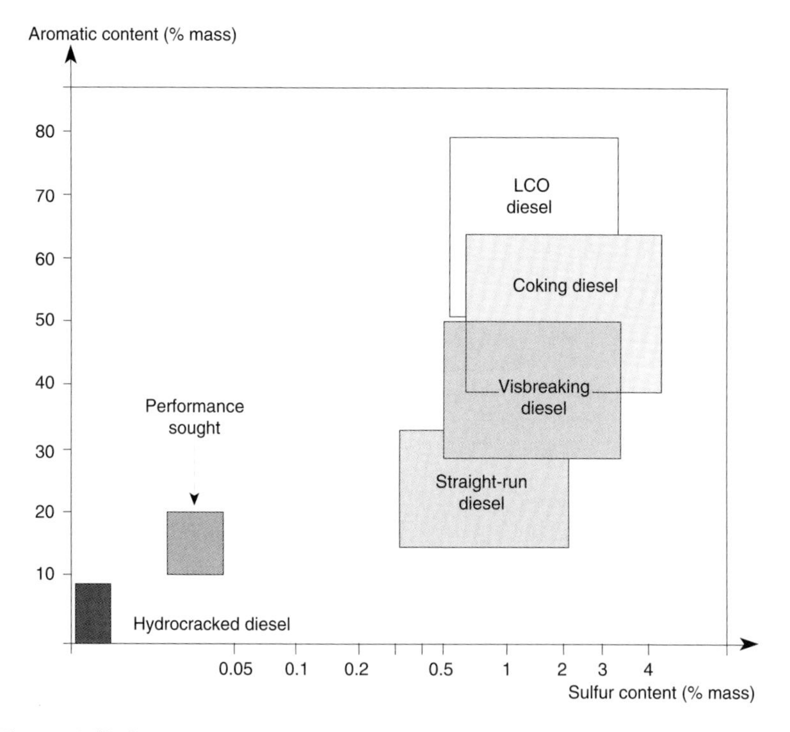

Figure 4.53 *Status of some medium cuts with respect to sulfur and aromatic content.*

The status of some medium cuts with respect to sulfur and aromatic content is shown in Fig. 4.53. It shows the excellent properties of hydrocracked products and it shows the need for the supplementary treatment of other fractions.

4.11.1 Hydrocracking

The hydrocracking process has already been described (see 2.2.2.2c). This section simply includes the reminder that it is a **very-flexible process** with respect to feedstocks (light or heavy vacuum-distilled diesel, LCO, deasphalted vacuum residue) and the distribution of products (kerosene, diesel fuel, lubricant basestocks). Table 4.17 lists some examples of the characteristics of the hydrocracked diesel fuels. In all cases, the sulfur and nitrogen contents are very low (a few ppm) and the aromatic concentration is about 10%. The cloud point is very acceptable at −10°C or lower. Finally, the cetane rating is between 55 and 65. The lower values (55) coincide with the relatively high naphthene content.

Table 4.17 *Examples of the composition and characteristics of hydrocracked diesel fuels.*

Operating conditions and product characteristics	Diesel fuels					Deasphalted vacuum residue
	Light vacuum distillate	Heavy vacuum distillate (HVGO)			70% HVGO 30% LCO	
Type of operation						
Hydrogen pressure	Medium	Low	High	High	High	High
Number of stages	1	1	1	2	1	1
Diesel fuel yield (% mass)	72	22	28	45	38	27
Product characteristics						
Density (kg/dm³)	0.810	0.835	0.825	0.825	0.825	0.827
Sulfur content (ppm)	< 1	< 5	< 5	< 5	< 5	< 5
Nitrogen content (ppm)	< 1	< 5	< 5	< 5	< 5	< 5
Aromatic content (% mass)	9	13	11	12	12	12
Cetane rating	55	55	64	63	60	63
Cloud point (°C)	−12	−18	−23	−12	−15	−19

Hence, the technical appeal of hydrocracking is undeniable. The counter point is the **high cost**, which includes equipment investment (high pressures) and operating costs (hydrogen consumption).

4.11.2 Hydrotreatments

These processes consist of severe **hydrogenation** of the **aromatic constituents** present in diesel fuel (Henirich et al. 1994). At the same time, an almost complete desulfurization and denitrification occurs.

Table 4.18 lists the operating conditions required to obtain aromatic contents of 20% and 10% for various types of feedstocks: SR diesel fuel, LCO, or a blend of the two. As indicated by the data, the hydrogenation of LCO is a difficult process, which is confirmed by the data in Table 4.19.

Table 4.18 *Hydrodearomatization of various diesel fuel cuts*

	Characteristics	Diesel cuts					
		Straight Run (SR)		80% Straight Run 20% LCO		LCO	
Feedstock	Aromatics (% mass)	26.7		39.2		78.4	
Operating variables	Hydrogen partial pressure (bar)	40	65	55	80	110	140
	GHSV (h^{-1})	Ref.	Ref.	Ref.	Ref.	0.5 · Ref.	0.5 · Ref.
	Hydrogen consumption (m^3/m^3)	61	90	116	135	375	415
Products	Sultur (ppm)	25	10	22	20	< 20	< 20
	Nitrogen (ppm)	< 1	< 1	< 1	< 1	< 1	< 1
	Aromatics (% mass)	20	10	20	10	20	10
	Cetane rating	60	63	55	56	41	41

Hydrotreatments can lead to cetane ratings greater than 60, if the feedstock has a high SR diesel-fuel content. Conversely, severe hydrotreatment of LCO does not result in cetane ratings greater than a range of 40 to 42 (in exceptional cases, 45 to 50 can be achieved with very-active catalysts). This is explained by the high output of **naphthenes**. In fact, hydrogenation stops at this stage and it is extremely difficult to **open the naphthenic rings**, which would lead to more inflammable structures.

The detailed changes in composition and characteristics that result from severe hydrotreatment are listed in Table 4.20.

Table 4.19 *Example of the hydrotreatment of LCO. Changes in the characteristics of hydrotreated diesel fuel as a function of operating conditions.*

Operating conditions	Feedstock				
	(1)	(2)	(3)	(4)	(5)
Hydrogen pressure (bar)	40	60	100	100	100
Temperature (°C)	360	360	360	380	380
GHSV (h^{-1})	2	2	2	2	0.5
Chemical consumption of hydrogen (m^3/m^3)	180	230	270	290	410
(% mass)	1.8	2.3	2.7	2.9	4.1
Hydrodesulfurization rate (% mass)	93.0	96.4	98.3	99	99
Hydrodenitrification rate (% mass)	75.0	98.3	99.5	99.5	99.5

Characteristics	Products				
	(1)	(2)	(3)	(4)	(5)
Density (kg/dm^3)	0.912	0.907	0.896	0.886	0.862
Viscosity at 20°C (mm^2/s)	4.1	3.8	3.67	3.58	3.44
Sulfur (% mass)	0.18	0.09	0.04	0.02	0.003
Total nitrogen (ppm)	100	10	3	2	1
Measured cetane number	22	24	27	29	40
Composition by mass spectrometry					
Paraffins (% mass)	10.5	12.5	12.3	11.0	8.0
Naphthenes (% mass)	9.5	12.5	22.3	35.0	66.0
Aromatics (% mass)	80.0	75.0	65.4	54.0	26
consisting of Monoaromatics	55	55	55.2	45.7	23
Polyaromatics*	12	6.0	4.6	4.9	1.7

* Other aromatics are diaromatics and benzothiophenes.

4.11.3 Non-conventional means

This section describes three other possible means of obtaining diesel fuels that are free of both sulfur and aromatics. The processes are oligomerization, Fischer-Tropsch synthesis (Chapter 2), and the production of various oxygenated organic compounds. Tables 4.21 and 4.22 list the characteristics of some products obtained using these processes.

Table 4.20 *Comparison of characteristics for a diesel fuel before and after hydrotreatment.*

Characteristics	Feedstock 80% SR 20% LCO	Products	
		P: 50 bar CSV: 1 h^{-1}	P: 75 bar CSV: 0.5 h^{-1}
Density at 15°C (kg/dm^3)	0.862	0.838	0.827
Viscosity (mm^2/s)			
20°C	5.55	5.12	4.90
50°C	2.76	2.63	2.54
Distillation (°C)			
IP	214	213	212
50%	288	278	275
90%	332	324	324
FP	353	345	347
Sulfur content (ppm)	11 600	22	4
Cetane rating	49	54	60
Chemical composition (% mass)			
Paraffins	36.5	36.9	41.5
Naphthenes	24.3	37.7	51.8
Monoaromatics	14.2	20.2	6.0
Di-aromatics	15.4	4.5	0.7
Tri-aromatics	1.8	0.4	0.0
Benzothiophenes	5.4	0.3	0.0
Dibenzothiophenes	2.4	0.0	0.0

Table 4.21 *Examples of diesel fuels from oligomerization and Fischer-Tropsch synthesis.*

Characteristics	Type of diesel fuel	
	Diesel fuel from oligomerization	Diesel fuel from Fischer-Tropsch
Density at 15°C (kg/dm^3)	0.781	0.761
Viscosity at 20°C (mm^2/s)	2.55	2.56
Distillation (°C)		
IP	156	205
FP	342	270
Sulfur content (ppm)	0.6	1.2
Nitrogen content (ppm)	0.25	0.5
Cetane rating	35	83
Chemical composition (% mass)		
Paraffins	85.6	100
Uncondensed naphthenes	9.5	0.0
Condensed naphthenes	4.9	0.0
Aromatics	0.0	0.0
Cloud point (°C)	< -48	–9

Table 4.22 *Characteristics of some oxygenated organic compounds appropriate for eventual use as diesel fuels.*

Name	Chemical formula	Density at 20°C (kg/dm³)	Boiling point (°C)	Cetane rating
Dimethylether	CH_3-O-CH_3	0.630	<0	50
Dimethoxymethane	$CH_3-O-CH_2-O-CH_3$	0.861	42	52**
Di-*n*-pentylether	$C_5H_{11}-O-C_5H_{11}$	0.787	187	109**
Di-*n*-pentoxymethane	$C_5H_{11}-O-CH_2-O-C_5H_{11}$	0.841	218	97**
Dimethylglycol	$CH_3-O-CH_2-CH_2-O-CH_3$	0.870	85	53
Dimethyldiglycol	$CH_3-O-CH_2-CH_2-O-CH_2-CH_2-O-CH_3$	0.950	162	>70
Ethyldiglycol*	$CH_3-CH_2-O-CH_2-CH_2-O-CH_2-CH_2-OH$	0.990	202	41
Ethyltriglycol*	$CH_3-CH_2-O-CH_2-CH_2-O-CH_2-CH_2-O-CH_2-CH_2-OH$	1.021	255	>70
Butyl-glycol acetate	$CH_3-COO-CH_2-CH_2-O-(CH_2)_3-CH_3$	0.942	156	41

* Low solubility in diesel fuel.
** Behavior when blended.

4.11.3.1 Oligomerization

The feedstock for oligomerization consists of light olefins (C_3, C_4) that are **dimerized and trimerised** by heterogeneous catalysis. The resulting products, which have a high medium olefin content (C_8, C_{12}), are then hydrotreated and separated into different fractions: gasoline, kerosene, and diesel fuel. The density of the diesel fuel cut is between 0.780 and 0.820 kg/dm³; the cetane rating is low (35 to 45) due to the presence of highly-branched isoparaffins, but the pour point is excellent (below –30°C).

4.11.3.2 Fischer-Tropsch synthesis

From all points of view, Fischer-Tropsch synthesis delivers a remarkable diesel fuel fraction: exclusive paraffinic composition with predominantly linear or lightly-branched chains, high cetane rating (80 to 85), and cloud point below –10°C.

As indicated previously, Fischer-Tropsch synthesis cannot become economically viable unless there is major increase in crude oil cost (on the order of $25/bbl).

4.11.3.3 Oxygenated organic compounds

As yet, there are no **oxygenated** constituents for diesel fuel that act in the same capacity as the alcohols (methanol, ethanol) or ethers (MTBE, ETBE) that are often added to gasoline.

However, this could be an interesting avenue to pursue for improving both the cetane rating and exhaust emissions (see 5.13.3).

Among the products that have been mentioned, the primary one is **dimethylether** (DME) (Sorenson 1995), which is easily created by the dehydrogenation of methanol:

$$2\,CH_3OH \rightleftarrows CH_3-O-CH_3 + H_2O$$

DME is a gas at ambient temperature and pressure, but it can be liquefied under low pressures of about 5 bar. It burns easily in a diesel engine and it produces low emissions, due to a cetane rating of about 50 and the presence of oxygen within the chemical structure.

Other oxygenated products that are potential candidates for mixing with diesel fuel have been proposed and tested (Pecci et al. 1991; Giavazzi et al. 1991) . Worthy of mention are di-*n*-pentylether (DNPE), di-*n*-pentoxymethane (DNPM), dimethoxymethane (DMM) and dibutoxymethane (DBM). Other compounds like ether, polyether, ether-alcohol, ester (acetate), which are described in summary in Table 4.22, have also been considered.

The **means of obtaining them are complex** and they involve, for the most part, an alcohol reaction with ethylene oxide. The production costs are **much higher** than those of conventional refining. In addition, most of these products are not very soluble in diesel fuel.

4.12 Characteristics of the finished products

Despite the establishment of relatively strict specifications, the composition and characteristics of commercial diesel fuel can vary considerably, which leaves some flexibility for the refiner during formulation.

To summarize, there are three major types of diesel fuel available in the world today, with the following characteristics:

- **European diesel fuels**, which are desulfurized to 0.05%, have moderate cold-temperature behavior (a pour point of about 0 to –8°C), and they usually have an aromatic content of 20 to 30%.
- **American diesel fuels** (United States and Canada) also have a low sulfur content (0.05%), but they have a low aromatic content (for example, less than 10% in California). They have very-good low-temperature behavior (cloud point of about –15°C) and a low final distillation point (330 to 340°C).
- **Special diesel fuels**, which are sold in the Scandinavian areas of Europe (Sweden, Norway, ...), have very-low sulfur content (10 to 50 ppm) and low aromatic content (5 to 20%). Their use is recommended because they produce very-low emissions. The distillation interval of these fuels is narrow (final point less than 310°C) and their cloud point is very low (–25°C), although the cetane rating is less than 50; in fact, these products are very close to **jet fuel** in a number of ways.

Table 4.23 shows examples of the characteristics of the three types of diesel fuels described above.

It is clear that the more specialized a fuel becomes, the more difficult it is to obtain it in **large quantities**. Therefore, in those parts of the world where diesel fuel consumption is high (Western Europe in particular), a compromise must be established in the medium term between the market demands for **quality** and **quantity**. In this regard, any massive increase in diesel fuel production, while maintaining high levels of desulfurization and dearomatization, would require a huge refining effort.

Table 4.23 *Diesel fuel characteristics from various countries.*

Characteristics	Type of diesel fuel				
	Swedish		American	Californian	European
	Classe 1	Classe 2			
Density at 15°C (kg/dm³)	0.813	0.818	0.842	0.835	0.838
Sulfur (ppm)	1.3	5.9	310	190	470
Aromatics (% vol)	3.9	16.5	10	8	30
Cetane rating	52.3	50.9	46.1	49.4	51.2
Distillation temperature (°C)					
Initial point	187	185	175	171	181
T 95	280	286			
Final point			340	351	356

Bibliography
(Volumes 1 and 2)

AEGPL (Association Européenne des Gaz de Pétrole Liquéfiés) (1994), "Assemblée générale. Statistiques 1993", Sevilla.

Anderson, C.L., O.A. Uyehara and P.S. Myers (1982), "An in-situ determination of the thermal properties of combustion chamber deposits". *SAE 820071.*

Andrews, J.W., P.H. Bigeard, A. Billon, R. Boulet, J.-C. Guibet and G. Heinrich (1989), "Future processing requirements? Heavier jet fuels, higher diesel fuel quality". *NPRA Annual meeting,* San Francisco.

Anon. (1990), "Moteurs Diesel". *Éditions Techniques pour l'Automobile et l'Industrie.*

Antos, G.J., A.M. Aitani and J.M. Parera (1994), *Catalytic naphta reforming.* Editions M. Dekker, New York.

Asian Autotechnical Report (1993), "Engine: Toyota's new generation D-4 engine", Vol. 206, pp. 24-25.

Asian Autotechnical Report (1994), "Engine: Mitsubishi's V6, 2,5-l lean burn engine", Vol. 206, pp. 19-20.

Asian Autotechnical Report (1995), "Mitsubishi's V6, 2,5-l lean burn engine", AI Publishing Company, Vol. 206.

ASTM (1958), *Knocking characteristics of pure hydrocarbons.*

ASTM D4865 (1995), "Standard guide for generation and dissipation of static electricity in petroleum fuel systems". *Annual Book of ASTM Standards.*

Aumont, B., A. Jaecker-Voirol, B. Martin and G. Toupance (1996), "Tests of some reduction hypothesis made in photochemical mechanisms". *Atmospheric Environment,* Vol. 30, No. 12, pp. 2061-2077.

Automotive Engineer (1993), "Engine-control-ignition/fuel injection management, petrol and diesel" Vol. 18, No.1.

Barker, D.A., L.M. Gibbs and E.D. Steinke (1988), "The development and proposed implementation of the ASTM Driveability Index for motor gasoline". *SAE 881668.*

Barnes, G.K., J.P. Liddy and E.G. Marshall (1987), "The ignition quality of residual fuels". *CIMAC paper 25,* Warsaw.

Batts, B.D. and A. Zuhdan Fathoni (1991), "A literature review on fuel stability studies with particular emphasis on diesel oil". *American Chemical Society, Energy & Fuels,* Vol. 5, No. 1.

Baudino, J.H. and F.L. Voelz (1993), "Emissions testing of three Illinois E 85 demonstration fleet vehicles". *10th International Symposium on Alcohol fuels,* Colorado Springs, Co.

Benson, J.D. and P.A. Yaccarino (1986), "The effects of fuel composition and additives on multiport fuel injector deposits". *SAE 861533.*

Benson, J.D., H.A. Bigley and J.L. Keller (1971), "Passenger car driveability in cool weather". *SAE 710138.*

Berlowitz, P.J. and H.S. Homan (1991), "Hardware effects on intake valve deposits". *SAE 912381.*

Bertoli, C., N. Del Giacomo, B. Iorio and M.V. Prati (1992), "Initial results on the impact of automotive diesel oil on unregulated emissions in DI light duty diesel engine". *SAE 922189.*

Bertrand, B. and C. Delarue (1992), "Véhicules électriques hybrides" *Programme de Recherche et de Développement pour l'Innovation et la Technologie dans les Transports Terrestres (PREDIT),* Éditions Paradigme, Caen.

Betts, W.E., S.A. Floysand and F. Kvinge (1992), "The influence of diesel fuel properties on particulate emissions in European cars". *SAE 922190.*

Beurdouche, P. and L. Monteiro (1992), "Fuel quality label: the three-year experience of french car manufacturers". *11th European Automotive Symposium AGELFI,* Sorrento.

Bidault, M. and P. Fléchon (1987), "Influence de la qualité du gazole sur les performances et les émissions des moteurs de véhicules industriels". Congrès International *"Moteurs Diesel pour Véhicules Automobiles et Utilitaires",* Société des ingénieurs de l'automobile, Lyon.

Bigeard, P.H., A. Espinosa and P. Marion (1994), "L'hydrocraquage: une solution simple pour produire des distillats moyens de haute qualité". *Revue de l'Institut Français du Pétrole,* Vol. 49, No. 5, p. 529.

Billon, A., J. Bousquet and J. Rossarie (1988), "HYVAHL® F and T processes for high conversion and deep refining of residues". *National Petroleum Refiners Association (NPRA), paper No. AM-88-62, Annual Meeting,* San Antonio, TX.

Bogdanoff, M.A. and S. Hada (1991), "Overview of on-board diagnostic systems used on 1991 California vehicles". *SAE 912433.*

Bosch (1985), "Pompe d'injection distributrice type VE". *Cahier Technique,* Éditions Delta Press, p. 13.

Bosch (1985), "Système combiné d'allumage et d'injection d'essence à régulation lambda-motronic". *Cahier Technique.* Éditions Delta Press.

Bosch (1987), "Mémento de technologie Automobile". *Cahier Technique,* Éditions Delta Press, pp. 400-403.

Bosch (1991), "Panorama de la technique d'injection diesel". *Cahier Technique,* Éditions Delta Press, pp. 22-26.

Bosch (1991), "Système électronique d'injection d'essence à régulation de richesse-mono-jetronic". *Cahier Technique,* Éditions Delta Press.

Brabetz, L., M. Siedentrap and G. Schmitz (1991), "The Siemens alcohol fuel sensor. Concept and results". *9th International Symposium on Alcohol Fuels, ISAF,* Florence.

Breitwieser, K., J. Henschel and G. Arnold (1993), "Der neue Vierventilmotor mit 1,6 l Hubraum von Opel". *MTZ Motortechnische Zeitschrift,* 54.

Brinkman, N.D. et al. (1983), "Effect of fuel volatility on driveability at low and intermediate ambient temperatures". *SAE 830593.*

Brioult, R. (1980), *Le moteur à essence.* Éditions ETAI.

Broadbent, S. (1994), "Experiences with commissioning and operation of biogas treatment plants for vehicle refuelling: Lille and Tours". *International Conference and Exhibition on Natural Gas Vehicles,* Toronto.

Burk, P.L. et al. (1995), "Cold start hydrocarbon emissions control". *SAE 950410.*

Bush, K.C., G.J. Germane and G.L. Hers (1985), "Improved utilization of nitromethane as an internal combustion engine fuel". *SAE 851130.*

Callison, J.C. (1987), "Octane number requirements of vehicles at high altitude". *SAE 872160.*

Caprotti, R., C. Bovington, W.J. Fowler, J. and M.G. Taylor (1992), "Additive technology as a way to improve diesel fuel technology". *SAE 922183.*

Carter, S., J. Heenann and B. Williamson (1992), "The GFI system-Prototype to product". *3rd International Conference on Natural gas Vehicles,* Göteborg.

Carter, W. (1994), "Development of ozone reactivity scales for volatile organic compounds". *Journal of the Air & Waste Management Association,* Vol. 44, p. 881.

CCPCS (Commission Consultative pour la Production des Carburants de Substitution) (1992). *Rapports des groupes de travail.* Ministère français de l'industrie, Paris

CEC (1981), "Tentative test method n° CEC F03 T81. Evaluation of gasolines with respect to maintenance of carburettor cleanliness". *Coordinating European Council (CEC).*

Cedra, C. and D. Gauthier (1990), *"Les moteurs diesel – Technologie et fonctionnement".* CEMAGREF, Paris.

CFBP (Comité Français du Butane et du Propane) (1987), "GPL (Gaz de Pétrole Liquéfiés) – Documentation Technique".

CFBP (Comité Français du Butane et du Propane) (1995), "L'industrie française des GPL Rapport annuel".

Charbonnier, M.A. and M. Andres (1993), "A comparative study of gasoline and diesel passenger car emissions under similar conditions of use". *SAE 930779.*

Chatin, L. (1992), "Using ETBE as a gasoline blending component. ElF-Aquitaine's Experience". *EFOA – Fifth Conference,* Bruxelles.

Chaumette, P. (1998), "Gas to liquid conversion - Basic features and competitors". *Pétrole et Techniques,* No. 415, Paris.

Chauvel, A., G. Lefebvre and L. Castex (1985), *Procédés de pétrochimie – Caractéristiques techniques et économiques.* Cours de l'ENSPM. Éditions Technip, Paris.

Choate, P.J. and J.C. Edwards (1993), "Relationship between combustion chamber deposits, fuel composition and combustion chamber deposit structure". *SAE 932812.*

Clark, G.H. (1988), *Industrial and marine fuels,* Butterworths, London.

Claude, J.-M., D. Herrier and J.-C. Guibet (1988), "Détermination du délai d'inflammation par voie optique sur moteur diesel d'automobile". *9th European Automotive Symposium AGELFI,* Bordeaux.

Connangle, J.N., P. Le Bitoux and B. Martel (1993), "Les Stations GNV: état de l'art". *Revue Gaz d'aujourd'hui,* p. 423.

Cotte, H. (1995), "Développement d'une technique spectroradiométrique pour la détermination expérimentale des fréquences de photolyse troposphérique: application au bilan photostationnaire de radicaux libres lors des campagnes Field VOC". *Thèse de Doctorat,* Paris VII.

Cox, F.W. (1979), "Physical properties of gasoline-alcohol blends". *US Department of Energy,* Bartlesville, OK.

CRC (1976), "Driveability performance of late model passenger cars at high ambient temperatures". *Coordinating Research Council (CRC),* No. 490.

CRC (1978), "Driveability performance of 1977 passenger cars at intermediate temperatures". *Coordinating Research Council (CRC),* No. 499.

CRC (1983), *Handbook of Aviation Fuel Properties. Coordinating Research Council (CRC),* No. 530.

Daly, D.T. *et al.* (1993), " A diesel regeneration system using a copper fuel additive". *SAE 930131.*

Damin, B., A. Faure, J. Denis, B. Sillion, P. Claudy and J.M. Letoffe (1986), "New additives for diesel fuels: cloud-point depressants". *SAE 861527.*

Dautray, R. et al. (1995), *Environnement et IFP.* Publication de l'Institut Français du Pétrole, Paris.

De Boer, C.D., J. Stokes, M.A. and T.H. Lake (1993) "Advanced gasoline combustion systems for fuel economy and emissions". *Institution of Mechanical Engineers (IME),* London.

De Soete, G. (1976), *Aspects fondamentaux de la combustion en phase gazeuse.* Éditions Technip. Paris.

Decker, G., H. Heinrich, U. Kammann and D. Steinke (1991), "The Volkswagen multi-fuel concept. A concept for flexible methanol/gasoline operation, development and results". *9th International Symposium on Alcohol Fuels, ISAF,* Florence.

Decker, G., H. Heinrich, M. Kröll and H. Loek (1993), "Field experience and progress of Volkswagen's multi-fuel vehicles". *10th International Symposium on Alcohol Fuels,* Colorado Springs, Co.

Degobert, P. (1995), *Automobile and pollution.* Éditions Technip, Paris.

Degobert, P. (1994), "Pollution atmosphérique". *Techniques de l'Ingénieur,* B 2710 – B 2714, Paris.

Demel, H., D. Stock and R. Bauder (1991), "2,51. Audi Turbodieselmotor mit Direckteinspritzung -Schadstoffarm nach Anlage 43-". *MTZ Motortechnische Zeitschrift 52,* p. 9.

Dementhon, J.B., B. Martin, P. Richards, M. Rush, D. Williams, L. Bergonzini and E. Morelli (1995), "Novel additive for particulate trap regeneration". *SAE 952355.*

Derwent, R. and M.E. Jenkin (1990), "Hydrocarbons and the long range transport of ozone and PAN across Europe". *AERE Report,* R 13816.

Desaulty, M. (1994), *La combustion appliquée aux moteurs à flux continu. Cours ENSPM,* Paris.

Descales, B., D. Lambert and A. Martens (1989), "Détermination des nombres d'octane RON et MON des essences par la technique proche infrarouge". *Pétrole et Techniques,* No. 349.

Dessus, B. (1993), "L'auto-condamnation: un exercice de perspective mondiale à long terme pour l'automobile". *Les cahiers du Clip No. 1,* Paris.

Diederichs, F. (1988), "A study of the influence of engine oil consumption, lead and fuel detergent additives on the engine ORI". *9th European Automotive Symposium AGELFI,* Bordeaux.

Dockery, D.W., J. Schwartz and J.D. Spengler (1992), "Air pollution and daily mortality: associations with particulates and acid aerosols". *Environmental Research,* 59(2), pp. 362-373.

Dockery, D.W. and C.A. Pope (1994), "Acute respiratory effects of particulate air pollution". *Annual Review of Public Healith,* 15, pp. 107-132.

Douaud, A. and P. Eyzat (1978), "Four-octane-number method for predicting the antiknock behavior of fuels and engines". *SAE 780080.*

Douaud, A. (1980). "Systèmes d'allumage asservi au cliquetis – Principe et perspectives d'économie d'énergie". *Ingénieurs de l'Automobile,* Paris.

Douaud, A (1981), "Analyse et optimisation du rendement à charge partielle des moteurs à 4 temps". *Pétrole et Techniques,* No. 281, Paris.

Douaud, A. (1994), "Engines and fuels for tomorrow". *SIA International Congress: The spark ignition engine: what challenges for the year 2000 ?* Paris.

Dumon, R., J.-C. Guibet and Y. Portas (1984), *Le méthanol: réalités et perspectives.* Éditions Masson, Paris.

Dupont, P. and J.-C. Griesemann (1995), "LPG liquid injection demonstration vehicles". *European Conference and Exhibition on LPG: a Clean and Efficient Motor Fuel,* Maastricht.

Durand, J.P., Y. Boscher and N. Petroff (1987), "Automatic gas chromatographic determination of gasoline components. Application to octane number determination". *Journal of Chromatography,* No. 395, p. 229.

Duret, P. and J.F. Moreau (1990), "Reduction of pollutant emissions of the IAPAC two-stroke engine with compressed air assisted fuel injection". *SAE 900801.*

Duret, P. (1993), "A new generation of two-stroke engines for the future", *IFP International Seminar.* Éditions Technip, Paris.

Ebert, L.B. (1985), *Chemistry of engine combustion deposits: Literature review.* Plenum Press.

Eckbreth, A.C. (1988), *Laser diagnostics for combustion temperature and species.* Editors A. K. Gupta, D.G. Lilley, Abacus Press.

Eilers, J., S.A. Posthuma and S.T. Sie (1990), "The Shell Middle Distillate Synthesis Process (SMDS)". *Catalysis Letters* 7, pp. 253-270.

Elitchegaray, C. (1990), "Problèmes liés à l'ozone troposphérique: effet de serre, pluies acides...". *Pollution Atmosphérique,* pp. 427-430.

Elsbett, L. (1983), "Alternative fuels on a small high speed turbocharged D.I. diesel engine". *SAE 830556.*

Eltinge, R.S. (1968), "Fuel-air ratio and distribution from exhaust gas analysis". *SAE 680114.*

Enga, B.E. (1990), "Development of a simplified diesel particulate filter. Regeneration system for transit buses". *SAE 900326.*

EPA (Environmental Protection Agency) (1993), "Regulation of fuels and fuel additives: standards for reformulated and conventional gasoline". 40 CFR part 80.

Ethyl (1983), "Ethyl MMT – Manganese octane improver for leaded and unleaded gasolines". *Ethyl International,* Petroleum Chemicals Division Publication.

Eyzat, P. and J.-C. Guibet (1968), "A new look at nitrogen oxides formation in internal combustion engines". *SAE 680124.*

Falk, K. (1983), "The development of a european cold weather driveability test procedure for motor vehicles with spark ignition engines". *SAE 831754.*

Faure, E. and X. Montagne (1997), "Octane requirement increase: assessment methodology and influence of engine running parameters". *SAE 972933.*

Fiskaa, G., K. Langnes, O. Toff and G. Ostvold (1985), "Some aspects on utilizing modern marine diesel fuels". *CIMAC,* Oslo.

Forti, L., X. Montagne, P. Marez and J.P. Pouille (1994), "The particulate number: a diesel engine test method to characterize a fuel's tendency to form particulates". *SAE 942021.*

Fortnagel, M., P. Moser and W. Pütz (1993), "Die neuen Vierventil-Dieselmotoren von Mercedes-Benz Entwicklung von Verbrennung und Abgasreinigungsystem", *MTZ Motortechnische Zeitschrift 54,* p. 392.

Foy, H. (1994), *La suralimentation appliquée aux turbocompreseurs.* Éditions ETAI, Paris.

Gairing, M. (1986), "Zur Qualitaet der Ottokraftsoffe aus den Sicht der Automobilindustrie: Vermeidung von Ablagerundgen auf Einlassventilen". *Mineroelrundshau,* Vol. 34, No. 11.

Gatellier, B., J. Trapy, D. Herrier, J.M. Quelin and F. Galliot (1992), "Hydrocarbon emissions of SI engines as influenced by fuel absorption-desorption in oil films". *SAE 920095.*

Gautier, M.M. (1930), "Nouveaux essais comparatifs des huiles végétales et minérales dans les moteurs diesel". *La Revue des Combustibles Liquides,* Paris.

Germane, G.J. (1985), "A technical review of racing fuels". *SAE 852129.*

GFC (1995), "Méthode d'essais d'agrément de conduite à chaud (véhicules à injection)". *Groupement Français de Coordination,* Paris.

Giavazzi, F., D. Terna, R. Patrini, F. Ancillotti, G.C. Pecci, R. Tréré and M. Benelli (1991), "Oxygenated diesel fuels. Part 2 -Practical aspects of their use". *9th International Symposium on Alcohol Fuels, ISAF, Firenze.*

Girard, C., J.-C. Guibet, A. Billon and X. Montagne (1993), "Technico-economic aspects and environmental impact of gas-oil desulphurization". *International Congress: state of the art. Potential.* SIA, Lyon, France.

Glatz, H.R. (1987), "The historic development, the political background and the future perspectives of motor emission control and emission control regulation in Europe". IMChE. *Conference "Vehicle emissions and their impact on European air quality",* paper C 358/87, London.

Glavincevski, B., O.L. Gülder and L. Gardner (1984), "Cetane number estimation of diesel fuels from carbon type structural composition". *SAE 841341.*

Golovoy, A. and J. Braslaw (1983), "On board storage and home refueling options for natural gas vehicles". *SAE 830382.*

Goodacre, C. (1958), "Les antidétonants, en particulier le plomb tétraéthyle. Son passé, son présent, son avenir". *Congrès de la Société des Ingénieurs de l'Automobile* (SIA), Paris.

Goodacre, C.L., D. Foord and M. Hedde (1961), "Comparaison des effets antidétonants du PTE, du PTM et des produits intermédiaires", *publications AFTP,* Paris.

Gorse, R.A. *et al.* (1992), "The effects of methanol/gasoline blends on automobile emissions". *SAE 920327.*

Goulley, E. (1987), "Synfuel – Le procédé Mobil et la production industrielle de carburants en Nouvelle-Zélande". *Pétroles et Techniques,* No. 332, pp. 19-23.

Graham, J. (1993), "Assessment of potential health risks of MTBE – oxygenated gasolines". *10th International Symposium on Alcohol Fuels,* Colorado Springs, Co.

Griesemann, J.-C., P. Dupont, J.F. Moreau, A. Leclair et P. Krieckaert (1994), "Injection de GPL liquide monopoint et multipoint". *Congrès International S.I.A. — Le moteur à Allumage Commandé: Quels Challenges pour l'an 2000?* Paris.

Groeneveld, R. (1995), "Auto LPG: global review and criteria for success". *European Conference and Exhibition on LPG: a Clean and Efficient Motor Fuel,* Maastricht.

Guibet, J.-C. and B. Martin (1987), *Carburants et moteurs.* Éditions Technip, Paris.

Guibet, J.-C. and A. Duval (1972), "New aspects of pre-ignition in european automotive engines". *SAE 720114.*

Guibet, J.-C., B. Martin, X. Montagne, A. Petit and A. Douaud (1991), "Gasoline composition effects on exhaust emissions". *13th World Petroleum Congress, Buenos Aires.*

Guichaoua, J.L. (1983), "L'évolution des lubrifiants pour moteurs". *Pétrole et Techniques,* No. 297, Paris.

Gulati, S. (1991), "Dynamic fatigue data for cordierite ceramic wall-flow diesel filters". *SAE 910135.*

Hadded, O., J. Stokes, T.H. Lake and K.J. Pendlebury (1994), "The ultra low emissions high fuel economy gasoline vehicle of the future". *International Congress SIA: The Spark Ignition Engine: what Challenges for the Year 2000?* Paris.

Harrington, J.A. (1982). "Water addition to gasoline. Effect on combustion, emissions, performance and knock". *SAE 820314.*

Hatamura, K., T. Gotô, M. Chôshi, H. Shimuzu and H. Abe (1993), "Development of Miller cycle gasoline engine". *Jidôsha Gÿutsukai Gakujutsu Kôenkai Maezurishû,* 935, pp. 33-36, (in Japanese).

Hatano, K., K. Iida, H. Higashi and S. Murata (1994), "Multi-mode variable valve timing engine". *Automotive Engineering,* pp. 111-114.

Haycok, R.F. and R.G.F. Thatcher (1994), "Fuel additives and the environment". *GEFIC,* Bruxelles.

Hayes, T.K., R.A. White and J.E. Peters (1992), "The in-situ measurement of the thermal diffusivity of combustion chamber deposits in spark ignition engines". *SAE 920513.*

Heimrich, M. (1995), "Demonstration of lean NO_x catalytic conversion technology on a heavy-duty diesel engine". *SAE 952491.*

Heinrich, G., M. Valais, M. Passot and B. Chapotel (1991), "Mutations of world refining: Challenge and answers". *13e congrès mondial du pétrole,* Buenos Aires, Vol. 3, pp. 189-198.

Heinrich, G., S. Kasztelan and L. Kerdraon (1994), "Diesel fuel upgrading: hydroprocessing for deep desulphurization and/or aromatics saturation". *IFP Publication, Industrial Direction,* Paris.

Heinze, P., R.C. Hutcheson et al. (1994), "A review of analytical methods for the quantification of aromatics in diesel fuels". *CONCAWE,* report No. 94/58, Bruxelles.

Heinze, P., C.J.S. Bartlett, W.E. Betts, M. Booth, P. Gadd, F. Giavazzi, H. Guttmann, D.E. Hall, R.F. Mayers, R. Mercogliano and R.G. Hutcheson, "Diesel fuel emissions performance with oxidation catalyst equipped diesel passenger vehicles -Part 1-", *CONCAWE,* report No. 94/55.

Hennico, A., A. Billon, P.H. Bigeard and J.P. Périès (1993), "IFP's new flexible hydrocracking process combines maximum conversion with production of high viscosity, high VI lube stocks". *Revue de l'Institut Français du Pétrole,* Vol. 48, No. 2, p. 127.

Heywood, J.B. (1988), *Internal combustion engine fundamentals.* Mc Graw Hill Book Company, New York.

Hitomi, M., J. Sasaki, K. Hatamura and Y. Yano (1995), "Mechanism of improving fuel efficiency by Miller cycle and its future prospect". *SAE 950974.*

Hollemans, B. and P. Schmal (1992), "Light and heavy duty engine equipment developments within the Netherlands for gaseous fuel usage". *3rd International Conference on Natural Gas Vehicles,* Göteborg.

Hollemans, B. (1995), "Regulated and non regulated emissions of a commercially attractive LPG vehicle". *European Conference and Exhibition on LPG: a Clean and Efficient Motor Fuel,* Maastricht.

Hong, H., T. Krepec and R.M.H Cheng (1993), "Optimization of electronally controlled injectors for direct injection of natural gas in diesel engines". *SAE 930928.*

Horie, K., K. Nishizawa, T. Ogawa, S. Akazaki and K. Miura, (1992) "The development of a high fuel economy and high performance four-valve lean burn engine". *SAE 920455.*

Houben, M. and G. Lepperhoff (1990), "Untersuchungen zur Rußbilding während der diesel motorischen Verbrennung". *MTZ,* Vol. 51, pp. XI-XVI.

Houghton, J. (1994), "Transport and the environment". *Royal Commission on Environmental Pollution, 18th Report,* London.

Houiller, C., J.N. Hanley, P. Laurent, M. Pohle, J. Marcucci, H. Grangette and B. Poitevin (1994), "Récupération des vapeurs en station-service. Stage II". *Pétrole et Techniques,* No. 385, Paris.

Hublin, M. (1983), "L'automobile et les moteurs de demain. Perspectives d'évolution". *Pétrole et Techniques,* No. 297, Paris.

Hublin, M. and J.F. Bondoux (1995), "OEM application of LPG in light duty vehicles", *European Conference and Exhibition on LPG: a Clean and Efficient Motor Fuel,* Maastricht.

Huls, T.A. (1973), "Evolution of federal light-duty mass emissions regulations". *SAE 730554.*

Iida, N. (1994), "Combustion analysis of methanol-fueled Active Thermo-Atmosphere Combustion (ATAC) engine using a spectroscopic observation". *SAE 940684.*

IEA (International Energy Agency) (1994), "Refining and environmental implications of increased use of diesel-engined passenger cars". Paris.

Ingham, M.C., J.A. Best and L.J. Painter (1986), "Improved predictive equations for cetane number". *SAE 860250.*

Inoue, T., S. Matsushita, K. Nakanishi and H. Okano (1993), "Toyota lean combustion system. The third generation system". *SAE 930873.*

Ishbashi, Y. and Y. Tsushima (1993), "A trial for stabilizing combustion in two-stroke engines at part throttle operation". *IFP International Seminar on Two-Stroke Engines,* Paris.

Ishino, J. (1995), "In-cylinder direct injection gasoline engine", *News from Mitsubishi Motors,* No. 9511.

Ito, A. et al. (1993), "Study of SIC application to diesel particulate filter. Part 1: material development". *SAE 930361.*

Iwamoto, Y., Y. Danno, O. Hirako, T. Fukui and N. Murakami (1992), "The 1,5-liter vertical vortex engine". *SAE 920670.*

Jenkins (S.R.), R.G.M. Landells and J.W. Hadley (1993), "Diesel fuel lubricity development of a constant load scuffing test using the Ball On Cylinder Lubricity Evaluator (BOCLE)". *SAE 932691.*

Jess, V.A. and K. Hedden (1994), "Die Modell – raffinerie Deutschland – Ein Instrument für Prognosen des zukünftigen Energieverbrauchs der Deutschen Mineralöl Verarbeitung". *Erdöhl Erdgas Kohle,* 110 Jahrgang, Heft 11/12.

Jetter, J. (1995), "Honda ULEV technology", *World Conference on Reformulated Fuels and Refinery Processing,* San Francisco.

Joumard, R., L. Paturel, R. Vidon, J.P. Guitton, A.I. Saber and E. Combet (1990), "Émissions unitaires de polluants des véhicules légers". *Rapport INRETS,* No. 116, Lyon.

Joumard, R. Vidon, L. Paturel, C. Pruvost, P. Tassel, G. de Soete and A.I. Saber (1995), "Évolution des émissions de polluants des voitures particulières lors du départ à froid". *Rapport INRETS,* Lyon.

Kalghatgi, T.G. (1990), "Deposits in gasoline engines – A literature review". *SAE 902105.*

Kasztelan, S., N. Marchal, S. Kresman and A. Billon (1994), "Production of environmentally friendly middle distillates by deep sulphur and aromatics reduction". *14th World Petroleum Congress,* Stavanger.

Kelly, N.A. and R.F. Gunst (1990), "Response of ozone to changes in hydrocarbons and nitrogen oxides concentrations". *Atmospheric Environment,* Vol. 24 A, No. 12, pp. 2991-3005.

Kitagawa, J. (1992), "Improvement of pore size distribution of wall-flow type diesel particulate filter". *SAE 920144.*

Klimstra, J. (1990), "The dynamixer. A natural-gas carburettor system for lean-burn vehicle engines". *SAE 901498.*

Kono, S. (1995), "Development of the stratified charge and stable combustion method in DI gasoline engines". *SAE 950688.*

Lacey, P.J. and S.J. Lestz (1992), "Effect of low lubricity fuels on diesel injection pumps. Part 1: field performance. Part 2: laboratory evaluation". *SAE 920823 et 920824.*

Lahn, G.C., R.F. Bauman, B. Eisenberg and J.M. Hochman, (1992), "Development of advanced gas conversion technology". *Proceedings from the European Applied Research Conference on Natural Gas,* Eurogas 92, Trondheim.

Lamure, C. (1995), *Quelle automobile dans la ville*? Presses de l'École Nationale des Ponts et Chaussées, Paris.

Langen, P., M. Theissen, J. Mallog and R. Zielinski (1994), "Heated catalytic converter competing technologies to meet LEV emissions standards". *SAE 940470.*

Le Breton, M.D. (1984), "Hot and cold fuel volatility indexes of french cars. A study by the GFC volatility group". *SAE 841386.*

Leduc, P. and X. Montagne (1995), "Characterization and analysis of the phenomena that produce ORI". *SAE 952393.*

Leiker, M., K. Christoph, M. Rankl, M. Cartellieri and U. Pfeifer (1972), "Evaluation of antiknock properties of gaseous fuels by means of methane number and its practical application to gas engines". *ASME,* No. 72–DGP–4.

Lemaire, J. (1990), " Dépollution diesel; contribution d'un additif à base de cérium; un exemple concret: Athènes". *Congrès SIA,* Lyon.

Leplae, M. (1921), "Tracteur agricole actionné aux huiles végétales". *Bulletin des Matières Grasses,* Marseille.

Leprince, P. and C. Raimbault (1988), "Development in natural gas conversion to transport fuel". *Offshore Northern Seas Conference,* Stavanger.

Levins, P.L. (1974), "Chemical analysis of diesel exhaust odor species". *SAE 740216.*

Lévy, R.H. (1993), *Les biocarburants.* Ministère français de l'industrie, Paris.

Leyrer, J. et al. (1995), "Design aspects of lean catalysts for gasoline and diesel engine applications". *SAE 952495.*

Linan, A. and F.A. Williams (1993), "Autoignition in non premixed flow". *Integrated Diesel European Action, Final Report (for the period 01.05.89/30.06.93), Fundamentals aspects of combustion.* Oxford University Press, NY.

Lippmann, M. (1989), "Health effects of ozone. A critical review". *JAPCA,* Vol. 39, No. 5, pp. 672-695.

Lorusso, J.A., E.W. Kaiser and G.A. Lavoie (1981), "Quench layer contribution to exhaust hydrocarbons from a spark-ignited engine". *Combustion Science and Technology,* Vol. 25, pp. 121-125.

Lovi, A.L. and W. Carter (1990), "A method for evaluating atmospheric ozone impact of actual vehicle emissions". *SAE 900710.*

Lucas-Diesel (1992), "Principe de fonctionnement de la pompe d'injection DPS". *Manuel d'atelier.* Technical Service Department, Haddenham.

Lyn, W.T. (1976), Institution of Mechanical Engineers, London.

Maier, C.E., P. H. Bigeard, A. Billon and P. Dufresne (1988), “Boost middle distillate yield and quality with a new generation of hydrocracking catalyst”. *NPRA* paper No. AM 88-76, Annual meeting, San Antonio.

Malakar, J.J., J.B. Retzloff and L.M. Gibbs (1983), “Throttle body deposits. The CRC carburetor cleanliness test procedure”. *SAE 831708.*

Marcilly, C. and M. Bourgogne (1989), “FCC gasoline: what is behind octane”. *American Chemical Society.* Division of Petroleum Chemistry, Miami.

Martin, B., C. Bocard, J.P. Durand, P.H. Bigeard, J. Denis, M. Dorbon and C. Bernasconi (1990), “Long-term storage stability of diesel fuels. Effect of ageing on injector fouling. Stabilization by additives or hydrotreating”. *SAE 902174.*

Martin, B. and D. Herrier (1990), “Efficiency of fuel additive on diesel particulate trap regeneration”. *Institution of Mechanical Engineers (IME)*, Londres.

Matsuura, H., A. Kato, Y. Ajiki and I. Fujii (1994), “Research and development of dedicated compressed gas (CNG) passenger vehicle”. *4th International Conference on Natural Gas Vehicles,* Toronto.

Mauléon, J.L. et G. Heinrich (1994), “Le procédé R2R: la technologie du XXIème siècle”. *Revue de l’Institut Français du Pétrole,* Vol. 49, No. 5, p. 509.

Maxwell, J.B. (1977), *Data book on hydrocarbons.* R.E. Krieger Publishing Company, Malabar, Florida, USA.

Mc Arragher, J.S. (1990), “The effects of temperature and fuel volatility on evaporative emissions from european cars”. *Institution of Mechanical Engineers,* 394/028, London.

Mc Arragher, J.S., L.J. Clarke and H. Paesler (1993), “Prevention of valve-seat recessing in european markets”. *4th International Symposium on the Performance Evaluation of Automotive Fuels and Lubricants,* Birmingham.

Mc Arragher, J.S. *et al.* (1994), “Motor vehicle emission regulations and fuel specifications. 1994 update”. *CONCAWE,* report No. 4/94, Bruxelles.

Mc Murry, J. (1994), *Fundamentals of organic chemistry.* Brooks/Col Publishing Company, Pacific Grove, California.

Mc Rae, G.J. and J.H. Seinfeld (1983), “Development of a second-generation mathematical model for urban air pollution – Evaluation of model performance”. *Atmospheric Environment,* 17, 501.

Menrad, H., W. Lee and W. Bernhardt (1977), “Development of a pure methanol fuel car”. *SAE 770790.*

Mikkonen, S., R. Karlsson and J. Kivi (1988). “Intake valve sticking in some carburetor engines”. *SAE 881643.*

Milford, J.B., A.G. Russel and G.J. Mc Rae (1989), “A new approach to photochemical pollution control: implications of spatial patterns in pollutant responses to reductions in nitrogen oxides and reactive organic gases”, *Environment. Sci. Technol.,* 23, 1290.

Miller, R.H. (1947), “Supercharging and internal cooling cycle for high output”. *Trans. ASME,* 69.

Misumi, M. et al. (1990), “An experimental study of a low-pressure direct-injection stratified charge engine concept”. *SAE 900653.*

Miyaki, M., H. Fujisawa, A. Masuda and Y. Yamamoto (1991), “Development of new electronically controlled fuel injection system ECD-U2 for diesel engines”. *SAE 910252.*

Modetz, H.J. (1994), “LNG heavy duty truck demonstration program”. *International Conference on Natural Gas Vehicles,* Toronto.

Mohnen, V.A., W. Goldstein and W.C. Chang (1993), “Tropospheric ozone and climate change”. *Journal of the Air and Waste Management Association,* Vol. 43, pp. 1332-1344.

Monden, A. and Y. Honda (1993), "Toyota's new generation D-4 engine". *Asian Autotech Report,* AI Publishing Company, Vol. 180.

Montagne, X., D. Herrier and J.-C. Guibet (1987), "Fouling of automotive diesel injectors. Test procedure, influence of composition of diesel oil and additives". *SAE 872118.*

Montagne, X., R. Boulet and J.-C. Guibet (1991), "Relation between chemical composition and pollutant emissions from diesel engines". *13th World Petroleum Congress,* Buenos-Aires.

Mulawa, P., S. Cadle, D. Hilden (1992), "Emissions from gaseous fuel vehicles". *U.S. E.P.A. and A&WMA International Specialty Conference on Toxic Air Pollutants From Mobile Sources.*

Müller, U.C., N. Peters and A. Linan (1992), "Global kinetics for *n*-heptane ignition at high pressure". *Twenty-Fourth Symposium (International) on Combustion, The Combustion Institute.*

National Research Council (1991), "Rethinking the ozone problem in urban and regional air pollution". *National Academy Press,* Washington, D.C.

Needham, J.R., C.H. Such and A.J. Nicol (1993), "Fuel efficient and green. The future heavy-duty diesel". *Institution of Mechanical Engineers (IME),* London.

Newbery, P.J., T.A.C. Davies and K.M. Chomse (1984), "Heavier residual fuels for marine diesel engines". *6th International Motorship Conference,* London.

Newcomb, T.P. and R.T. Spurr (1988), *A technical history of the motor car.* Editions Adam Hilger, Bristol and New York.

Nikanjan, M. and E. Burk (1994), "Diesel fuel lubricity additive study". *SAE 942014.*

Nivière, P. (1993), *Cours de chimie organique. Fonctions et mécanismes réactionnels.* Éditions Eyrolles, Paris.

Nocca, J.L., A. Forestière and J. Cosyns (1993), "IFP's new technologies for reformulated gasolines". *NPRA Meeting,* San Antonio.

O'Connor, C.T. and M. Kojima (1990), "Alkene oligomerization" *Catalysis today,* Vol. 6, No. 3.

Okasoe, H. et al. (1993), "Study of SIC application to diesel particulate filter. Part 2: engine test results". *SAE 930361.*

Olree, R.M. and D.L. Lenane (1984), "Diesel combustion cetane number effects". *SAE 840108.*

Onishi, S., S.H. Jo, K. Shoda, P.D. Jo and S. Kato (1979), "Active Thermo-Atmosphere Combustion (ATAC) – A new combustion process for internal combustion engine". *SAE 790051.*

Orselli, J. (1992), *Énergies nouvelles pour l'automobile.* Éditions Paradigme, Caen.

Owen, K. and R.G.M. Landells (1989), "Precombustion fuel additives". *Gasoline and diesel fuel additives.* Editions J. Wiley and Sons.

Owen, K. and T. Coley (1990), *Automotive Fuels Handbook.* Editions SAE International.

Owen, K. and T. Coley (1995), *Automotive Fuels Reference Book.* Editions SAE International.

Pande, S.G. and D.R. Hardy (1990), "A practical evaluation of published cetane numbers". *Fuel,* Vol. 69, pp. 437-442.

Panico, R. and J.C. Richer (1994), *Nomenclature UICPA des composés organiques.* Éditions Masson, Paris.

Papachristos, M. and J.J. Russel (1992), "Gasoline additives – Their influence in controling octane requirement increase with particular reference to the Renault F2N engine". *SIA 92049.*

Papachristos, M., A. Raath and M. Trainor (1994), "Effects of gasoline detergent additives on formation of intake valve and combustion chamber deposits". *International Congress SIA,* Paris.

Paramins (1994), "Worldwide 1994-winter diesel fuel quality survey", *Exxon Chemical Limited.*

Parry, S., S. Miller and L. Walker (1993), "Detroit Diesel Corporation alcohol engines and field experience". *10th International Symposium on Alcohol Fuels,* Colorado Springs, Co.

Pattas, K.N. *et al.* (1989), "The effect of exhaust throttling on the diesel engine operation characteristics and thermal loading". *SAE 890399.*

Pearson, J.K., K.H. Reders and V.M. Tertois (1985), "The correlation of consumer and chassis dynamometer cold weather driveability". *CEC Symposium,* Paper EF4, Wolfsburg.

Pecci, G.C., M.G. Clerici, F. Giavazzi, F. Ancillotti, M. Marchionna and R. Patrini (1991), "Oxygenated diesel fuels. Part 1 -Structure and properties correlation". *9th International Symposium on Alcohol Fuels, ISAF,* Firenze.

Pedley, J.F. and R.W. Hiley (1988), "Investigation of sediment precursors present in cracked gas oil". *3rd International Conference on Stability and Handling of Liquid Fuels.* London.

Peters, B.D. and R.F. Stebar (1976), "Water gasoline fuels. Their effects on spark ignition engine emissions and performances". *SAE 760547.*

Petit, A. and X. Montagne (1993), "Effects of the gasoline composition on exhaust emissions of regulated and speciated pollutants". *SAE 932681.*

Peyla, R.J. (1994), "Deposit control additives for future gasolines – A global perspective". *27th ISATA Meeting.* Paper 94 EN 068, Aachen.

Pinchon, P. (1989), "Three dimensional modeling of combustion in a prechamber diesel engine". *SAE 890666.*

Pinchon, Ph. (1993), *La combustion dans les moteurs d'automobile. Progrès de la modélisation et de l'approche expérimentale. Colloque du Groupement Scientifique Moteurs (GSM).* Éditions Technip, Paris.

Prigent, M. and G. de Soete (1989), "Nitrous oxide N_2O in engines exhaust gases. A first appraisal of catalyst impact". *SAE 890492.*

Prigent, M., B. Martin and J.-C. Guibet (1990) "Engine bench evaluation of gasoline composition effect on pollutants conversion rate by a three-way catalyst". *SAE 900153.*

Prigent, M., C. Gauthier and S. Tauzin (1992), "Pots catalytiques, influence de la répétition de démarrages à froid sur l'activité des catalyseurs 3 voies et évolution en fonction du temps de l'activité des catalyseurs d'oxydation pour moteurs diesel. *PREDIT. Colloque à mi-parcours,* Versailles, Éditions Paradigme, Caen.

Raimbault, C., P. Abrassart and M. Espeillac (1994), "L'hydrogène: bilan actuel et perspectives en raffinerie". *Pétrole et Techniques,* Paris, No. 390.

Raynal, B. (1975), "Les techniques d'analyse de gaz et leurs applications dans le domaine des moteurs thermiques". *Revue Générale de thermique,* No. 162-163.

Renault, F. (1984), "Consommation pendant la mise en action des automobiles". *GFC, Journée d'Études Carburants-Moteurs,* Paris.

Rinolfi, R., R. Imarisio and R. Buratti (1995), "The potentials of a new common rail diesel fuel injection system for the next generation of DI diesel engines". *VDI Berichte,* Germany.

Rojey, A. et al. (1994), *Le gaz naturel – Production, traitement et transport.* Éditions Technip, Paris.

Rose, J.W. and J.R. Cooper (1977), *Technical Data on Fuel.* The British National Committee of the World Energy Conference, Editions Scottish Academic Press, Edimbourg.

Roumegoux, J.-P. (1994), "Calcul des émissions unitaires de polluants des véhicules utilitaires". *3rd International Symposium: Transport and Air Pollution. INRETS,* Avignon.

Russak, S. and P. Naon (1994), "Service experience with NGV – compressors in heavy duty applications. Importance for high reliability and availability". *4th International Conference on Natural Gas Vehicles,* Toronto.

Russel, M.F. and R. Haworth (1985), "Combustion noise from high speed direct injection diesel engines". *SAE 850973.*

Ryan, T.W. and S.S. Lestz (1980), "The laminar burning velocity of iso-octane, *n*-heptane, methanol, methane, and propane at elevated temperature and pressure in the presence of diluent". *SAE 800103.*

Ryden, C. and R. Berg (1991), "Extensive tests on ethanol operated buses fitted with diesel engines". *9th International Symposium on Alcohol Fuels,* Firenze.

Rytler, E.R. and A. Solbakken (1990), "Statoil's GMD, gas to middle distillate process". *Proceedings from the European Applied Research Conference on Natural Gas,* Eurogas 90, Trondheim.

Saber, A., L. Paturel, J. Jarosz, J. Subtil and M. Martin-Bouger (1993), "Instrumentation for analysis of trace-level PAHs by shpol'skii high resolution spectrofluorometry". *13th International Symposium on Polynuclear Aromatic Hydrocarbons.* Gordon and Breach Science Publications, Philadelphia, USA.

Salles, J., J. Janischewski, A. Jaecker-Voirol and B. Martin (1996), "Mobile source emission inventory model – Application to Paris area". *Atmospheric Environment,* Vol. 30, No. 12, pp. 1965-1975.

Sanger, R.P. et al. (1995), "Motor vehicle emission regulations and fuel specifications in Europe and the United States. 1995 update". *CONCAWE,* report No. 5/95, Bruxelles.

Schäpertons, H. et al. (1991), "VW's gasoline direct injection (GDI) research engine". *SAE 910054.*

Schneider, W., M. Stöckli, T. Lutz and M. Eberle, (1993), "Hochdruckeinspritzung und Abgasrecirculation im Kleinen, Schnellaufenden Dieselmotor mit Direkter Einspritzung", *MTZ Motortechnische Zeitschrift 54,* p. 588.

Schrepfer, M.W., C.A. Stanky and R.J. Arnold (1983), "Middle distillate stability, time for reassessment". *National Petroleum Refiners Association, (NPRA), Fuels and Lubricants Meeting,* Houston.

Schreyer, P., K.W. Starke, J. Thomas and S. Crema (1993). "Effect of multifunctional fuel additives on octane number requirement of internal combustion engines". *SAE 932813.*

Schuetzle, D. and J.M. Perez (1983), "Factors influencing the emissions of nitrated polynuclear aromatic hydrocarbons (nitro PAH) from diesel engines". *JAPCA,* 33, pp. 751-755.

Seinfeld, J. (1988), "Ozone air quality models – A critical review". *JAPCA,* Vol. 38, No. 5, pp. 615-645.

Senden, M.M.G. (1998), "The Shell Middle Distillate Synthesis Process: commercial plant experience and outlook into the future". *Pétrole et Techniques,* No. 415, Paris.

Shiratori, A. and K. Saitoh (1991), "Fuel property requirements for multiport fuel injector deposit cleanliness". *SAE 912380.*

Short, G.D., J.M. Betton and L.C. Antonio (1991), "Global experience with ignition – improved alcohol fuels in heavy duty diesel engines". *9th International Symposium on Alcohol Fuels,* Firenze.

Sirtori, S., P. Garibaldi and F.A. Vincenzetto (1974), "Prediction of the combustion properties of gasolines from the analysis of their composition", *SAE paper No. 741058,* International Automobile Engineering and Manufacturing Meeting, Toronto, Ontario.

Smith, D.A. and S.R. Ahern (1993), "Developments in the Orbital ultra low emissions vehicle". International Seminar: *A New Generation of Two-Stroke Engines for the Future?* Éditions Technip, Paris.

Socha, S., J. Thompson, D. Thompson and P. Weber (1994), "Optimization of extruded electrically heated catalysts". *SAE 940468.*

Sofiprotéol (1995), *Bilan technique du diester.* Proléa, Paris.

Sorenson, S.C. and S.E. Mikkelsen (1995), "Performance and emissions of a 0.273 litter direct injection diesel engine fuelled with neat dimethyl ether". *SAE 950064.*

Spiegel, L. et al. (1992), "Mixture formation and combustion in a spark ignition engine with direct fuel injection". *SAE 920521.*

Spikes, H.A., K. Meyer, C. Bovington, R. Caprotti and K. Krieger (1994), "Development of a laboratory test to predict lubricity properties of diesel fuels and its application to the development of highly refined diesel fuels". *International Symposium,* Esslingen (Germany).

Spindt, R.S. (1965),"Air-fuel ratio from exhaust gas analysis". *SAE 650507.*

Springer, K.J. and L.R. Smith (1993), "Experiments with MTBE-100 as an automobile fuel". *10th International Symposium on Alcohol Fuels,* Colorado Springs, Co.

Steeg, H. (1994), *Véhicules électriques: technolo*

gie, performances et perspectives. Agence Internationale de l'Énergie, Éditions Technip, Paris.

Stern, R., J.-C. Guibet and J. Graille (1983), "Les huiles végétales et leurs dérivés: carburants de substitution (analyse critique). *Revue de l'Institut Français du Pétrole,* Vol. 38, No. 1, pp. 121-136.

Strobl, W. and W. Peschlea (1986), "Liquid hydrogen as a fuel of the future for individual transport". *Hydrogen Energy Progress VI, Proceedings of the 6th WHEC,* Vienna.

Takatsuka, T., Y. Wada, H. Suzuki, S. Komatsu and Y. Morimura (1992), "Deep desulfurization of diesel fuel and its color degradation". *Sekiyu Gakkaishi.* Vol. 35, No. 2.

Taniguchi, B.Y. *et al.* (1986), "Injector deposits. The tip of intake system deposit problems". *SAE 861534.*

Tiedema, P. and L. Wolters (1990), "Recent developments in gas/air mixers and in micro processor air/fuel ratio control systems". *Institution of Mechanical Engineers.*

Tims, J.M. (1983), "Benzene emissions from passenger cars". *CONCAWE,* report No. 12183, The Hague.

Toupance, G. (1988), "L'ozone dans la basse troposphère – Théorie et pratique". *Pollution Atmosphérique,* pp. 32-42, Paris.

Trapy, J. (1985), "Modélisation d'un coefficient d'échanges thermiques spécifique au moteur à allumage commandé et prenant en compte les fonctionnements avec cliquetis". *Journées Internationales "Le Moteur d'Automobile à Allumage Commandé dans le Nouveau Contexte Européen",* Aachen.

Tucker, R.F., R.J. Stradling, P.E. Wolveridge, K.J. Rivers and A. Ubbens (1994), "The lubricity of deeply hydrogenated diesel fuels. The swedish experience". *SAE 942016.*

Tupa, R.C. and D.E. Koehler (1986), "Gasoline port fuel injectors – Keep clean/clean up with additives". *SAE 861536.*

Tupa, R.C. and C.J. Dorer (1986), "Gasoline and diesel fuel additives for performance/distribution quality". *SAE 861179.*

Unzelman, G.H., E.J. Forster and A.M. Burns (1972), "Are there substitutes for lead antiknocks"? *Midyear Meeting of American Petroleum Institute,* San Francisco.

Unzelman, G.H. (1984), “Diesel fuel demand. A challenge to quality”. *Petroleum Review.*

Urbanic, J.E. (1989), “Factors affecting the design and breakthrough performance of evaporative loss control systems for current and future emissions standards”. *SAE 890621.*

Van der Weide, J., J. de Haas, J.A.N. van Ling, J. de Rijke and M. van der Steen (1994), “CNG city bus engine with optimized part-load efficiency, high mean effective pressure and low emissions”. *4th International Conference on Natural Gas Vehicles,* Toronto.

Van Tiggelen, A. et al. (1968), *Oxydations et combustions.* Éditions Technip, Paris.

Von Fersen, O. (1986), *Ein Jahrhundert Automobil technik.* VDI Verlag, Düsseldorf.

Wang, J.C. and D.J. Reynolds (1994), “The lubricity requirement of low sulphur diesel fuels”. *SAE 942015.*

Warnatz, J. (1981), “The structure of laminar alkane-, alkene-, and acetylene flames”. *Eighteenth Symposium (International) on Combustion, The Combustion Institute,* p. 369.

Warnatz, J. (1984), “Chemistry of high temperature, combustion of alkane up to octane”, *Twentieth Symposium (International) on Combustion, The Combustion Institute,* p. 845.

Waters, P.E. and S.D. Haddad (1989), “Some developments of the effect of fuel and other parameters on diesel engine noise control”. *SAE 890131.*

Wauquier, J.-P. (1995), *Crude Oil – Petroleum Products – Process Flowsheets.* Éditions Technip, Paris.

Waynick, J.A. (1994), “Evaluation of commercial stability additives in middle distillate fuels”. *5th International Conference on Stability and Handling of Liquid Fuels,* Rotterdam.

Weick, L. (1998), “Gas to liquids technology: current developments and strategic implications for the energy industry”. *Pétrole et Techniques,* No. 415, Paris.

Weismann, J. (1970), *Carburants et combustibles pour moteurs à combustion interne.* Éditions Technip, Paris.

Whitney, K. and B. Bailey (1994), “Determination of combustion products from alternative fuels, Part 1: LPG and CNG combustion products”. *SAE 941903.*

Williams, G.R., F. Lagarde and D.D. Hornbeck (1971), “Étude des paramètres carburant affectant le cliquetis à haute vitesse des moteurs à combustion interne”. *Ingénieurs de l’Automobile.* Paris.

Wolff, J.P. (1968), *Manuel d’analyse des corps gras.* Édition Azoulay, Paris.

Woschni, G. (1967), “A universally applicable equation for the instantaneous heat transfert coefficient in the internal combustion engine”. *SAE 670931.*

Young, M.B. (1981), “Cyclic dispersion in the homogeneous-charge spark-ignition engine. A literature survey”. *SAE 810020.*

Yule, W. (1983), “Blood lead concentrations in school age children, intelligence, attainment and behaviour”. *Annual Conference of the British Psychological Society,* University of York, England.

Zingarelli, J.A. (1994), “LNG as a fuel for severe service heavy duty trucks”. *International Conference on Natural Gas Vehicles,* Toronto.

Index

(Volumes 1 and 2)

Each word or expression is followed by the page numbers were it occurs. The pages shown in boldface type cover topics that are extensively reviewed and expanded upon.

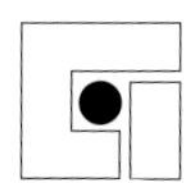

COMPOGRAVURE
IMPRESSION, RELIURE
IMPRIMERIE CHIRAT
42540 ST-JUST-LA-PENDUE
AOÛT 1999
DÉPÔT LÉGAL 1999 N° 5191
N° D'ÉDITEUR 1006

IMPRIMÉ EN FRANCE